Technical Writing

TECHNICAL WRITING:
Method, Application, and Management

Alice I. Philbin, Ph.D.
John W. Presley, Ph.D.

Delmar Publishers Inc.®

NOTICE TO THE READER

Cover photo by Tom Carney Photography

Delmar Staff
Associate Editor: Cynthia Haller
Editing Manager: Barbara A. Christie
Project Editor: Christopher Chien
Production Coordinator: Larry Main
Design Coordinator: Susan C. Mathews

For information, address Delmar Publishers Inc.
2 Computer Drive West, Box 15-015
Albany, New York 12212

Printed in the United States of America
Published simultaneously in Canada
by Nelson Canada,
A Division of International Thomson Limited

10   9   8   7   6   5   4   3   2   1

Library of Congress Cataloging-in-Publication Data
Philbin, Alice.
    Technical writing: method, application, and management/Alice
I. Philbin, John W. Presley.
        p.   cm.
    Includes index.
    ISBN 0-8273-2685-8 (pbk.)   ISBN 0-8273-2686-6 (Instructor's
guide)
    1. Technical writing.   I. Presley, John W.   II. Title.
T11.P44 1989
808'.066—dc19   88-39715
                                                        CIP

Contents

PART TWO   APPLICATION

PART THREE   MANAGEMENT

Preface

Technical Writing: Method, Application, and Management speaks to two audiences — students beginning to explore technical communication as a major field of study and students learning technical communication as a personal or degree requirement for a particular technical or scientific major. Typically, we teach classes composed of both groups, and they have guided us in selecting the content and special features of this book.

The content of *Technical Writing* is divided into three sections.

Part One (Chapters 1–7) addresses the writing process as the basis of technical writing. Each chapter discusses a particular phase of the writing process for use in technical writing. Thus, the organization chapter (Chapter 5) reviews not only the basics of outlining, but it concentrates specifically on the outlines that technical writers use such as decimal and headline outlines. Each chapter of this section features a process diagram to show the objective or topic as part of the writing process.

Part Two (Chapters 8–16) treats the specific types of technical communications you may draft for college or for your occupation. Since professionals, managers, and technicians need to complete various kinds of technical communications on the job, we supply checksheets, outlines, and other guides in these chapters to show how to apply the general strategies of the writing process to specific assignments.

Part Three (Chapters 17 and 18) introduces you to the world of corporate technical writing by describing how a publications group works and by introducing cost estimating.

Overall, *Technical Writing: Method, Application, and Management* prepares students for the corporate culture in which they will collaborate with associates to plan, budget, draft, and produce technically oriented publications.

In addition to comprehensive content coverage, *Technical Writing* contains several special features:

Case Studies. Each chapter starts with a case study of a technical communication problem that is work-related. The chapter then develops the writing problems outlined in the case study as we explain how to apply knowledge to communications in a given situation. The case studies are drawn from the experiences of students who work as they attend college and our graduates in industry; thus, as a collection, the cases answer many questions students may have about writing for occupational purposes.

Varieties of Writing. The cases and the examples in the chapters expose students to the vast collection of documents that can be called "technical communication" including computer manuals, office procedures manuals, proposals, science lab reports, various types of correspondence, and other technical documents. There are as many kinds of technical communications as there are occupations, and new needs and requirements will emerge during an individual's career. By studying examples from the various occupations, students can be prepared for a career of continuous growth and learning about the types of writing they will need to do well to excel in their professions.

Students' Writing. As you skim this book, you'll notice that we include passages written by students. We also show how the students revised and edited their work. We have selected case histories and passages from a variety of students: young undergraduates, working adult undergraduates, parents who attend college, graduate students, and occupational specialists who attend school to retrain or upgrade. All of these people share writing problems that are universally experienced, and including their work allows other students to see how these problems can be approached.

ACKNOWLEDGEMENTS

Many individuals have contributed to this book, but we wish to thank particularly Cindy Haller, Karen Hawkins, and Chris Chien, the editors who guided us through the review, revision, and production process.

Because our text has developed in response to our students, many individuals have contributed. We are grateful to the students of Bowling Green State University who tested the materials in their classes. Our graduate students in English 574, Professional Editing, reviewed Chapter 16; we appreciate their suggestions for revisions. Special thanks go, also, to the students in English 388, Introductory Technical Writing and English 488, Technical Writing, who tested all the materials in Sections One and Two. We appreciate the reviews of Xiao Duan Yang, Qi Quan Wang, and Xue-mei Zhang, who examined the text for cultural biases. Finally, we appreciate the assistance of Bonnie Fink with content reviews of Chapters 8 and 9.

ABOUT THE AUTHORS

Alice I. Philbin has taught technical communication, composition, and literature to students in high school, college, graduate, and post-secondary, occupational programs. A former technical journal editor, Dr. Philbin developed and taught on-site technical communication and career development courses for all the branches of the United States military services. During her tenure at Bowling Green State University, Dr. Philbin has maintained her consulting practice with businesses, taught in the undergraduate and graduate scientific and technical communication programs, and serves currently as Director of Graduate Studies for the Department of English.

John W. Presley has taught, and written about, writing at every level, from basic to advanced, for several colleges and universities since 1970. He has taught business and technical writing in the private sector as well, and has written or edited a half-dozen textbooks on reading and writing. Dr. Presley is currently Professor of Business Administration at Augusta College, where he also works in the Academic Affairs Office. In addition to teaching graduate courses in communications, he supervises faculty development and curricular revision. Firmly convinced that writing varies only by audience and purpose, Dr. Presley is also a published poet.

PART ONE

Method

CHAPTER 1

Definitions of Technical Communication

Good technical writing requires an understanding of the ways technical professionals communicate. In reading this chapter, you will learn how successful technical professionals:

- Define technical communication.
- Discuss basic terms they use.
- List characteristics of good technical communication.
- Describe major types of technical communication used in business, industry, and other workplaces.

CASE STUDIES   Technical Writing and Work

Many workers and students need help with their technical writing in order to reach goals or to sustain job success. Because technical communicators work in many professions, the following random sample shows the diversity and complexity of their tasks.

Louise Mitchell—an operations manager who must compile a lengthy feasibility report for her manager, a corporate vice-president.

Bob Andrews and **Ann Middleton**—two individualistic employees who must together prepare for a client a series of recommendations for purchasing two corporate airplanes.

Steve and **Elaine Miska**—owners of a small construction company who must use the library's resources to diversify and expand their business.

Bill Dorsey—a technical writer for a consulting firm who must continually win new contracts to expand the firm's business. To maintain a client list, Bill writes successful proposals and feasibility studies.

Sue Chin—a college senior majoring in environmental science. She must learn to organize a report on methods of radon detection in the home.

Laura Heberlein—a successful business major in college. For a required technical writing course, she is learning to revise and edit her technical communication.

Pat Pryer—a gifted programmer who has developed a software package for scheduling team sports. She must design appropriate illustrations for the product's users.

John Sanchez—a technical communication student who must complete a set of assembly instructions during his internship. Although John writes well, he must illustrate his instructions so users can assemble their furniture from his drawings, not his text.

Randy Goetz and **Linda Harris**—job applicants. Randy, a recent college graduate, must find his first job. His sister Linda, an experienced technician, must reposition herself in the job market.

Joe Rachiele—a veteran and owner-operator of a small electronics business. His business requires extensive paperwork and communication with suppliers and customers, so he needs a work file of form letters and memoranda.

Carlos Andrade—a training evaluator who writes for his livelihood. As a consultant, he writes progress reports, closeout reports, and multivolume studies of his clients' training programs.

Ved Krishnan—an accountant who understands accounting software packages. He must write procedures for his relatives and business associates that explain how a computer disk operating system packages a file for easy, systematic access.

Ted Wilson—a successful agricultural journalist. He must conquer his fear of public speaking to promote the Society for Technical Communication among college students and local technical writers.

Maryann Hasama—a graduating technical communication major. Despite excellent writing skills, she is unprepared to work in a corporation. To succeed, Maryann must learn how corporate writers publish and distribute their work.

Jeff Beetham—a young technical writer recently promoted to his first management position because of his excellent writing in waste management proposals. Jeff must now learn to plan, schedule, and estimate the costs of his employees' work.

Each of these 18 technical communicators has a special communication need. This book analyzes the various communication problems these writers face. While examining these and other writers' drafts, you will meet still more technical communicators.

DEFINITION OF TECHNICAL WRITING

The first step is to define technical communication. Is it technical writing? Or is that term too exclusive? Is there, for example, technical speech as well? Where are examples of technical communication found? Is technical communication on television or is technical communication only a written medium?

We can answer these questions neither easily nor quickly. Lengthy works elaborate the philosophy and psychology of rhetoric, expository writing and its origins, occupational and business English, interpersonal communication, journalism, and mass communication. Technical communication combines ideas and practices from all of these fields and borrows techniques from such diverse disciplines as training and development and public relations. Although no one chapter of this book covers the antecedents of technical communication in all of these areas, we can outline some basic terms, some theoretical bases, and certain applications to your education and future employment.

Some Terms and Assumptions

Before reading on about the applications of technical communication, you need to know some of the discipline's basic terms:

Technical communication—a universal expression that refers to the written documents, videotapes, slide shows and other illustrated learning aids, demonstrations, and electronic messages created by professional writers and designers. Because this field changes rapidly, "communication" is more generic than writing, and "technical" more comprehensive than "scientific." For example, a user manual for a word processing software package and a flight simulation program for training pilots are technical communications.

Technical writing—printed or electronically transmitted messages. These include instructions, manuals, procedures, statements of policies, articles for publication, and various other technology-related communications designed for publication. The magazine *Fine Woodworking*, with its numerous feature stories and regular columns on woodworking technology, is an example of technical writing.

Scientific communication—again, a general term for all the kinds of communication about various fields of science. The television program *Nova* is an outstanding example of scientific communication.

Scientific writing—printed or electronically transmitted messages, including laboratory reports, experimental results for publication, and scientific discussions of specific theories and applications, designed for public reading. A classic example is *Scientific American*.

Scientific and technical communication are often taught as one discipline. In fact, this book contains a sample lab report and several examples of writing in the applied sciences. But the sciences and technologies have so expanded, that the essential differences between the two fields are more noticeable. Thus, this text, though containing examples of scientific writing, emphasizes technical writing.

Listed below are more terms that will help you understand this overview of technical communication.

Technical journalism—communication about science and technology for the general public, transmitted by the mass media such as newspapers, magazines, and television. *Car and Driver* is a technical magazine.

Oral technical communication—spoken communication about science or technology, delivered to an audience either live or via videotape. The popular television program, *This Old House,* is a well-engineered example of oral technical communication.

Documentation—the generic term for industry-generated manuals, instructions, procedures, policies, and other technical communication. For example, a reference guide for a stereo system or an "owner's manual" for a car is a form of documentation.

Electronic publishing—the electronic transmission of coded manuscripts, via computers and telephone lines, to a printer who processes the electronic manuscript into its final printed form. Technical communicators write and edit, using software codes, to format their manuscripts. They then use a modem and telephone lines to transmit the electronic manuscript from their computers to the printer.

Communication on-line—the telecommunication of electronic data via computers. For example, both CompuServe and Genie use on-line communication.

Desktop publishing—the process that allows a writer with a computer to draft, edit, illustrate, and print the final version of a manuscript. Although not of offset printing's technical quality, desktop publishing does expand the writer's and publisher's production choices. A simpler form of desktop publishing involves producing complete documents with a software program and a compatible laser printer.

Keyboarding—the process of entering words, illustrations, and codes into a computer for writing, designing, and publishing documents. In a sense, keyboarding replaces typing as a way of accessing a manuscript. Anytime you compose at a computer, you keyboard.

Technical Communication as Profession and Process

Technical communication is both a process for and a profession of presenting information. Professional technical communicators, graduates of "technical communication" programs, work in industry, education, government, and on a contract or freelance basis as full-time documentors. Chapters 16 and 17 examine both technical communicators' roles in industry and a firm's costs for documenting its products. On the other hand, Chapters 2, 3, and 4 emphasize technical communication as a process that most people need in their professional or personal lives to help explain new products, define technical problems, or clarify pressing community issues. Thus, everyone, not just the professional technical communicator, needs to understand how to communicate about technology and science. Whether you write as an experienced professional, an occasional author, or only an informed reader, you should know technical communication's specific requirements.

A technical communication reports facts, processes, or results of processes in terms familiar to a well-defined audience. To accomplish this, the technical communicator often uses special or distinct forms to create a visual structure for the communication. Note that this definition focuses on the information, the audience, and the form. Once a technical communicator understands the information to be presented, the writer then conducts a thorough audience analysis to determine the appropriate form, or medium, of presentation. This audience analysis may involve interviews with product users, demographic research, and even re-

FIGURE 1.1   A Tree Diagram of the Communication Process Showing That Technical Communication and the Other Disciplines Share Certain Similarities.

A TREE DIAGRAM OF THE COMMUNICATION PROCESS

Can you relate technical communication to the other types of writing you have studied? Perhaps you see similarities between technical communications and certain types of journalism or exposition. Maybe you have learned with surprise that technical texts rely greatly on illustrations to clarify their explanations. This tree diagram shows that the various disciplines within the field of communication, though different, share common categories. In short, the processes of writing and speaking common to all communication create similarities of form as well.

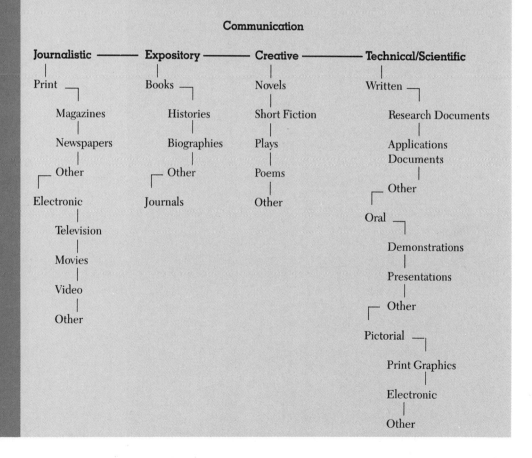

views of the company's historical users. With this data you can select the appropriate format and structure of the actual document. Figure 1.2 shows some of the types of technical communications this text describes, as well as the audience and form of each document.

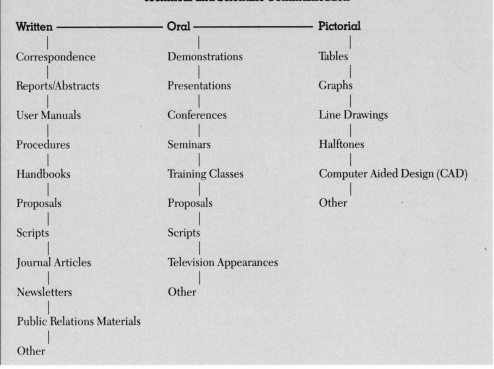

FIGURE 1.2 A Tree Diagram of the Three Categories Into Which Technical Communication Can Be Grouped.

A TREE DIAGRAM OF TECHNICAL AND SCIENTIFIC COMMUNICATION

Perhaps in your occupation you have discovered a form of technical communication that should be added to this classification drawing. Since the list is by no means complete, where would you place the forms of technical communication you have defined? Note again that the categories are not mutually exclusive; that is, a proposal may be both written and spoken.

Technical and Scientific Communication

Written	Oral	Pictorial
Correspondence	Demonstrations	Tables
Reports/Abstracts	Presentations	Graphs
User Manuals	Conferences	Line Drawings
Procedures	Seminars	Halftones
Handbooks	Training Classes	Computer Aided Design (CAD)
Proposals	Proposals	Other
Scripts	Scripts	
Journal Articles	Television Appearances	
Newsletters	Other	
Public Relations Materials		
Other		

Characteristics of Technical Writing

Stylistic and visual cues in a text can help identify technical writing.

1. Technical writing tends toward *objectivity* in nature and tone. A technical text usually presents facts directly through definition, description, example, and causal analysis.

Although difficult to define, objectivity is easily identified. For example, a science periodical may feature a story on the breakdown of steam generator tubes in American nuclear energy facilities and an anti-nuclear energy editorial. The feature story's text describes steam generator tube problems at several plants while its artwork and photographs illustrate the inside of a tube and its construction. On the other hand, the editorial angrily details the probable effects of nuclear plant breakdowns on the environment of future generations of human beings. The feature article is factual and objective. However, the editorial, which may be right, uses fact, conjecture, emotion, and other techniques to argue. Its tone is not objective.

2. Technical writing often follows careful specifications. Typically, a technical writer uses a conventional report form designed by either a director of communications or a company's technical writing supervisor. The form may tell the writer, based upon an audience analysis, where to state experiment results, how many sentences to use in describing cost implications for the company, and the average number of words to use per sentence. For example, writers may use Gunning's Fog Index (shown later in this text). Writers may also compose text in conformance with a firm's design standards. Industrial psychologists and technical writers have found that guidelines for arranging a manual's content and graphics can save time and money and help ensure clarity and uniformity of image in company-disseminated information.

3. Technical writing is usually graphic writing. Technical writers often work with documentation groups employing graphic designers. The writers and artists use drawings, halftones, graphs, and tables as frequently as the audience's comprehension level permits. Publications groups plan densely illustrated manuals, reference sheets, reports, and technical news stories to stimulate readers and to simplify verbal presentation, technology, and its applications.

4. Technical texts often exhibit a characteristic called "chunking." Chunking is the arrangement of text into visually separate sections emphasized by clear headings and subheadings. Borders and white spaces also show the audience the points of emphasis. These features help readers to retain new or possibly complicated information.

5. Technical writing is frequently numeric. Numbers, preferably Arabic numerals, appear often in the text. In complex technical writing for experts, numbers are presented to the last significant decimal point in complete lists, tables, and graphs.

6. Technical writing is often symbolic because mathematical and scientific symbols are frequently used (see Handbook for symbols). Equations, too, may appear, although appendices may contain the complete steps of a complicated formula.

7. Technical and scientific writing use a special vocabulary, words such as "input," "output," "software," "debit," "hypothesis," "task," "articulation," and "appliances." Attorneys, educators, engineers, scientists, and business persons all use words and figures of speech characteristic of their professions. If an audience is familiar with a specialized vocabulary, trade-specific words can create brevity and clarity.

8. Technical communication requires clarity because important outcomes such as human safety and equipment preservation may depend on accurate communication. Technical writers must check all facts for accuracy, include every step in a process, and be certain that a document communicates to its users.

FUNCTION OF TECHNICAL WRITING IN THE OFFICE

Many examples of technical writing can be found at work. A balance sheet from a spreadsheet program and a multivolume feasibility study of environmental costs are both examples of technical writing. For reference and convenience of definition, we list 15 kinds of technical writing:

- ☐ Correspondence.
- ☐ Reports, including summary reports, research studies, feasibility studies, needs analyses, and evaluations.
- ☐ Abstracts both for documentation and as introductory matter.
- ☐ Product specifications and updates.
- ☐ Procedures and instructions, both self-instruction sets and traditional instruction manuals.
- ☐ Handbooks and internal procedures statements for employees.
- ☐ Proposals for internal funding and prospective clients.
- ☐ Educational materials for in-house training and new product orientation.
- ☐ Graphic elements, charts, graphs, tables, and halftones.
- ☐ Audio-visual materials, including slide tape, videotape, and film presentations.
- ☐ Scripts for oral reports and company presentations.
- ☐ Documentation, either manual or computerized, for access to technical materials.
- ☐ Articles for publication in trade magazines and scientific journals.
- ☐ House organs and company newsletters.
- ☐ Public relations materials such as new product brochures and advertising copy.

A common misassumption students share about work is that a staff of technical writers is available to prepare technical materials for them. Major firms do employ groups of technical writers, and some companies add to their writing staff in proportion to the company's volume of such work. But even if a technician has access to trained technical writers, they may not be expert in a particular technical field. Technicians and experts must communicate to writers an understanding of the relationship between a company's new products and particular communication requirements.

Typically, an expert or a technician must first draft a document. Studies in large corporations indicate that up to 25 percent of a college graduate technician's time is spent writing, often in this pattern:

Calculations and data gathering	30%
Organizing materials for writing	25%
Writing reports and correspondence	20%
Communicating with supervisors about their work	15%
Meetings and oral reports	5%
Other	5%

The emphasis on writing differs with the company, the field the technician writes in, and the particular technician's expertise and seniority. Nevertheless, managers usually value communication skills. When evaluating employees' performance, industry executives always indicate that technical training is the most needed skill, and that the ability to communicate effectively is second. Communication skills become critical to promotion within a company. Success or failure may well hinge on technical writing ability, that is, the ability to inform supervisors of new projects and to persuade managers to support new product development. Technical writing, then, is a positive resource. An ambitious employee learns quickly that technical writing is a critical occupational survival skill that he or she will use at least one full working day a week.

SUMMARY

This chapter includes the definition, scope, and application of technical communication. On the basis of your reading, you should be able to:

- Identify technical writing at work, at a library, or at home.
- Identify the characteristics of technical writing: objectivity, chunking, specificity, graphics, numbers, symbols, occupational terminology, and clarity.
- Identify 15 types of technical writing that may be done on a job, either individually or with a team.
- Prepare as a professional or occupational specialist to spend at least one day a week composing technical writing related to an occupation and its place in society.

ASSIGNMENTS

1 Each profession and occupation offers a collection of technical communication. Select either your major or the area of your present employment to conduct this research:
   1.1 Collect and examine a range of publications in your chosen field. You may include television programs, videotapes, expert publications, journalism for the mass market, or specific works such as policy statements and guidelines for employees. After reviewing your research materials, prepare an informal oral report for your class that answers the questions listed here.
   1.2 How is your chosen profession changing? What are the current trends in practices and concerns voiced in your research materials?
   1.3 Does your occupation currently face ethical issues? For example, technical communicators are often concerned about the issue of confidentiality. What ethical or substantial professional issues can you identify after studying your materials?
   1.4 Is obsolescence an issue? Do the publications indicate a need for the continuing education in your field? What recommendations are made?
   1.5 Can you identify any other issues or topics of concern that your materials mention regularly? Briefly define and explain these issues.

2 The chapter lists 15 types of technical communication. Which types appear most often in your profession or occupation? To answer this question, broaden your research; con-

sider work-related materials as well as reading materials. In view of the needs your research identifies, which types of technical communication do you need to learn about?

3 Assuming neither that you have selected a major nor that you work in your chosen occupation, examine your environment for examples of technical communication. Consider, but do not limit yourself to:

Insurance policies
Instructions and manuals for appliances and tools
Institutional communications from schools and colleges
Textbooks and software documentation
Cookbooks
Agricultural and gardening supplies

Now that you have gathered your research materials, prepare for your class an informal oral report about the technical nature of your environment. Are you ahead, with, or behind the times? Which technical communication is most useful to you? Why? What is it about your favorite materials that you would like to imitate?

CHAPTER
2

Your Audience

As a technical writer you will need to know much about your audience. This chapter explains how to analyze your readers as an audience, and should teach you to:

- Define your purpose for preparing a document.
- Describe the primary and secondary audiences.
- Explain a traditional and a functional view of audience analysis.
- Determine whether tests should be used with the audience.
- Profile an audience.

CASE STUDY   Preparing to Persuade an Audience

Louise Mitchell, a personnel manager for International Avionics, Inc., must write a feasibility study for the vice-president. The report should recommend that International Avionics adopt a three-year plan to replace its health maintenance organization (HMO) with a return to a group health insurance policy.

Louise's report will necessarily address some sensitive issues for International Avionics. The HMO, introduced 10 years ago, should have saved the firm considerable money as the prices of group health insurance policies soared. Employees were initially reluctant to join the HMO, but later grew to expect its total preventive health care package. The HMO's subsequent popularity caused health care costs to soar beyond company estimates. To further complicate matters, during the trial period, the average age of International Avionics' employees has increased, creating a greater demand for comprehensive medical coverage.

Louise has a difficult assignment: the vice-president has commissioned the report to prepare the employees for the benefits change. To win the employees' support, she must clearly demonstrate how the increased costs make the HMO unaffordable. Thus, she must write to inform and to persuade.

To effectively inform and persuade her readers, Louise must understand several aspects of her audience. First, she must have a clear sense of purpose to communicate with the company's personnel. Second, she must clearly define her two audiences. Third, she must study her readers and clarify her purpose as their author. Finally, she must compose a profile of her readers so that she and the vice-president can avoid any unpleasant surprises as they prepare the final document. A flowchart shows the four-step process that Louise Mitchell uses to complete the pre-writing audience analysis.

STEP ONE—DECIDING YOUR PURPOSE

Deciding on your purpose is an essential step in the pre-writing process. To analyze an audience, consider your functional and rhetorical purposes.

Functional Purpose

The functional purpose refers to the document's content. Authors usually think first of their subject matter or topic, and then define the communication's purpose. What is your subject? For example, Louise's subject is the feasibility of replacing the HMO option with a group health insurance policy.

Your functional purpose needs to be clear. Louise must write a report with a very specific set of stylistic and organizational requirements: the third-person-singular form of narrative, three levels of headings, and a summary.

But if not assigned a report, she would have to ask, "What is my purpose as the writer?" To answer her question, she might compose such a list:

To show employees how to sign up for the new benefits.
To explain how an HMO will eventually bankrupt the company.

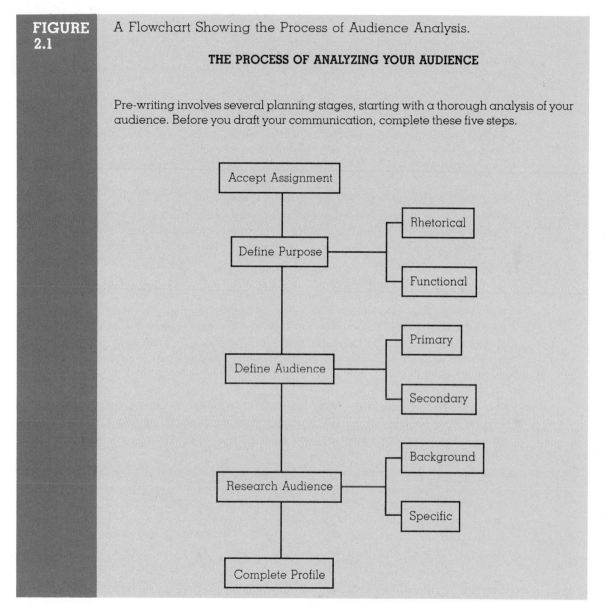

FIGURE 2.1 A Flowchart Showing the Process of Audience Analysis.

THE PROCESS OF ANALYZING YOUR AUDIENCE

Pre-writing involves several planning stages, starting with a thorough analysis of your audience. Before you draft your communication, complete these five steps.

To prove that International Avionics is concerned and timely in cases of work-related injuries.
To prevent employees from blaming the vice-president.
To educate employees about comparative merits of group health insurance and HMOs.
Some combination of these list items.

A document may have one or several purposes. Like Louise, you must identify your purpose before writing any type of document. Your understanding of your unique purpose will determine the document's length, word choice, rhetorical strategy, and many other factors.

Rhetorical Purpose

After defining a purpose, an author must then ask, "How will I tell my readers about my topic and purpose?" With such a question, you essentially ask, "Which rhetorical strategy should I use?" Although various techniques are available, we present five methods of communicating a message:

Describing a Process. Is your purpose to explain how to operate a battery-operated screwdriver? Perhaps instead you want to explain to employees the process they should use to estimate their benefits. Either way, you describe a procedure, how to perform a sequence of actions. Technical communicators probably use this rhetorical approach most often.

Describing a Mechanism. Perhaps you need, as Louise did, to explain how an HMO serves its members, including the services it provides and the hospitals that are participating. Maybe you need to explain how either a newspaper is printed or a new scanning device works. If you are explaining how the parts and whole of a system operate, and not a procedure, then you are describing a mechanism.

Defining an Object or Idea. Your assignment may call for introducing and explaining new terms and ideas. For example, Louise Mitchell's report may clarify terms such as "third-party payments," "liability," or "dependent." In each case, she must explain the word's meaning, the class of terms the word belongs to, and its specific use in an insurance or personnel context. Usually, an author uses description and explanation as techniques to enhance the rhetorical purpose of definition.

Comparing and Contrasting Topics. Louise Mitchell can use these techniques to demonstrate the benefits of group health insurance. She can contrast the expenses of the two programs as she lists the similar types of coverage both plans offer.

Arguing for a Case. You may try persuading the readers to adopt a point of view or plan of action. For example, you might prepare an argument against a local toxic waste facility, use financial tables to convince your supervisor that in-house publication is cost-effective, or, like Louise, try persuading readers that a different type of insurance coverage is better. Using inductive and deductive reasoning to arrange facts, you employ persuasion as a rhetorical strategy. (For a detailed discussion of rhetorical strategy, see Chapter 6.)

Before beginning a thorough study of any audience, you should decide which of these five approaches best suits your topic and purpose. Only after defining the *what*—your topic—and the *how*—your rhetorical strategy—can you focus on influencing your audience.

A Limitation of Your Purpose—Primary and Secondary Audiences

Finally, you must define your purpose for very specific readers, not every kind of reader. Your purpose will be clearer, your choice of material more specific, and your argument more appropriate if you plan for only two audiences, primary and secondary readers.

The *primary audience* of your document is the most important group of readers, not necessarily the majority. For example, the vice-president is Louise Mitchell's primary reader because she will approve it or suggest changes before the report is distributed. The primary audience consists of decision-makers, the people who will decide whether or not to follow the recommendations of your document.

The *secondary audience* is all the other document readers. In Louise Mitchell's case, these are the International Avionics employees. Corporate personnel who read a draft only for informational purposes are considered a secondary audience. On the other hand, manual users are a primary audience, the decision-makers, because they rely on the instructions to detail the use of the product. If the instructions are confusing or intimidating, they will return a product, hesitate to buy it again, and recommend purchasing a different one.

Despite current pressure to cut publications costs, you should compose a document that addresses no more than two audiences. You risk readers ignoring your document if it addresses too large an audience, and not their specific needs.

If you can afford to draft only one document, then what should you do?

1. Draft different sections for different readers. In a report, the executive summary can speak to the decision-makers, the body of the report to the technicians, and the tables in the appendixes to the engineers. Do not address everyone throughout the text.
2. Use visual cues to emphasize section differences so specific readers can easily find their own sections. Use tabs, markers, color-coding, and the table of contents to direct different segments of your audience.
3. In a preface explain how to use the document. Tell each audience which pages pertain to them.
4. Consider dividing a lengthy document into several shorter works that address specific audiences. For example, if you compose comprehensive documentation for a medical software package, recommend several manuals: a short reference manual for administrators, a reference sheet or short user's guide for nurses, and an advanced user's guide with hardware and software specifications for engineers and installers.

STEP TWO—DEFINING YOUR AUDIENCE

Defining an audience is not necessarily easy. First, you must ask, "What method should I use?" If from a business background, you may visualize your readers as various consumer groups. If you studied advertising or journalism, you may picture your audience as readers, listeners, and viewers of the media. Technology and technical communication majors may imagine product users. English majors might think more directly and simply of the reader.

These views are all valid because technical communication draws upon and contributes to so many disciplines. In this current phase of technical communication's development, consider and feel free to modify the two prevailing approaches to audience analysis. The first is sometimes called "traditional," because it was developed first. The second, the "functional" approach, emphasizes the reader as a user of documentation.

A Traditional Approach to Audience Analysis

For years, technical communicators learned to address combinations of three types of audiences: experts, technicians, and lay readers. The three-audience view offers a "traditional" and quite useful method of classification. Consider each group.

Experts. These readers have extensive formal training in highly specialized fields, sometimes holding Ph.D.'s or terminal master's degrees. Experts prefer reading for facts and for professional updating, often reading complicated tables and charts before the narrative. Typically, experts write and understand complex sentences, averaging 17 or more words, that use their profession's specialized jargon. Text for this audience should contain all relevant facts, even if only in footnotes and appendixes, and less background material. For brevity and reading ease, formulas, abbreviations, and equations should appear directly in the text.

Technicians. Technicians work in applied science and technology. While the expert studies why, the technician concentrates on how particular scientific laws operate. Technicians implement experts' decisions. Typically, a senior chemist with a Ph.D. in pharmacology is an expert. The lab supervisor with a bachelor's degree in chemistry is a technician who may construct apparatus to carry out the chemist's experiments. Pharamacists are experts; their assistants are technicians. Medical research doctors are experts.

Technicians may have two- or four-year degrees from community colleges, trade and industrial schools, or some United States military training. In many areas, such as the health professions, they obtain state licenses through examination. Technicians may prefer shorter sentences averaging 14 words or less. They understand basic formulas as they apply to the job at hand, and their field's common abbreviations and symbols. For technicians, give details about how to do the job but minimize theory unless your report presents material typically outside their expertise. Technicians, as all readers, read for many reasons, but you might assume that their technical reading is job-oriented.

Lay Readers. Everyone is a non-technician in some areas. A physicist expert in minimal surface theory may have no background in macroeconomics. A college professor with a photography hobby may not understand the physics of light. The ordinary driver usually doesn't understand how advanced automobile repairs are made.

Lay readers are most difficult to address since they may have no one common area of experience. Before selecting facts, organization patterns, and visual materials for nontechnicians, further segment your readership. Use company files to profile nontechnicians. Use library resources such as census information, or interview members of your audience. Try to segment your nontechnical audience by education and experience, because these factors determine a person's reading level and attitude toward a product or task.

A Functional Approach to Audience Analysis

Recent audience analysis pictures the reader as a document user. This view emphasizes why the reader needs documentation and what the reader expects from a manual or report. To understand this view, consider four points.

FIGURE 2.2   A Typical Audience Analysis Sheet.

AUDIENCE ANALYSIS WORKSHEET

1. Approximately ____ persons will read my document. This audience has ____ experts, ____ technicians, and ____ nontechnicians.

2. **Primary Audience**

   Group your audience by age and sex. State the approximate number of readers in each group.

   under 18 ____   19–35 ____   36–50 ____   51–65 ____   over 65 ____

   number of males ____   number of females ____

   Classify the primary audience by writing down the approximate number of persons who have trained in each of these categories.

   ____ high school only   ____ military training   ____ BA or BS degree

   ____ trade school   ____ associate degree   ____ post-grad.

   In about 20 words, state the attitude of the primary audience toward the document.

   What does this audience need to learn from the document?

3. **Secondary Audience**

   Group your audience by age and sex. State the approximate number of readers in each group.

   under 18 ____   19–35 ____   36–50 ____   51–65 ____   over 65 ____

   number of males ____   number of females ____

   Classify the primary audience by writing down the approximate number of persons who have trained in each of these categories.

   ____ high school only   ____ military training   ____ BA or BS degree

   ____ trade school   ____ associate degree   ____ post-grad.

   In about 20 words, state the attitude of the secondary audience toward the document.

   What does this audience need to learn from the document?

| FIGURE 2.3 | A Typical Audience Analysis Sheet (Functional Emphasis). |

AUDIENCE ANALYSIS WORKSHEET

Topic:_____

Your Purpose for Writing This Document (Select only one):

____ To inform   ____ To persuade   ____ To teach a procedure

Primary Users

Reasons for Using the Document (Select only one):

____ To gain knowledge   ____ To make a decision   ____ To learn a procedure

How well can the users read?
illiterate   advanced
_____ / _____ / _____ / _____ / _____ / _____ / _____
Extremely Moderately Slightly Neutral Slightly Moderately Extremely

How do the readers feel about the subject?
unfavorable   favorable
_____ / _____ / _____ / _____ / _____ / _____ / _____
Strongly Moderately Slightly Neutral Slightly Moderately Strongly

If you want to obtain a decision, how would you rate the readers?
opposed   disposed
_____ / _____ / _____ / _____ / _____ / _____ / _____
Strongly Moderately Slightly Neutral Slightly Moderately Strongly

If you write to teach the readers a procedure, how much do they already know about
the subject?
uninformed   informed
_____ / _____ / _____ / _____ / _____ / _____ / _____
Extremely Moderately Slightly Neutral Slightly Moderately Extremely

How will readers use the document? (Select one):
____ Leisure reading, minimal distractions
____ Work-related reading, routine distractions

Briefly describe the typical reader's environment, for example, a crowded workstation
or a comfortable, large desk.

FIGURE 2.3   Continued.

Secondary Users

Reasons for Using the Document (Select only one):

_____ To gain knowledge   _____ To make a decision   _____ To learn a procedure

How well can the users read?
illiterate   advanced
_____ / _____ / _____ / _____ / _____ / _____ / _____
Extremely Moderately Slightly   Neutral   Slightly   Moderately Extremely

How do the readers feel about the subject?
unfavorable   favorable
_____ / _____ / _____ / _____ / _____ / _____ / _____
Strongly   Moderately Slightly   Neutral   Slightly   Moderately Strongly

If you want to obtain a decision, how would you rate the readers?
opposed   disposed
_____ / _____ / _____ / _____ / _____ / _____ / _____
Strongly   Moderately Slightly   Neutral   Slightly   Moderately Strongly

If you write to teach the readers a procedure, how much do they already know about the subject?
uninformed   informed
_____ / _____ / _____ / _____ / _____ / _____ / _____
Extremely Moderately Slightly   Neutral   Slightly   Moderately Extremely

How will readers use the document? (Select one):
_____ Leisure reading, minimal distractions
_____ Work-related reading, routine distractions

Briefly describe the typical reader's environment, for example, a crowded workstation or a comfortable, large desk.

Special Considerations

Note here any special concerns or information related to either audience.

The User's Needs and Purposes. The functional view encourages you to clearly understand the reader's interest before preparing your documentation. Why will the reader use this document? Will he want to read it all? Will she wish to consult illustrations only? Will the reader use this document to install, or to maintain, a dishwasher? Your understanding of the reader's purpose will determine the type of document to produce.

Consider Louise Mitchell's case. Her primary user, the vice-president, will use the document to institute a policy change. Her secondary user, the employees, will use the document to understand how the change compares with the old policy. Now, how should she structure the report? This functional approach allows her, as the author, to think very explicitly about the needs of her two user groups.

The User's Ability to Read and Interpret. If users become your primary audience you should next consider their abilities to understand texts and drawings. How detailed can text be? Should you use polysyllabic words, or basic English? How do readers respond to sentence length? Can they comprehend long blocks of text? Are they visually oriented? Can they understand a diagram, or must it have an explanatory text and markers attached? Remember, in the functional view, your document is more than a text or a book; it is a tool to serve its users' needs. How can they best use it?

The User's Attitude. The worksheet for a functional analysis asks questions that define users' attitudes. For example, if users have little education, they may fear the change your product and its documentation represent. You need to know how the user feels toward your product. Will the user read in order to use a product? You will also want to know the user's time constraints, and whether the user purchased the product for work or recreation. These questions of attitude affect not only the project's content, but its design as well.

The User's Environment. Where will the typical user read your documentation? In the kitchen trying to install a garbage disposer? At a desk trying to select the best insurance policy? At an electronic workstation with limited reading space? Technical communicators are learning to design better documents by studying their users' environments. For example, which is better at a crowded workstation: a hefty reference manual or a pocket-sized reference guide? The pocket-sized communication is the more probable choice. A laminated reference card or a template an operater can place at the top of a keyboard? Again, the question is, "Which tool works best?"

STEP THREE—STUDYING YOUR AUDIENCE

Knowing an audience can earn an employer customer sales and praise, just as poor research can produce an industry-wide fiasco.

Before conducting any formal research, examine your current knowledge of the audience. For example, Louise Mitchell has previously prepared reports for the vice-president. She knows the types of arguments the vice-president wants to study, as well as the design and types of illustrations this crucial report needs to succeed.

If possible, obtain examples of any previous documents your audiences liked. Also, speak with others who prepared documents, designs, and products for these groups. Then examine two types of sources, background materials and user reviews.

Background Sources

Various types of library and commercial sources may help you to fully research your audience.

Proprietary Materials, sources your company owns, may be free. Before starting a project, check with your company's information specialists to obtain similar documentation. Perhaps the files contain writers' records of audience profiles, records of readers' reactions to test drafts, and suggestions for improving the product. Interview marketing and advertising personnel in your firm. Their research on consumers' attitudes and motivations may be useful. Explore this "local" resource in detail since technical writers neglect it most often.

FIGURE 2.4   A Collection of Research Sources for an Audience Analysis.

LEARNING ABOUT YOUR AUDIENCES

Where do professional writers, advertising copywriters, novelists, journalists, or scriptwriters research their viewers' and readers' preferences? Most professional writers use many resources to thoroughly study and analyze their audiences. We list only some here. Can you name others?

Proprietary Materials—A company's or writer's primary research materials, including surveys, interviews, product reports, and user tests.

Reference Materials—Library materials defining specific populations and their interests. Rely on U.S. government publications and the advertising and marketing literature available in business periodicals.

On-Line Information—Information on an audience's preferences available from direct marketing or advertising businesses.

Print Media—Articles in technical and scientific writing and journalism that discuss the requirements of various readers and product users.

Electronic Media—Videos and other audio-visual presentations prepared by your competitors. How do they communicate with their audiences? Do audiences respond well to the messages?

Popular Culture—Print, electronic, and other media such as comic books, cartoons, and posters that transmit information to the public. What styles and approaches seem to work best? Can you imitate them to reach your audience?

Reference Materials are obtainable at any local library through interlibrary loans and electronic media. Again, technical communicators tend to neglect these sources. Chapter 4 explains how to search your local library for audience analysis and consumer behavior sources.

On-Line Information is offered by both public and private databanks and research services. This information is expensive, but you can tailor your search to very specific audience segments. Before purchasing such a service, read the discussion in Chapter 4. Ask for a price list and the names of the databanks the service uses. If you have limited time and proprietary resources to target a very specific audience, on-line information may be a wise investment.

Print Media, specifically, publications of the Society for Technical Communication, the Association of Teachers of Technical Writing, and related interest groups can help you profile your audience. Consult indexes and abstracts that help locate specific sources.

Electronic Media, for example, your competitors' work, should be examined. Try to view videos, training tapes, and on-line documentation that your competitors designed for their current products. How do they speak to their users? Which approaches worked for them?

Popular Culture, such as television shows, movies, magazines, and books that your audience likes can help determine your approaches and rhetorical appeals. Study your audience's popular culture for the designs, retail displays, and other visual cues they prefer.

User Reviews

If possible, have users sample your documentation before publishing it. Your firm might practice a combination of these activities:

Internal review—a reading of your document by selected company employees to confirm its strengths and to suggest improvements. For more details on the conduct of reviews, see Chapter 7's discussion of revising and rewriting.

External review—consultants, experts in content or in publication, can comment on the manuscript and answer specific questions put to them.

Site testing—finished-looking copies of the document are reviewed by actual users and their supervisors, so documents can be used on the equipment, in the surroundings they are written for.

Focus groups—small groups of sample readers study and comment on the document. Writers interview them for suggestions and improvements.

Laboratory testing—testing under controlled conditions in a room designed especially for the purpose. Subjects agree to be videotaped so you can study how they react to the materials, how they handle and read the document, and when they appear puzzled or confused.

You may also test for specific features of documentation. For example, a comprehension test is designed to measure whether readers remember important text features. A performance test measures whether readers can do the task the manual describes. You may test for specific qualities in your documentation: ease of locating specific sections, clarity of specific passages, reaction to colors and designs, or any number of unique characteristics. Although testing is labor-intensive and expensive, it allows you to sample an actual audience. If planning a major collection of documentation, your plan should include testing.

STEP FOUR—CREATING AN AUDIENCE PROFILE

To create a useful audience profile, you need to complete a worksheet, visualize your idea as an object, and obtain supervisor or client approval for your defined audience.

Complete an Audience Analysis Worksheet

This chapter displays two worksheets to use when composing your primary and secondary audience profiles. The first worksheet uses traditional categories of experts, technicians, and lay readers. It asks you to consider your audience's age, sex, education, and attitude, because these factors greatly affect a reader's understanding of your topic. They can also influence the document's design. Older readers will need larger type, and sex may partially determine color preferences.

Traditionally, education is considered the most important of reader characteristics. Level and type of training often determine or predict readers' comprehension of large and difficult words, understanding of complex or abstract concepts, and attitude toward new problems and technologies. Note that this worksheet concentrates on the reader as a member of a category, primary or secondary, and that it asks about the reader's use of the document only after establishing categories. After completing this sheet, you will be able to classify your reader as expert, technician, or lay reader.

The second worksheet emphasizes the functional approach to audience analysis. First, it asks you to clarify your purpose. Next, it offers a set of choices you can make to classify the reader as a user. Note, too, that this worksheet asks you to visualize the reader's environment. Although this worksheet appears easier to complete than the first, both worksheets deliberately lead you through a series of objective choices to a written formulation of your ideas about readers.

Visualize the Idea as an Object or Product

Because you compose for real people, you should now "visualize," or imagine, your audience as real individuals and your manuscript as an actual product. To start, we pose some questions. How do your readers look? What are they wearing? How are they standing or sitting as they read your work? Do they look happy about the product? How do they sound? What questions do they ask about the product? What are they telling you about their interests in the product? What problems are they describing in relation to the product?

Now consider your project. Earlier in the chapter, you, as the author, stated your purpose. How does that purpose look? To assist you, we have included a list of illustrations that correspond to your rhetorical strategies. Now what do you see? Does your project consist of tables or graphs? Will schematics and diagrams be parts of your story? What else can you see? Text with illustrations elsewhere? Text and illustrations positioned to explain each other? Is your text in a book? A checksheet? A reference guide? In short, what type of product is your idea?

Obtain Supervisor Approval for Your Definition

Now that you are imagining your readers and product, remember your project's scope and limitations. We began by stating that you should write one text for no more than two audiences. Which will you address primarily and which secondarily? At this point, it is wise to obtain editorial approval, either from a supervisor if you work for a firm, or from a client if you are an independent contractor.

The method of obtaining approval should depend on your routine with your supervisor. In some cases, a detailed audience analysis worksheet serves as an audience profile. If this is the case, a short conference may serve as an explanation of the analysis. The supervisor can then sign and date the existing form to indicate agreement on the project's direction. For ex-

FIGURE 2.5   How to Visualize for Your Audience.

VISUALIZING THE IDEA FOR YOUR AUDIENCE

Sometimes you can write effectively by "sketching" or composing simple pictures of your idea before writing your draft. Then you can outline a text to accompany the illustrations. Here are some questions to ask about any technical writing assignment. As you answer the questions, list types of illustrations you might select. Can you think of any other types of visual approaches for your readers?

Are you describing a process? If so, consider
    flowcharts
    trees
    checksheets
    exploded views
    cutaways

Are you explaining measurements, trends, or results? If so, draw
    tables
    boxed or screened equations
    various graphs

Are you comparing several ideas, products, or systems? If so, use
    column comparisons
    word pictures
    comparative checklists

Are you describing a mechanism or system? If so, include
    parts lists
    specification sheets
    line drawings of specific views

Are you writing to persuade readers to make a purchase? If so, use
    colored illustrations
    spot color in your text
    tints for emphasis
    a mixture of line drawings and colored prints

ample, Louise Mitchell works directly for the vice-president; a more detailed profile would be wasteful.

If you are working on a new project, or if you are having difficulty limiting the scope of your audience, you should write a one- or two-page audience profile to define and describe your primary and secondary users. This document should explain your rationale for restricting the audience. The Checksheet for Audience Analysis will help you. The memo, letter, or other

FIGURE 2.6   A Checksheet for an Audience Analysis.

UNDERSTANDING YOUR AUDIENCE

Do you completely understand your audience's unique needs? What concerns, biases, and experiences affect their understanding of your document? Ask yourself these questions as you "visualize" your audience.

1. What do your readers know about this topic? How did they gain their knowledge: education, work, recreation?

2. How do the readers feel about this subject or product? Are they hostile or frightened, neutral, or strongly supportive? If they are hostile, how will you treat that bias?

3. In what context will the users read the document: in a crowded office with too little time; alone with no help interpreting the information; standing on a factory, garage, or kitchen floor with no place to spread out a manual; in or under an automobile; in a comfortable setting with plenty of time? What other contexts can you describe?

4. How much of this document will users read? Do you have data available for this question? Will readers look up one fact they need, or read about an entire system?

5. How well educated and how sophisticated are your readers? Most people—even educated ones—avoid reading. What types of appeals will work with your readers?

6. What do these people need to visualize? What do they need to see: Schematics, elevations, tables and flowcharts? What types of illustrations will help them most?

short document you design should answer the questions the checksheet poses. In fact, if you pretend that a supervisor or client is asking the checksheet questions you can create a simple document that clarifies your purpose and audience. Now during a conference, you can both design a communication that delivers the information users expect.

SUMMARY

In this chapter, you learned to define both your and your audience's purposes before drafting a technical document. Then you should complete the four step process of audience analysis, leading to an editor-approved profile of your audience. After reading this chapter, you should be able to:

■ State your functional purpose for a document.
■ State the rhetorical purpose for a document.
■ Limit the scope by defining your primary and secondary audiences.

■ Use the traditional categories to define your audience as expert, technician, or lay reader.

■ Use functional categories to define your audience as users with specific needs and preferences.

■ Research your audience cost-effectively, using a variety of traditional sources.

■ Suggest user tests and indicate a familiarity with the various types available.

■ Create an audience profile either by completing an audience analysis worksheet or by using the audience analysis checksheet to compose a short memo or letter for your supervisor's approval.

ASSIGNMENTS

1 List your writing assignments done during a typical workweek at home, school, or on the job. Using either audience analysis worksheet, describe the audiences. Have you considered your readers' needs? Do any of your typical writing assignments require revision to address their audiences more accurately?

2 Select an assignment from the collection above and prepare a list of suggestions about its approach to an audience. How could the writer use better materials to address the audience?

3 After examining some other textbooks, complete an audience analysis worksheet on what you think is a well-written chapter. Produce for your class a transparency of your worksheet. Be prepared to explain, in a five-minute oral report, why this chapter spoke effectively to its audience.

4 For this assignment, concentrate on your home or residence instead of classes. Examine a technical communication that you have dealt with recently, such as an appliance manual or a policy statement from a bank or insurance company. Perform an audience analysis using the worksheet for functional analysis. How well did the communication meet its purpose? Would you suggest revisions? Write a one-page memo to your instructor that discusses the communication and proposes any revisions.

5 How would you describe your classmates as an audience for technical communication? Can you fill out an audience analysis worksheet for this class?

CHAPTER
3

Planning for Writing

Good technical writing proceeds from a plan. This chapter explains how to plan technical writing projects. After reading it, you will understand how to:

- Pre-write your project.
- Organize your work.
- Write your report.
- Produce your report.

CASE STUDY   Planning an Aircraft Purchase Study

Florida Fruit and Imports, Inc. (FFI), specializes in sales and delivery of citrus and exotic fruit. Its directors recently commissioned a feasibility study. To enhance their firm's competitiveness in the exotic fruit market, the directors are considering the purchase of one or two cargo airplanes. The planes would enable them to ship their products directly from Central America to major supermarket warehouses in the United States.

The company hired Aviation Consultants, Inc. (ACI), to analyze the costs and to make recommendations. ACI assigned Bob Andrews, a retired military pilot and licensed aircraft mechanic, and Ann Middleton, a research specialist, to the Florida Fruit project.

As a team, they studied Florida Fruit and Imports, researched transportation options, and composed a feasibility study. Their systematic approach involved four phases: pre-writing, organizing, writing, and producing. In their pre-writing phase, Ann and Bob planned their illustrations; designed proper charts, graphs, and maps; and delivered a professional report by the contractual date.

COMPREHENSIVE PLANNING DEFINED

Ann Middleton's and Bob Andrews' task, the composition of a feasibility study, is one any technical writer can expect, completion of a lengthy document. This chapter explains how comprehensive planning helps define this task and improve time management skills.

Comprehensive planning is a process by which the technical writer defines the scope of the activities needed to accomplish a goal or complete a writing project. As Figure 3.1 indicates, comprehensive planning is more than outlining, researching, and writing a report. Comprehensive planning, the logistics of technical writing, requires use of a timeline so that the writer, mixing production and editorial decisions, develops and completes the project on time.

To the comprehensive planner, a written report is the result of outlining; researching; writing; editing; selecting typefaces, illustrations, and paperstock; packaging; distributing; and more. The comprehensive planner combines these editorial and production decisions on a time line in order to deliver a finished, coherent package. The drawings explain the text that the typefaces enhance. Planning comprehensively, the author divides the writing process into four phases:

☐ Pre-writing
☐ Organizing
☐ Writing
☐ Producing

The complexity or number of steps within each phase depends, among other things, on report length. A two-page status report may have one or two steps within each phase. A 32-page report may have 5 or even 10 steps within each phase. Figure 3.1, The Process of Comprehensive Planning, shows the process used to create an Aircraft Purchase Study.

PRE-WRITING

Pre-writing includes most activities that define the project and limit its scope. When Ann and Bob received the Aircraft Purchase Study assignment, they performed six distinct pre-writing steps, listed as action statements.

Assess Capabilities

In this first step, the writers briefly assessed their own capabilities and assigned tasks. Ann would design and develop all mathematical figures, graphs, and cost analyses. Bob, a former pilot, would research all suitable aircraft, establish comparison points among the airplanes, and gather computer data for analysis in Ann's research department. They assigned the writing based on each writer's capabilities: Bob the plane descriptions and other transportation options, and Ann all explanations of research methods and computations.

They then conducted an informal assessment of their organization's capabilities. With the company's adequate research library, they could gather and verify all their data to enhance the report's persuasive style. The company also had in-house printing and binding capabilities. Ann and Bob discussed with their printer how to print the report, how to produce illustrations, what paperstock to use, and how to bind and present the report. The two authors also decided to design all artwork with graphs, simple maps, and tables. This art could then be reproduced on slides for presentation at Florida Fruit and Import's annual meeting.

Gather Information

After dividing the work and scheduling assessment meetings, each writer worked independently. Bob called the aircraft companies under consideration to arrange site visits and obtain specifications. He listed anticipated problems in recommending specific aircraft. For example, all firms indicated that delivery of a new aircraft would take six months to two years. Besides listing all problems that might limit his recommendations, Bob also listed methods he would use to derive calculations and data for Ann. Ann, on the other hand, had simpler tasks during this stage. She listed the fruit company's preferred financing strategies, selected an appropriate analysis procedure, and awaited Bob's data on his selected aircraft's capabilities.

During this preliminary data-gathering stage, particularly when working with a co-author or a group, it may be helpful to "brainstorm" the topic with other colleagues. Technical writers are seldom asked to write from imagination. Usually, they write on subjects familiar to them through training or experience, or they write after gathering information from a variety of sources: technicians or experts, library research, surveys, interviews, observation, and data analyses.

However, "brainstorming"—talking over a topic with associates to generate a list of every possible facet or avenue of approach—is invaluable. Brainstorming ensures that no profitable evidence or proof is overlooked.

If you work alone on an assignment, take time before you begin to list every subtopic imaginable, every possible pitfall, every source of information. Don't forget that writing these reminders is itself a form of investigation. Reminding yourself to check experienced pilots for their impressions may remind you to check experienced ground crews as well, so you can include any unusual maintenance or repair problems associated with the airplanes.

FIGURE 3.1 A Flowchart Shows the Steps Involved in Planning for a Report.

THE PROCESS OF COMPREHENSIVE PLANNING

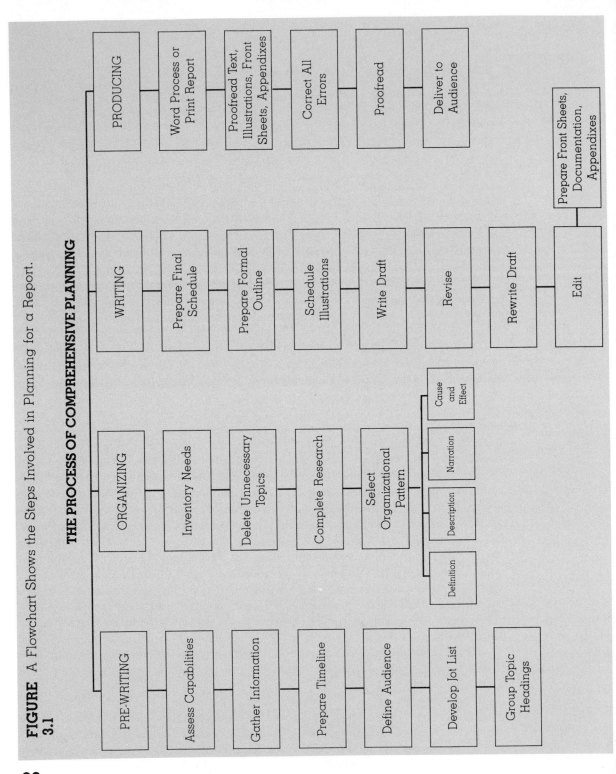

Prepare Timeline

Bob and Ann scheduled the writing and production of the report at their second conference. Since they had 30 days to return to FFI's executives, they created a tight schedule. First, they listed milestones, or major deadlines, then grouped lesser deadlines within each milestone. The initial list was in reverse chronological order. Note that the list of summary deadlines includes preparing to write, composing the report, editing, and production. Preparation of a time line allows a report writer to think through the whole report preparation process.

A timeline is helpful when a report with illustrations must be delivered on a tight schedule, and absolutely necessary when several reports are in preparation simultaneously. We recommend using a wall calendar, or a tack or magnetic board for overlapping deadlines. The wall calendar permits scheduling long-term projects and last-minute requirements. Also, when multiple authors share report writing, a calendar of detailed deadlines and milestones apprises each writer of the group's progress and specifies each writer's responsibilities. Although researching, organizing, and writing are time-consuming, so is the actual production of the report. Remember, most late reports are so because writers failed to anticipate adequate production time once the report is written. A timeline apprises a writer of the "lead

FIGURE 3.2	A List of Summary Deadlines Shows How to Plan the Writing of a Report.

SCHEDULE FOR AIRCRAFT PURCHASE STUDY

July 31	Present finished report to executives at headquarters. Deliver bound copies of report to the company.
July 29	Travel to FFI's headquarters to set up meeting—allow one extra day in case of delay or emergency.
July 27	Deliver all copies of bound report to Ann's office. Deliver all audio-visual aids to Ann's office.
July 22	Complete final proofreading of report—last chance to correct errors.
July 18	Proofread typed manuscript and illustrations.
July 15	Finish manuscript text and deliver to word processor for final preparation.
July 10	Finish text for all illustrations, complete list of illustrations, and deliver all materials to graphics or artist for production. Begin to write report.
July 9	Complete organization phase of report at final planning meeting.
July 5	Prepare timeline.
July 1	Assess capabilities, scope of report, anticipated problems.

time" for each production stage and permits a writing team to assist should production slowdown occur at any phase.

Define Audience

Audience analysis should occur during the pre-writing process because clearly defining an audience saves research, rewriting, and illustration time. Bob and Ann used the audience analysis techniques described in Chapter 2. Their readers are no more than 20 North and Central American males, many with Masters of Business Administration degrees from American universities. Most know the fruit and grocery brokerage businesses, but little of aviation and aircraft. Bob's and Ann's audience analysis determined their writing style: relatively complex sentences, numerous figures, and detailed costs analyses.

Develop Jot List

During this phase, gather any other ideas that have come to mind since the earlier "brainstorming." At their second conference Ann and Bob added such topics as the company's image, its managers' insights into their operations, and FFI's needs assessments and projected growth. Their list contained 50 items and ideas, categorized by headings and related ideas. After this session, their preliminary research began.

Group Topic Headings

To save time in your library, develop a list of topic headings at your second pre-planning session. In Ann and Bob's Aircraft Purchase Study, several types of topic headings evolved: airplanes the writers had in mind, concerns and problems Ann and Bob wanted to address, and process descriptions such as maintenance, fueling requirements, and distance. Grouping ideas narrowed their focus and allowed them to pinpoint specific information quickly.

Perform Preliminary Research

During this step, Ann and Bob read general information about their topic. They noted references to cite and eliminated those that were too detailed, general or outdated. They sought aviation experts for suggestions. Remember, your first research session is an opportunity to compile a working bibliography of written materials, experts, site visits, and other sources. More detailed material will be extracted during the next phase of your comprehensive plan, the organization phase.

ORGANIZING

Because it is so critical to a project's success, we devote Chapter 5 to methods of organizing technical writing. For now, though, organization is that phase of the comprehensive plan involving the decisions the author must make before completing the preliminary research and before starting the actual writing.

| FIGURE 3.3 | A Collection of Grouped Topic Headings Shows How Ann and Bob Thought Through the Problem of Their Topic. |

GROUPED TOPIC HEADINGS

Group 1—Airplanes and Their Characteristics
    manufacturers—Boeing, McDonnell Douglas, Grumman, Shorts, others
    loading facilities including palletized, bins, direct loading
    refrigeration facilities—freezer deck, refrigerator deck
    cargo door installation and servicing
    ground facilities including landing sites
    maintenance
    air conditioning
    availability and scheduling

Group 2—Process and Concerns to be Studied and Explained
    ground transportation and air costs compared
    time saved
    safety factors
    feasible delivery dates
    maintenance requirements—costs, facilities
    inventory control
    refrigeration
    cargo capabilities
    expansion of domestic travel
    expansion of international travel
    cost recovery
    purchase options
    rental options

Inventory Needs

During this stage, a writer lists those items, materials, and processes necessary to finish a report. For example, Aircraft Purchase Study included maps of selected ground and air routes and specification drawings of cargo areas. The graphics department received a list of all non-tabular illustrations so they could develop appropriate diagrams and charts. To fit the report's trim size, lengthy tables were composed and their allotted space in the final report planned. The final cost of the manual production of all these computer results, too, was planned. This inventory is an author's last chance to discover production problems and to contract for special equipment or outside professional skills.

Delete Unnecessary Topics

At some stage, an experienced writer usually eliminates about half the brainstorming ideas from the pre-writing phases. For example, in Aircraft Purchase Study, Bob and Ann decided to recommend purchasing two airplanes designed for both domestic and intercontinental shipping. They discarded any unrelated ideas, facts, or notes. You may find in your own writing that having defined your audience's needs, you want to delete irrelevant, unnecessary topics or details. Your financial officer, for example, probably does not need a review of leasing contracts.

Complete Research

Plan for a second research phase, more directed and specific than your first. Bob made site visits to inspect production and determine once again the availability of all airplanes under consideration. Ann began computer analyses of Bob's statistics for ground maintenance, flight distances, and delivery of perishable fruit. These figures were then verified in the ACI's library and with the aircraft companies.

Group-Related Topics

At this point, Ann and Bob began outlining their text. Formal outlines of reports should be dictated by rules of formal report organization, but a report's "internal outline" should begin to take shape at this stage. The internal outline of their aircraft purchase report required Ann and Bob to (1) discuss the fruit company's present travel situation; (2) analyze the company's predicted travel needs for the next 10 years; (3) explain their analysis of the most suitable planes' characteristics; and (4) show figures indicating which two airplanes to purchase. Ann and Bob then wrote a thesis statement to introduce and summarize their main ideas.

Select Organizational Pattern

Four patterns of writing are generally available for technical reports: definition, description, narration, and cause-and-effect reasoning. After analyzing their outline, Ann and Bob indicated their preferred method for each section. By doing so, they assured overall continuity of tone and similarity of style. In their report, Ann and Bob decided on a narrative pattern wherever possible for discussing Florida Fruit and Import's past needs and current problems. They chose definition and description for evaluating the selected airplanes because these two patterns simplify summary tables and graphs for the reader.

DRAFTING THE REPORT

Writing is more than typing a draft. Writers must also compose for illustrations, prepare front sheets, and documentation, and compile appendixes as needed. Because writing is central to every phase of a project's production, your first draft should begin with preparation of a final schedule.

Prepare Final Schedule

Figure 3.2 shows that the authors of Aircraft Purchase Study allowed 5 days for drafting the entire report, and 17 for writing, proofing, printing, proofing, and delivery. When the writing phase begins, compare your real progress with your timeline's estimate. Have you run into problems obtaining information? If so, can you write "around" the missing information and allow for its last-minute insertion? Consider all probable difficulties at this point so you can adjust your schedule. Make certain you have a back-up production system in case your planned system becomes unavailable. You should have an alternate word processor and print shop available if necessary, so you can deliver your report on time.

Prepare Formal Outline

A formal topic outline eventually becomes part of a table of contents. Each section of your formal outline indicates a major discussion in your paper. Subsections provide the details and descriptions necessary for a clear explanation of that section's main idea. The outline for Aircraft Purchase Study follows:

Introduction (descriptive abstract)	International Flight Needs
Summary (informative abstract)	Local Flight Needs
General Discussion	Recommendations
Recommendations	General Transportation
Aircraft Description	Aviation
Discussions	Conclusion
General Discussion	

Note, this outline uses generic names for parts of technical reports as titles for report sections. You may find that a client—or an instructor—prefers more specific category labels. These decisions depend upon the reason why you prepare an outline and the degree of formality desired in both an outline and table of contents. In the aircraft study, because the fruit company wanted to ensure confidentiality, the authors worked from a general, more formal outline, helping keep their information secure.

Schedule Illustrations

Note that their calendar requires Bob and Ann to complete the text for all illustrations and to draw up a list of illustrations on the first scheduled day of writing. Assuming your research is thorough and your understanding of the audience complete, you should plan for illustrations as you planned your text's length and detail of presentation. Writers in large companies with graphics departments will enlist experienced graphic artists. Other writers, and probably most students, will prepare their own graphics. Therefore, learn enough about graphics, then, to either prepare them efficiently yourself or to communicate knowledgeably with graphic artists. (For more details, see Chapter 8.)

For planning purposes, you should define an illustration as any words, lines, or numbers not arranged line for line in text or in rows and columns. For example, Figure 3.1 is an illustration because it is a process chart which uses boxes and arrows to indicate development of an objective's conclusion. Boxes and arrows require an artist, graphics specialist, or a word processor equipped with these limited graphic functions. Aircraft Purchase Study required 16 appendixes with graphs, 8 charts in text, 3 graphs in text, and 2 maps. For each item, Ann and Bob prepared a sheet of plain paper with all that drawing's text and numbers. Next they attached a second sheet with a rough sketch of the drawing. The general rule is the more illustrations, the earlier you must schedule them. Preparing your test list and sketch for each illustration immediately after completing your outline allows the art department to illustrate, while you continue to write text.

WRITING THE DRAFT

Writing your report should be relatively simple using your detailed outline as a starting point. If you do not know your writing speed, research may help. Studies indicate that professional writers, such as journalists or technical writers, average about 1,000 words per hour, or roughly four pages of double-spaced, typed text. With an outline in hand, and notes organized for reference, you should write 500 words per hour, or two pages of double-spaced typed text. This is about the average speed for an intermediate level, college writer.

At this point in the writing process, many experienced writers find that careful planning and meticulous work leave them with a virtually complete draft already in note form. Some experienced writers use large note cards for individual paragraphs, sentences, or ideas gathered from their research. For these writers, organizing the report means sequencing their cards suitably for their audience. Producing the first draft means revising these notes into a coherent sequence, linking sentences, providing transitions, and revising sentences for logic. To keep this draft physically manageable, headings and subheadings can be written at the top of a card, or on a separate card inserted into the sequence at the proper place. Some writers are so proficient with this technique that a typist can eventually work directly from their cards.

Other writers prefer a technique called a "free-write." A free-write involves producing a draft as quickly as possible, without consulting notes or checking correctness. Simply sit and start writing as quickly and as long as possible. Rely on memory, imagination, and perhaps a quick once-over of your notes before beginning. Later, when you correct your writing, you can insert information from your notes more precisely. Experienced writers often use this technique for short writing assignments requiring a minimum of research, or for drafting a troublesome assignment. The resulting "zero draft" often provides an approach or a sequence suitable for a more complete first draft.

There are, in fact, an infinite number of ways to produce writing. A writer may be able to produce a final manuscript of a short memo in one draft, but too many people expect the same perfection from drafts of longer assignments. Your final version should have at least one—or better, several—initial drafts. A distinguishing characteristic of good writing, no matter how it was produced, is that your audience sees it only after several revisions.

The methods recommended here for pre-writing, producing drafts, revising, and editing are only some of the ways to organize the writing process. But if you are a relatively inexperienced writer, try them. Then experiment with other methods until settling on your own comfortable writing process.

Producing your first draft can be more pleasant if you plan ahead. First, practice thinking with a keyboard so you can type first drafts. Use double or triple vertical spacing and 1½-inch top, side, and bottom margins to leave room for your revisions. On a word processor, double-check that your draft is safely stored in memory for later editing and rewriting. If you use a typewriter without storage functions, use carbons so you can produce two copies of your first draft. Carbons are still more cost-efficient than copying machines—especially since your first draft need not be a finished report. An extra copy of your first draft saves time in rewriting since you can use the extra copy for "cut and paste" revisions.

If you cannot "think" at a keyboard, consider learning. A keyboard is faster and less demanding than a pen or pencil. If you write your drafts longhand, write clearly and legibly on every other line of a sheet of lined paper. You will need the space for revision. Write on only one side of a sheet. If a draft is handwritten, have it typed or type it yourself, before the revision stage. Typed copy is much easier to rewrite and edit.

Try to write quickly and easily without worrying too much about grammar, spelling, or usage. Don't stop to correct typos. You can make corrections after completing a first draft. If you forget to add information from your notes or calculations to the report, add them later, in your second draft. Instead, mark your manuscript with a title indicating the missing information and continue writing. Again, the object of the first draft is a working "hard" copy, not a perfect report, that you will later revise, edit, and rearrange for your final report.

While preparing a draft, skip a line or two wherever you intend to insert an illustration. In the space, type "Insert Chart 3," or "Insert Figure 2," whatever you call your rough illustrations. It is important to note all illustrations in your rough draft because you may not see finished illustrations until the graphics department needs them proofread. You need to keep your text and your placement of illustrations "tied" together. This problem is avoidable by placing extensive illustrations in appendixes, as in Aircraft Purchase Study.

First, revise for spelling, punctuation, and word choice. Then review the entire manuscript for mistakes in verb tense agreement, noun pronoun agreement, and other points of literacy. On your third review, examine the content for accuracy, logical presentation, and style. You may decide to add or delete more material. Whole sentences can be omitted, and most tightened. Revision can take as long as writing and, in fact, takes up to 45% of the time professionals allot to the entire process. If you are an inexperienced writer, a good rule of thumb is to spend half your time writing the first draft and half revising it. If possible, have a competent person assist with revision. If you co-author a report, remember that co-authors customarily revise or comment on each other's work. This process saves rewriting time and usually improves the finished report's quality.

Rewrite Draft

If you had to make only grammar, spelling, and punctuation revisions, this step may be unnecessary. However, most authors make stylistic and sequence changes, or make last-minute insertions of new or previously unavailable information. For example, in Aircraft Purchase Study, the information and calculations from site visits were unavailable when the first draft was written. These results significantly affected the draft, and an entire section, almost a second draft, had to be written. If you write a second draft, its style and tone should approach those intended for the published report. Work quickly but concentrate on style and integration of all changes from the first draft. Decide where illustrations will appear and mark these places in the text. If possible, note the lines of vertical space the figure will occupy.

Edit

Again, read your report three times:

first reading—spelling, punctuation, literacy;
second reading—accuracy, logic, style, or readability;
third reading—inclusion of all last-minute material and changes.

Editing is so critical that we devote a handbook section to a discussion of its essential skills.

Remember, the steps of writing—drafting, revising, editing, proofreading—are not necessarily linear. They may neither always occur in the order listed here, nor as separate steps for every writer. Don't hesitate to add an entire sentence, even an entire paragraph if you must during the later editing stages. But remember, doing so may alter your production schedule. Many famous writers, for example, James Joyce, continued revising even while the printer was setting type. His works are still not in his intended final forms. Our model, then, is just that, a model of the process.

Prepare Front Sheets

The Aircraft Purchase Study had eight front sheets and a cover letter. Writers often forget that the front of the report, a reader's introduction to a work, is psychologically strategic to the report's success. As with illustrations, prepare samples of the title page, distribution page, table of contents, list of charts, and list of appendixes for your typist or printer to follow. The front sheets package your report, so show your printer exactly how things should look. The writer should also prepare the documentation in complete detail for the typist or printer. All references should be arranged neatly on paper so that the printer or typist must only prepare them according to the writer's instructions.

PRODUCTION

In the Summary Deadline for the Aircraft Purchase Study, the authors planned to produce the finished report in 12 days. Production accounts for nearly half the scheduled time because transportation of the manuscript between the printer and writer for proofing, correcting, and printing may require whole days. As more people become involved in manufacturing a report, more time is needed. If a weekend interrupts the schedule, two days should disappear from the timeline. Also, the print or graphics shop may experience broken equipment or slow-downs due to an employee's illness. The writer may have no knowledge of such situations.

Reports are late because writers fail to observe necessary amenities as part of the planning process. First, meet the person who will supervise or print or type your report. Second, specify and include as a deadline a report's usual lead time. Third, tell the supervisor what the whole job will be. From the printer's point of view, a 32-page report of straight text, no illustrations, and a few columns of data differs substantially from a 32-page report with approximately 25 illustrations. The former requires a competent keyboardist, a person to duplicate the report, and another to bind it. The latter requires an artist or computer graphics, or perhaps both.

The printer must understand your needs and your schedule, so all functions of production can be provided as you requested. Finally, you can assist in production by attaching to the draft a separate sheet of paper with written instructions for the printer or typist. If your instructions appear on a report page, the keyboardist may accidentally set your instructions as part of the report. Your written instructions should always specify the mutually agreed upon date when you can expect the finished report.

Type Report

During the first conference with your printer, you need to decide whether to type or print your report. Reports of 50 pages or less are generally typed, and their illustrations use the same or a similar typeface for continuity. When a report approximates monograph length (75 to 100 pages), consider typesetting or offset printing. Again, limited distribution reports such as Aircraft Purchase Study are usually typed, while for cost-effectiveness and appearance, mass distribution reports are printed. Before final production, a professional typist or printer should supply you with a copy to proofread.

Correct All Errors

Do not assume that the proofreading done in the graphics shop or the word processing center will catch all errors. You, as the writer, must proofread every letter you have written. Even though others may assist in preparation, the responsibility is the writer's. This responsibility includes figures in both text and illustrations, words in illustrations, tables, appendixes, references and front sheets. You should proofread your report three times, once for accuracy of meaning, once for errors in type and space, and once for any errors missed in the first two readings. If someone else can also proofread, let them. It may be useful to read the text aloud, or even sentences one at a time, backwards from the end of your text. Both techniques help you to spot possible errors. Make a list of errors and of any corrections to be inserted and return both the list and your report to the printer or typist for revisions. Be sure to keep a copy of your error list for yourself.

Proofread

Upon your next receipt of the document, first proofread for the errors on your list. Next, slowly read the whole report, checking for missed or new errors. Return the report to the printer or typist with a written statement that the report is ready for duplication.

Deliver to Audience

Because this planning stage is so obvious, it is often one of the most overlooked. The authors of the Aircraft Purchase Study delivered their report in person, not only because the company wanted a traditional presentation at a business meeting, but also because they wanted to be certain that the final copies arrived in excellent condition.

Because of a preoccupation with production, delivery of the report may seem unimportant to the writer. But to the client, your audience, delivery is one of the most important aspects of the job since it initiates the client's contact with your company's product. Discuss report packaging with your printer or a good graphic artist. If the United States Postal Service delivers the copies, certain requirements must be met. Still, an attractive cover and binding that complement your letter of transmittal can enhance your firm's image. You simply cannot allow delivery of a damaged or dirty report. Good paper and competent wrapping and shipping can eliminate these problems.

CHARACTERISTICS OF COMPREHENSIVELY PLANNED REPORTS

Each technical report has a life of its own, created by the audience's special needs and the writer's subject matter. However, certain traits distinguish a well-planned report; it

☐ Arrives by the due date.
☐ Arrives attractively packaged and undamaged by shipment.

☐ Invites the client to read because the front pages are neat.
☐ Files easily because the title is readable.
☐ Contains lists of contents and illustrations in the front pages.
☐ Contains a factual summary at the start.
☐ Reveals a plan of organization through the table of contents.
☐ Uses visible sequence markers in text so readers can find and read only the sections that are of interest to them.

PLANNING FOR SHORT WRITTEN DOCUMENTS

This chapter emphasizes comprehensive planning for long reports because they contain many of the types of problems encountered daily in smaller technical writing projects. In condensed steps, the comprehensive planning model also applies to shorter forms.

You can use the planning model for short documents such as letters, memos, short progress reports, short proposals, public relations materials, and short articles for house organs or professional journals. Planning removes the mystery from writing and the use of planning statements like Figures 3.1 and 3.2 can serve as useful guidelines when you supervise a writing project.

SUMMARY

In this chapter, you learned about comprehensive planning by studying a case in which two authors collaborated to prepare, deliver and present a 32-page report in 30 days. Based on your reading, you should understand:

■ How to assess your own capabilities as an author or co-author of a technical report. What capabilities do you have? What capabilities do you need?
■ How to assess your college's or business library's capabilities. What services do they offer?
■ How to prepare a timeline or schedule similar to Figure 3.2 for a report you may write.
■ How to perform preliminary research on a report topic and to compile a list of sources you will consult to research the report.
■ How to compile a jot list of topics your report might consider.
■ How to read a technical report in your area of interest and list the problems you think the authors encountered in the pre-writing and organizing stages.
■ How to evaluate in your own words a technical report in your area of interest.

ASSIGNMENTS

1 Make an error list while proofreading the following paragraph. Proofread it once by yourself, then again with a classmate. Which method is more thorough?

You should not assume that the proofrreading done in th graphic shop or the word processing center will catch all errors. You as the writer must proofread ever letter you

FIGURE 3.4 A Flowchart Shows How to Plan for Short Documents.

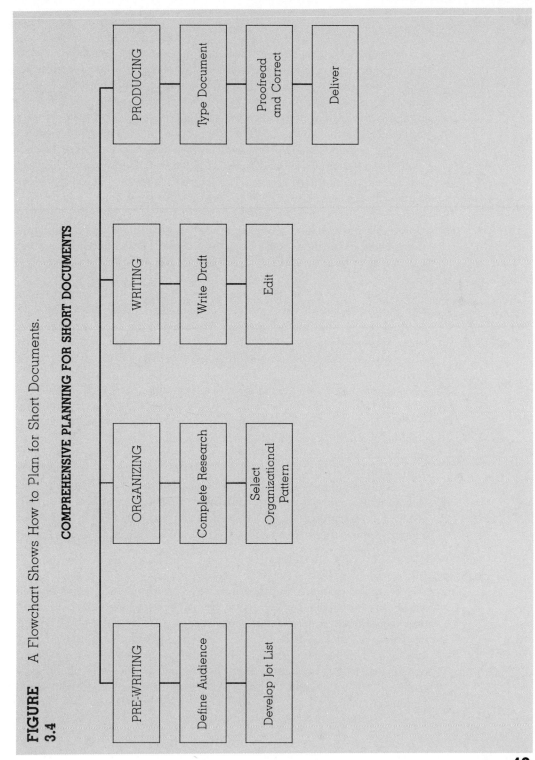

COMPREHENSIVE PLANNING FOR SHORT DOCUMENTS

PRE-WRITING
- Define Audience
- Develop Jot List

ORGANIZING
- Complete Research
- Select Organizational Pattern

WRITING
- Write Draft
- Edit

PRODUCING
- Type Document
- Proofread and Correct
- Deliver

have written. Even though others may assist preparation, the responsibility is the writers'. this responsbility includes figures in both text and illustations, words in illustrations, tables, appendixes, references, and front sheets. You should proofread you report there times, once for accuracy of meaning, once for errors in type nd space, and once to cathc any errors you any have missed on the frist two readings. If someone else is avalable to proofread it after you finished, let them proofread it too. You may find it useful to read the text a loud, or even to read the sentences one at a time, backward from the end of you text ; Both these techniques help focus you attention on possibel errors. Make a list of errors and of any correction to be isnerted, and return both the list and you report to the printer or tpyist for repair. Keep a copy of you error list fro youself.

2   Browse through your library's collection of corporate annual reports, the sample reports in this textbook, or those your instructor provided. Which include illustrations in the text? Which place illustrations at the end, in appendixes? Which seems the better approach?

3   Draft a short memo to your instructor. Assume that one of the reports you examined, one with the illustrations in the appendixes, is actually a preliminary draft. You have now decided that the final draft will run the illustrations in the text. Your memo should detail where you would insert the illustrations. Or you may photocopy the report and indicate, with revision comments between the lines or in the margins, where you would insert each illustration.

4   Assume you are the director of the writing team that produced the report you chose for #3 above. How long do you think the writers worked on the report? Find the publication or transmittal date of the report. Working backward from that date, create a timeline that includes all the steps you think were necessary, and allow what you think is a reasonable time for completing each step. Does the complexity of the task surprise you after reading this chapter? Draft a memo presenting your timeline to your instructor. Assume your instructor is the supervisor who gave you this assignment.

5   Here are some informal abstracts students wrote to explain and obtain an instructor's approval of the oral presentations they would later give in class. Using the techniques in this chapter, improve these drafts and submit the revisions to your instructor for approval.

OPPORTUNITIES FOR BLACKS IN TEHNICAL WRITING

Technical writint is a vastly growing and promising feild which few black college students persue. Technicel writing is in the process of conveying scientic and business-oriented infromation in a cleer and concise manner   This field would purmit Black colege studnets with strong literey or scientific backgrounds to use thier expertize toward writing such materials as goverment documents pamphlets, instructiion manuals and newsletters. The presnatation discribes a detailed process of applying for study in technical communication. Infromation regarding types of programs and potential employer is included.

THE MINI TRUCK IN AMERICA; AN AUTOMOTVE CULT IS BORN

The first compact pick-up trucks reeched Amerca's shores in the late 1950's and despite there lowely beginings (early sales were pathitcally slow), have consistenly increased in popularity as a mini-pickup "cult rose amony U.S. car buyers. This presnatation will provide historicel perspectives on the trucks themiselves the manufacturer's marketing stratagies and the "cultists" who drive them. The discusion will feature technical informatiaon on the changes the trucks have undergone over the coarse of the last 20 years, and lite hearted commentary on the "cutlists" who buy them, and on the manufacturers" various marketing strategies. The truck enthusaists will gain a historical perspective on their vehicles (or find out what make, model, and year they own,) and general automotve enthusaists will enjoy learning about these economical 'off beat' and fun little vehicles.

DEALING WITH THE PRESENT; ERASING SEXISM FROM COLLEGE TEXTBOOKS

The editor/writer needs to be aware of sexist language a method of dominatiaon and discrimination as archaic as quil pens, this presentation discussses how language expression illustrate thinking, ergo if sexist language is spoken sexist loanguate is though. The presnatation explains methods of dealing with sexist language in industry and other ares, apllying these methods to colleg textbook publication

CHAPTER 4

Information Gathering

This chapter introduces methods of collecting information so that you can write successful technical reports. The chapter focuses on these information sources:

- Personal experience.
- Library materials.
- Informative interviews.
- Questionnaires.
- Letters of inquiry.
- Organizational records.
- Site observation.

CASE STUDY   An Assigned Research Topic

Steven and Elaine Miska own and operate a small construction company with a successful record in concrete and insulation installations. Operating in suburban areas of a fairly large city, they have built over the last five years a substantial client list, including light industry and government contracts. However, expecting a slowdown in the light industry and residential markets, they must seek new prospects.

Over the last few months, they have monitored the progress of a county ordinance aimed at controlling noise pollution in their city. During the rapid growth in the 1960s and 70s, the ordinance permitted light industry and expressway construction near residential neighborhoods. When this bill becomes law, the Miskas realize that a new market for sound-buffer walls along highways and industrial areas, as well as new markets for insulation against noise, will be created. They have been developing package plans for retrofitting industrial buildings with a new kind of insulation. Their experience in concrete construction may help them land contracts with developers who will be required to provide buffer walls before building along highways. Steven and Elaine want to survey their previous clients' needs and interest in both these possibilities.

The Miskas may be eligible for government loans to cover some of the startup costs. To qualify, they must write a statement of interest that includes a client (market) survey as well as a documented statement of the community's need for their products. Although they cannot afford a consulting firm, the Miskas have access to a good suburban library and their district community college library. They have asked you to help them research the general topic of noise pollution, to relate it to the local legislation and their clients' needs, and to document these needs. Relying on your research, you will compose a feasibility report that will also be a part of the Miskas' loan application.

The Miskas, tough survivors in a very competitive construction business, must now become successful researchers. They need to prove to their audience — a loan officer — that not only their experience and intuition but also their facts, research, and analysis of published data will enable them to prosper.

Many technical writing students, and even some professional writers, seem to have difficulty using library resources. This chapter, then, focuses on checking your topic's suitability and on getting the most from your college or public library resources. Other resources, such as interviews and questionnaires, are important but highly evolved information-gathering tools. We recommend that beginning and intermediate technical writers avoid these methods at first, until you can receive advanced training in survey and interviewing methodology.

To the beginning researcher, the library offers many advantages. The most important is its publications, available either for free or a minimal charge. The information can enhance your knowledge of a subject or meet your employer's or client's knowledge needs for a given subject. When you use the library to complete research projects on time, you become a cost-efficient evaluator of existing literature that can be used in your employer's research, development, or sales.

CHOOSING YOUR OWN TOPIC

In the working world, people usually write reports about their work, write proposals to secure more work, or update work-related records. Only in a writing class do we ask people to select a topic for a research paper or a feasibility study. Students sometimes say that selecting a topic is difficult or impossible. Often they feel compelled to choose a topic that interests the teacher, ignoring their own interests and needs.

These concerns are real. As a writer, you must make the topic as interesting as possible, and always consider the reader's preconceptions. Also, that initial feeling of incompetence is sometimes overwhelming. Just remember, you learn more when you write about a project. Every writer starts out somewhat ignorant about his or her topic. Through research and reading, the writer prepares to report on the subject at hand. This chapter illustrates the topic problem two ways: by supplying a list of issues to investigate and develop into report topics and by illustrating how a topic might be assigned at work. If you need a topic for your paper, choose one from our list and develop it according to the guidelines in the next section.

 FIGURE 4.1   The Process of Research.

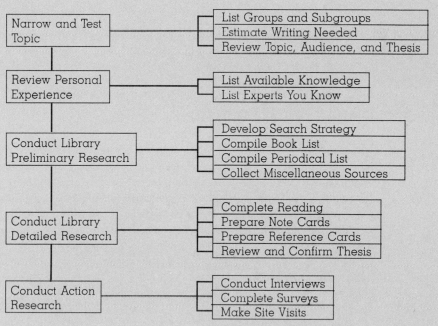

YOUR RESEARCH STRATEGY

The process of research follows the five clear steps shown in this flowchart. Use this diagram to check your research process, and modify the sequence of tasks to suit your requirements.

Narrow and Test Topic	List Groups and Subgroups
	Estimate Writing Needed
	Review Topic, Audience, and Thesis

| Review Personal Experience | List Available Knowledge |
| | List Experts You Know |

Conduct Library Preliminary Research	Develop Search Strategy
	Compile Book List
	Compile Periodical List
	Collect Miscellaneous Sources

Conduct Library Detailed Research	Complete Reading
	Prepare Note Cards
	Prepare Reference Cards
	Review and Confirm Thesis

Conduct Action Research	Conduct Interviews
	Complete Surveys
	Make Site Visits

| **FIGURE 4.2** | A List of Some Contemporary Issues in Technology and Science—Possible Topics for Your Research and Paper. |

ISSUE TOPICS FOR LIBRARY RESEARCH

☐ Nuclear energy—cost-effective energy option?
☐ Genetic engineering—discuss one aspect—legal, biological, medical
☐ Pollution control—toxic waste sites, acid rain
☐ Business outlook in—a field you are studying or would like to study
☐ Alternate technologies for automobiles—propane, hydrogen, and others
☐ Home computer purchase—justify your choice of one computer, defend its cost, and design its payback schedule
☐ Alternate construction technology—design and report on a family home made from adobe, designed for solar, or an alternate construction method
☐ The advantages of passive over active solar heat
☐ Population engineering—the concept, the problems
☐ The census and its annual revision—one implication of the data
☐ The next generation of electronic equipment in—office machines, stereo equipment, television
☐ Cable television or satellite dish—which is a better buy for the homeowner?
☐ Nuclear medicine—developments of the last five years
☐ Occupations of the future—define and explain one emerging occupation
☐ Agriculture as a business—what can farmers in the U.S. do to stay in business?
☐ Corporate mergers—business of the future or anti-trust violators?
☐ The physiology of aging—what is the latest research? What can be done to keep the elderly healthy?
☐ Personal investments—design an investment package for a two-income, childless couple
☐ Risk management—what benefits will employers of the future offer?

NARROWING AND TESTING YOUR TOPIC

We now need to condense the topics listed in the first section to a 15- to 20-page paper. Narrowing a topic to your specifications is part art and part training. Your topic must be broad enough to interest a well-defined audience, yet specific enough that you can provide enough detailed and pertinent facts and research materials to have it engage your audience. For example, you may decide that pollution is a good topic. Rather than start in the library, your own list of facts about pollution can serve as step one in narrowing your topic. This first list might emerge:

Pollution:	**Ideas:**
air	chemical
water	high altitudes
soil	who pollutes most?

industrial	noise—cars, trucks, loud music, lawnmowers
urban	Do the data vary? How many people die from pollution? Relationship between pollution from automobiles using leaded gas and cancer? Unleaded gas?
rural	How does the local factory get away with violating the noise ordinance? What do fertilizers and pesticides do to the water?
local	Bad smelling air when the window is open late at night
	120 dB—painful, 130 dB—damaging

To narrow the topic, list your related thoughts and feelings, as well as any facts or ideas that come to mind. The list can be the free-form jot sheet of whatever comes to mind or a rough set of labeled columns. You can then use the classification process to narrow your topic. To do so, the writer lists, for example, pollution as the central topic, then creates the groups of ideas relevant to the topic.

Central Topic:	**Groups of Locations:**	**Subgroups**
pollution	water	water—oceans
		—lakes
		—rivers
		—groundwater
	air	air—noise
		—odors (chemical)
		—dangerous chemicals (odor)
		—visual (upper atmosphere
		—chemicals in it)
	soil	soil—chemical dumps (synergism)
		—agricultural fertilizer
		—pesticides
		—herbicides
		—nuclear waste disposal
		—chemical and experimental weapons sites

In step one of narrowing the topic, list the known "parts" of pollution. There are probably other parts of the universe (pollution), but those parts may be unknown at the beginning of the writing process. Step two requires the writer to study the jot list and estimate how much can be written about each item. With a jot list on as general and popular a topic as pollution, we see immediately that volumes have been written on each group and subgroup of the topic. We can estimate that the topic is easily researched but its subgroups are still too large for a short paper, and that general information may be unsuitable or boring to an audience.

Steps one and two are matters of training. The writer learns to list a topic's major items, groups, and subgroups. Step three is personal because the writer rejects all but one subgroup in order to make the topic more manageable. The choice can be arbitrary, but in both the writing classroom and the workplace, the writer's audience—its needs, biases, capacity for interest — determine the choice of topic. In this case, let's say our writer chooses to write on noise pollution and to reject all other subgroups as topics.

Now the matter of preference enters. We will write about noise pollution in our town of 30,000 people, Somewhere, Ohio. After examining its sources, we will propose methods of controlling noise pollution.

Thus in step three, we limited the topic to a subcategory of pollution (noise) and partially defined it by deciding to write about noise pollution in one town only. We do not yet have a thesis because we must now do a final check of resources before determining the topic's suitability for a technical paper.

Step four requires the writer to check that the library has enough resources for the adequate research, composition, and documentation of the paper. This step is critical. If the library does not have adequate materials, the writer cannot complete the assignment on time. Typically, step four is where an inexperienced writer may decide too quickly on a topic without checking the available resources. Then closer to the deadline, the inexperienced writer realizes he/she does not have enough information and must abandon the project, travel to distant libraries, or wait for material to arrive by interlibrary loan. Always check the library before you commit to a topic.

The library, according to one rule of thumb, should have at least one recent book for every two pages of text you plan to write. Thus, for a 10- to 15-page paper, we need five to seven recent books (published within the last five years) on noise pollution. We also need articles for the most recent information; here the test should be one article for each page of text, or, in our case, 10 to 15 noise pollution articles. The books and articles must be recent publications, because science and technology change rapidly. The only exception to the five-year rule is when a definitive text remains unchanged in subsequent publication, such as, for example, Charles Darwin's *Origin of Species*, published in 1859.

If you checked the university library and found adequate books and articles, the topic is acceptable. Let's review the selection process.

Step One   List all your ideas, facts, and questions about the proposed area of interest for a paper.

Step Two   Treat your area of interest as the universe; write it down. Then write down all possible sets and subcategories of the universe. Estimate whether the topic is manageable at this point or whether you must list more subcategories until the topic is small enough for the assigned length.

Step Three   Choose the subcategory that you like or that interests or concerns your audience to research and discuss in a paper.

Step Four   Go to the library you will use for research purposes and be certain it has the books, periodicals, pamphlets, microfiche, and other research items you will need to adequately cover the topic.

You are now ready to start your paper using the scheduling techniques outlined in Chapter 3. A word of caution is useful here. Some writers think that before beginning research, you should formulate your narrowed topic into a thesis, that is, a statement of attitude toward a topic. We, however, do not recommend a thesis at this stage, unless you have chosen a topic about which you are already very knowledgeable. For example, if you are an engineering student and have taken a physics course that has treated aspects of noise pollution, you may have chosen your thesis when you selected your topic. Steve and Elaine Miska, for example, may know the subject of noise control well enough to have established their thesis before setting foot in the library. If you are new to the topic, though, or if you are assigned a topic, read generally on the subject before deciding your approach. Since this text has a whole section on formulating a thesis, consult Chapter 5 when you are ready to define your thesis.

Most writers agree that there are five to seven ways of gathering information for most research/technical papers: personal experience, library research, informative interviews, questionnaires, letters of inquiry, organizational records, and site observations.

PERSONAL EXPERIENCE

Writers often overlook this method of information collection, although they should not. Your research or technical report cannot turn into a personal essay, but it should reflect your personal experience or competence in the subject. After choosing a topic, list your ideas, facts, and opinions. Ask yourself how you know these facts. Is it through a previous course? If so, check your course notes. They may save you time in collecting references for your research. Perhaps your knowledge of, or interest in, the topic comes from your work experience in the field. If so, what records have you of that experience?

Researching from personal experience is similar to the personal resources survey described in Chapter 3. List everything you know about the topic, then list every person you know who knows something about the topic. For example, our list on noise pollution might look like this.

Researchers divide noise pollution into two kinds: damaging and irritating.
There is some controversy over these distinctions because they overlap. Irritating noise, such as noise from power tools, can be damaging if one is exposed to it regularly.
Music can also have the same effect, even though the listener may actually enjoy the music without realizing that the volume is destroying his or her hearing.
Government regulations exist to protect workers from damaging noise at work, but these regulations are widely ignored.
There is noise pollution at the local tape factory. Are the workers properly protected? Allen Meadors used to work there; call him.
Is a plant tour available for the public? Check.
Noise levels are high on this street because it is an alternate truck route. Call the city to see if we can get measurements.
Call Dr. Harty in the audiology department. Arrange an interview—maybe he has a complete list of local noise polluters.

Our list, although rather diverse, has helped to start the project. For example, the personal experience list unearthed Meadors and Harty, two acquaintances who will assist us and maybe save us time, because their research materials are probably more current than the library's.

LIBRARY RESEARCH

Library research can be simple because all libraries are alike. They use the same classification methods, even though they may use the different call numbers of the Dewey Decimal or the Library of Congress systems. After learning to use one library, you can function successfully in any library in the United States. We have some initial pointers to save time:

1. Do not use the time-consuming and discouraging "freestyle enter-and-search" method. Instead, enter the library and pick up every bit of free information available on this particular library. This information is usually available in distribution stands near the main circulation desk or at the circulation desk itself. Librarians spend hours writing one-sheet documents such as "Using the Central Card Catalog," "Biographical Information — Selected References in the Undergraduate Library," and other such free resources. These aids will teach you anything you want to know about a library.
2. Get a map of the library if you are working in a large university or city library. Book locations, by subject area, are usually posted in signs near the central card catalog. Find where your materials are located and if a card catalog is available for just that floor or location.
3. Choose your location. If working on an unfamiliar topic in a large library, work first in the undergraduate library, then in the more specialized engineering or technology library. If the topic is new to you, work from general to specific information so you understand new material as you find it and do not become overloaded.
4. Start your work by looking up basic books in your area of research, then move to periodical or journal articles, monographs, vertical file, computer information, and other sources. The general works may yield further valuable bibliographical references that will assist exploration of the topic. In the same vein, start your reading with a well-recognized encyclopedia such as the *Encyclopedia Britannica*. Do not use a major encyclopedia as a source for a citation of authority, but you can get an excellent reading and research list at the end of the encyclopedia article. Since encyclopedia articles are written by recognized authorities, you will probably save time by using a recent encyclopedia article as your starting point. All encyclopedia articles are unsigned but initialed at the end. To identify and read more of the author's specialized work, look up the initials in Volume I. Many professional writers begin research projects by checking what a first-rate encyclopedia includes on the subject.
5. Get to know a librarian who will assist you. Librarians are highly skilled researchers who help writers with all sorts of arcane and exciting subjects. Too often students, particularly beginners, fail to approach the librarian with a set of answerable questions. Before working with a librarian, write out your questions so you can obtain all the help you need during the consultation.

If you expect to use the library regularly, and most students will do so at least once a week throughout their undergraduate careers, invest in a basic guide that tells you everything

FIGURE
4.3

A Process Diagram Shows the Six Steps of a Search Strategy.

A SEARCH STRATEGY FOR WRITERS

If you systematically research your materials, you can find exactly what a library has to offer. This approach allows you to cross-check the listings, so you have checked all the major sources on your topic.

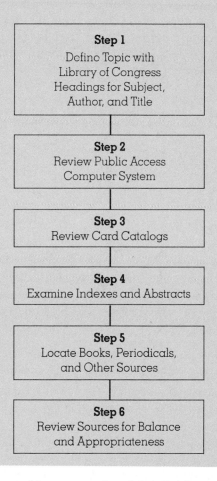

Step 1
Define Topic with
Library of Congress
Headings for Subject,
Author, and Title

Step 2
Review Public Access
Computer System

Step 3
Review Card Catalogs

Step 4
Examine Indexes and Abstracts

Step 5
Locate Books, Periodicals,
and Other Sources

Step 6
Review Sources for Balance
and Appropriateness

you could possibly ask about a library. Several such handbooks exist; ask your librarian which is best for your library. Perhaps the best-known is Jean K. Gates's *Guide to the Use of Books and Libraries, 5th ed.* (New York: McGraw Hill, 1983).

In this chapter, we describe the basic steps of library research, using the basic aids available in every library: the library's basic classification system, the two parts of a card catalog, indexes to periodicals and newspapers, dictionaries and encyclopedias, government documents, computerized bibliographic databases, and vertical files. Of course, the larger the library, the more options available. Researchers should be sure, by consulting librarians or the library's own descriptions of its resources, they know, for example, which indexes and specialized dictionaries are available, and the nature and location of any other specialized resources relevant to your project. Remember, our lists are basic; you should add your library's resources to these lists as you discover and use them.

Finding Your Way in the Library

Libraries in the United States conform to either the Dewey Decimal or the Library of Congress classification systems. The local system is really not that important since you need to know how to use both systems. Public libraries tend to be organized by Dewey Decimal, university libraries by Library of Congress. The Dewey Decimal system uses numbers for classification and storage.

Summary of Dewey Decimal Classification System

000:	General Works	500:	Pure Science
100:	Philosophy and Related Fields	600:	Technology and Applied Science
200:	Religion	700:	Arts
300:	Social Sciences	800:	Literature
400:	Languages	900:	History, Geography, and Biography

Colons appear after these three-digit numbers because subsets of each category are created by adding numbers to the right of the colon. As a field expands and new materials appear, they can be classified within the existing numbers structure. A more detailed listing of Dewey Decimal classifications is available in any good library handbook.

The Library of Congress classification system is alphabetical by 20 basic categories.

Summary of Library of Congress Classification System

A	General Works	M	Music
B	Philosophy, Psychology, Religion	N	Fine Arts
C	History, Related Subjects	P	Languages and Literature
D	History and Topography—not including the U.S.	Q	Sciences
		R	Medicine
E	American History	S	Agriculture, Plant and Animal Husbandry
F	American History		
G	Geography, Anthropology	T	Technology
H	Social Science	U	Military Science
J	Political Science	V	Naval Science
K	Law	Z	Bibliography and Library Science
L	Education		

Any library in the United States will use one of the two systems. Familiarize yourself with the numbers and letters relevant to your areas of research, so you can quickly use your library's classification system.

Using the Card Catalog

All books appear in the card catalog by author, title, or subject. If you know the author's name, locate the book in the author-title catalog with the first initial of the author's last name. Remember, an "author" may be a person (or several persons if the book has multiple authors or an editor), an organization, a conference name, a government agency, or, in the absence of a listed author, whoever commissioned the book.

If you know the book title but not the author's name, again use the author-title catalog. Locate the book under the first word of the book title. Do not consider "a," "an," or "the" as first words.

If you simply want to look for books about a given subject, use the subject catalog. Most libraries have standardized lists of their subject headings available, so you can locate your material quickly. Using the subject catalog is the least efficient method of locating a book—unless you simply want an overall view of the library's books in your area. What the subject catalog does give, though, is a list of good cross-references to check under other headings. For example, we found *Occupational Hearing Conservation* and the *Handbook of Noise Control* listed in the subject catalog. *Occupational Hearing Conservation* was listed under the subject heading "noise control," but the *Handbook of Noise Control* appeared under the subject heading "handbooks."

Figures 4.4 through 4.6 show the cards from the author-title and subject catalogs for these two books. Each card has seven points of reference on it:

① **Call number**—either Library of Congress or Dewey Decimal in the upper left-hand corner. The call number tells you where to find the book in the library.

② **Author line**—the first typed or printed line on the card. It states the author's full name and sometimes his or her birth date.

③ **Title line**—gives the title in full, including an editor if appropriate, as well as edition number, place of publication, publisher, and date of publication.

④ **Collation information**—tells the book length, the number of illustrations, trim size, and whether or not the book is part of a series.

⑤ **Bibliography line**—below the collation. This line states whether or not the book has a bibliography or any indexes (very useful to the researcher).

⑥ **Subject headings**—near bottom of card. Lists subjects under which the book may be found in the subject catalog.

⑦ **Classification information**—the first line of the card, listing internal classification data such as the date of the card code and library personnel code numbers.

Proper use of the card catalog can save time as you research a topic. For example, the publication date on the card catalog permits you to determine the book's usefulness. If older than five years in a technical field, the book may be obsolete unless it is an important work in the field. Also, if the book has no illustrations, it may not meet your specifications. Much of your evaluation can occur right at the card catalog, thus cutting research time.

FIGURE 4.4	Author Card.

①TD
    892   ②Harris, Cyril M., 1917-
    H37   ③   Handbook of noise control / edited by
    1979   Cyril M. Harris, 2d ed. New York:
          McGraw-Hill, [1979]
      ④   xiii 724 p. in various pagings:
      ill. ; 24 cm.
    ⑤   Includes bibliography and index.

    ⑥1. Noise control.   1. Title

⑦ICarbs   SOIUsc   78-6764

If your library has closed stacks — and many research libraries do — you will not have access to every book of interest. You must then use the card catalog or a computer terminal to determine a book's worth before ordering it at the circulation desk. If your library does not have open stacks, allow more time for card catalog work. Make a complete note of any text that may interest you so that you later have all the information needed to fill out an order form. You must order books by call number, author, title, and sometimes by publisher in a closed stacks library, so always check your accuracy in recording card catalog entries. Finally, as you record references, note on your reference card whether the book looks useful, unusable, arcane, or interesting in any other way. You may forget this information when you get to the circulation desk, unless you annotated your cards.

Locating and Using Periodicals

A typical undergraduate or small public library stocks about 500 periodicals. Large research libraries stock tens of thousands of magazines and journals, so knowing how to find information in periodicals is essential. Usually, since important new research, and just about all primary research, appears in journals before books, you must consult periodicals for the latest available information. Five steps will take you quickly through the maze of indexes you need to use the periodicals wisely:

1. Select an appropriate index.
2. Search the index subject list.
3. Compile a list of the articles listed in the index that may be useful in your research.
4. Consult the library's periodical list (usually kept at the circulation desk) to ascertain that the library subscribes to the publications you need.
5. Go to the shelf or to the microfiche to find the periodical.

A list of useful indexes may make this task less intimidating. An annotated list is available in most library handbooks. You can find indexes in the reference section. Following is a list of useful Indexes and Bibliographies.

FIGURE 4.5	Title Cards.

Handbook of noise control

TD
892   Harris, Cyril M., 1917-
H37   Handbook of noise control / edited by
1979   Cyril M. Harris, 2d ed. New York:
McGraw-Hill, [1979]
                xiii 724 p. in various pagings:
ill. ; 24 cm.
                Includes bibliography and index.

1. Noise control.   1. Title

ICarbs   SOIUsc   78-6764

SCI
RF   Occupational hearing conservation /
293.5   edited by Maurice H. Miller, Carol A.
.O33   Silverman.—Englewood Cliffs, N.J.
1984   : Prentice-Hall, c1984.
xvii, 270 p. : ill. ; 24 cm.
Includes bibliographical references
and index.
        ISBN 0-13-629386-7

1. Deafness, Noise induced—
Prevention. 2. Industrial noise.
3. Noise control—United States.
4. Noise control—Law and legislation—
United States. I. Miller, Maurice H.
II. Silverman, Carol A. (Carol Ann),
1953–

OBgU   BGUUnt   83-3393

Applied Science and Technology Index
Biological and Agricultural Index
Business Periodicals Index
Chemical Abstracts
Cumulative Book Index (always called CBI)
Education Index
Engineering Index
Guide to Reference Material for Science and Technology
Monthly Catalog of United States Government Publications

FIGURE 4.6	Subject Cards.

NOISE CONTROL

```
TD
892      Harris, Cyril M., 1917-
H37              Handbook of noise control / edited by
1979     Cyril M. Harris, 2d ed. New York:
         McGraw-Hill, [1979]
                 xiii 724 p. in various pagings:
         ill.  ; 24 cm.
                 Includes bibliography and index.


         1.  Noise control.    1.  Title


ICarbs                       SOIUsc                    78-6764
```

NOISE CONTROL—UNITED STATES.

```
SCI
RF        Occupational hearing conservation /
293.5         edited by Maurice H. Miller, Carol A.
.O33          Silverman.—Englewood Cliffs, N.J.
1984          : Prentice-Hall, c1984.
              xvii, 270 p. : ill. ; 24 cm.
              Includes bibliographical references
          and index.
              ISBN 0-13-629386-7


OBgU                        BGUUsc                    83-3393
```

Music Index
Readers' Guide to Periodical Literature
Sources of Business Information

Useful Newspaper Indexes.

The New York Times Index
The Wall Street Journal Index

**FIGURE
4.7**

Listing from *Applied Science and Technology Index* for "Noise."

Noise
> See *also*
Ear—Injuries
Ear—Protection
Noise control
Radio communication—Interference
Radio communication—Interference elimination
Sound pressure
> See *also* subdivision Noise under various subjects
Noise and vigilance: an evaluative review. H.S. Koelega and J.-A. Brinkman. bibl(p479-81) *Hum Factors* 28:465-81 Ag '86

Standards

Measurement and control of noise: comments on the present and solutions for the future. W.R. Thornton. *Am Ind Hyg Assoc J* 47:683-5 N '86

Recommended maximum valve noise levels. A.C. Fagerlund. bibl *InTech* 33:47-50 N '86

Suggested standard hearing protector attenuation function relating to the NRR. G.L. Cluff. bibl *Am Ind Hyg Assoc J* 47:776-8 D '86

Noise control

"Ducting" as a pneumatic noise control technique. J.C. Beck and others. bibl *Am Ind Hyg Assoc J* 48:28-31 Ja '87

Floating floors for impact and airborne noise control. W.W. Fearon. bibl diag *Sound Vib* 20:20-2 O '86

Measurement and control of noise: comments on the present and solutions for the future. W.R. Thornton. *Am Ind Hyg Assoc J* 47:683-5 N '86

Program for combining sound pressure levels in noise. O.F. Offodile. *Ind Eng* 19:22-4 Ja '87

Reducing and preventing noise in control valves. M. Burton and W.R. Faas. il diags *Plant Eng* 40:53-6 N 13 '86

Saudi petrochemical plant noise control. A. Gharabegian and J.E. Peat. bibl il diags *J Environ Eng* 112:1026-40 D '86

Noise measurement See Sound—Measurement
Noise prevention See Noise control

From the *Applied Science & Technology Index* (March 1987), p. 410. Copyright © 1987 by the H.W. Wilson Company. Reproduced by permission of the publisher.

Newspaper indexes are grouped with the major indexes because a newspaper index allows the researcher to pinpoint the exact date an event was covered as a news story. You need only one good newspaper index to look up a topic in several newspapers because libraries file newspapers by date. Several other good indexes of newspapers are available, but the two listed here are the most comprehensive.

As does the card catalog, indexes require flexibility in your approach to a topic in order to locate all the related information. For example, we looked up pollution in the *Applied*

Science and Technology Index, thinking it would list noise as a subcategory of pollution. Because *Applied Science and Technology Index* started many years ago, its categories are well established and, in some cases, old-fashioned. There were no listings for noise under "pollution" and "noise pollution." We also found a list of related topics, half of which seemed related to noise pollution.

Applied Science and Technology Index is a useful index to start periodicals research, because it is the largest of its kind. Several other indexes, such as the *Readers' Guide to Periodical Literature*, have the same format. The *Applied Science and Technology Index* lists entries by article title, author(s), publication title, volume, pages, and date. Thus in Figure 4.7, an entry under noise standards is an article entitled "Measurement and Control of Noise: Comments on the Present and Solutions for the Future," written by W.R. Thornton. It was published in *American Industrial Hygiene Association Journal* in volume 47, pages 683–685, November 1986. These statements are all abbreviated in the entry, but all abbreviations are explained in the front of the index. Note that the index lists periodicals by volume. Check to make sure that your library collects the periodicals listed under your topic in the indexes.

Other Library Sources

Every library has a vertical file, a collection of pamphlets, brochures, unpublished reports, graphic materials, and other odd-sized pieces that are not bound and shelved. The vertical file is usually near or behind the circulation desk, with its items listed by subject, title, and author.

Dictionaries are also useful sources because they provide uncontested definitions of terms you may use in your paper. Like encyclopedias, they also offer basic, authoritative, but short articles on diverse subjects. Some useful dictionaries are:

Harris, C.M., ed. *Dictionary of Architecture and Construction.*
McGraw-Hill Dictionary of Scientific and Technical Terms.
Monkhouse, F.J., ed. *A Dictionary of Geography.*

Atlases are useful for checking demographic and geographic data. Some basic sources available on the reference shelf of most libraries are:

Hammond Medallion World Atlas.
National Geographic Atlas of the World.
Rand McNally New Cosmopolitan World Atlas.
The Times Atlas of the World.
U.S. Geological Survey. *The National Atlas of the United States of America.*

Statistical sources are always needed in business and technical communications. Every library has its own United States Government documents section. Learn what kinds of documents your library carries, and ask the librarians to order any you need. Three essential lists of United States data are listed here:

Monthly Catalog of United States Government Publications. USGPO Superintendent of Documents. (This publication lists all documents published by U.S. agencies, and it also introduces the student to the U.S. government document classification system.)

United States Bureau of the Census. *County and City Data Book* (First issued in 1952 and issued every five years since then.)
United States Bureau of the Census. *Statistical Abstract of the United States.* (Issued annually.)

Computers have changed the nature of documentation. Each library has its own computer systems and searches. You may ask a librarian for a list of computer searches and retrieval systems. By saving space and printing costs, computer storage has allowed libraries to maintain large collections of published materials. Since in the future, libraries will increasingly rely on computers to store and index information, learn to access this information now.
There are five typical computer systems most libraries either own or can obtain:

Information Retrieval Systems

Chemical Industry Notes—abstracts and lists recent significant events and results in the chemical industry. Selections are from business and chemical trade periodicals and journals.
Educational Research and Information Service—abstracts, indexes, and prints microfiches and articles from every area of education. Known everywhere as ERIC, the service is classified by levels of education and disciplines.
National Technical Information Service—NTIS for short, a United States government service offering searches of more than 600,000 government documents. You can purchase a packaged topic search or an individualized search.
Scisearch—published by the Institute for Scientific Information, indexes interdisciplinary materials on science and technology from the Institute and other sources.
Social Scisearch—also published by the Institute for Scientific Information, lists interdisciplinary research not only from the Institute, but also from major social science journals.

Some university libraries may subscribe to 250 computerized bibliography databases. Be sure you check with a librarian to see what services are available locally. A very specific service may include your topic. Most search and storage systems are available in microfiche for a user fee of $3 to $20 per list, depending upon the number of variables listed in any given request. For another fee, you can usually obtain "hard," that is, paper copies of these materials.
Computer indexing and storage save the researcher the tedious, repetitious task of finding, copying, and alphabetizing every bit of basic information needed for a complete bibliography. The computer search also saves the researcher the demanding task of gathering every publication fact for a complete list of references and a bibliography. The searches are usually quite accurate in their citations.
However, computer searches do not replace your own thorough search of materials. Like all other indexes, computerized information retrieval systems are language or "variable" bound. For example, to order a computer search, you list the "variables" or key words the librarian-operator will use to sort through the computerized list of information. The computer can also cross-tabulate, that is, search a certain number of topics related to or contained within the set of variables you list. But that is all the computer can do. Since the computer is a tool, not a thinking, discerning researcher, it inevitably misses important articles that can fill out

FIGURE 4.8
Sample Screens Show Part of a Typical Computer Search for Research Sources in a Library.

HOW A COMPUTER SEARCHES FOR INFORMATION

Your library's computers may use the Library of Congress subject titles to list books. At a university library we used a terminal dedicated to searches to find books about noise pollution. An example of its menus follows. Although no two libraries have the exact same systems, they usually look like these two screens:

Screen 1

```
Public Access Catalog

Subject:   Noise Pollution


5. Subject

S   Noise Pollution


Choices


1. Noise pollution
2. Noise pollution—Bibliography
3. Noise pollution—Congresses
4. Noise pollution—Costs
5. Noise pollution—European Economic Community Countries—Public Opinion
```

We selected item four by typing the number 4. This menu then appeared:

Screen 2

```
4. Costs

Search: S Noise Pollution

Total Matches


1. Author:   Walters, A.A.
   Title:    Noise & prices/ A.A. Walters
   Date:     1975
```

Under category 4, noise pollution costs, the computer found only one book in the library's collection. Perhaps next we should try item 2, noise pollution bibliography, so we can cross-check the computer's list with the library's card catalog.

weak spots in your report. The computer program can neither judge a piece of research's worth, nor perceive subtle but unlabeled relationships among apparently unrelated topics.

For example, we can feed the key words "noise" and "pollution" to our search system and receive thousands of items listed in the literature. But the computer may not catalog every article that is relevant to your unique topic. Computers can provide excellent basic and lengthy reference lists, but you yourself must search for the latest material and for all the sub-variables that will produce one or two useful, topic-related articles (see Figure 4.8). The next generation of computers could become better literature searchers than humans, but for now as the researcher and author of your report, you must creatively coordinate your computer and manual searching.

Recording and Processing Research Information

Setting up a notation system is essential to the success of a research or technical paper. Well-organized notes enable the writer to produce a paper entirely from note cards — without repeated trips to the library to double-check references. The researcher must: set up bibliography or reference cards, decide what to write on reference or note cards, and appropriately detail the note cards so the information neither "dies," nor loses its meaning between the library and the typewriter.

Although professional researchers have their own gimmicks for annotating references, most library users agree that you should use $3'' \times 5''$ index cards to store your reference list only twice — when you note your source during library research and when you compile the alphabetical reference list at the end of a report.

Reference Cards. When you skim through the card catalog and the other research sources, record access numbers on scrap paper. After deciding a reference may be useful, fill out a reference card. Beginning writers sometimes forget reference cards, thinking they can "pull the references together at the end" of the paper. Instead, fill out a reference card at the beginning of your research. It is always easier to toss a card than to return to the library for an un-named source. Reference cards should all have the same format, such as:

1. Author (last name, first name, middle initial — use editor if there is one).
2. Full title, including edition (if more than one edition exists).
3. Place of publication.
4. Name of publisher.
5. Date of publication.
6. Useful extras your reference list may need.
    A. Volume number if the book is part of a multivolume series.
    B. Access number — either Dewey Decimal, Library of Congress or other in pencil, so that if you make several trips to the library, you can go right to your source each time without looking up its location.
    C. Any other reference information you might want later.

A typical $3'' \times 5''$ reference card should look like ours for Occupational Hearing Conservation (see Figure 4.9).

Note that the reference card contains all the information needed to write a complete reference list item. Although reference list regulations vary, this card contains the facts

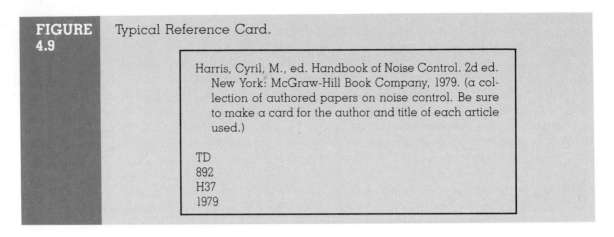

**FIGURE
4.9**
Typical Reference Card.

Harris, Cyril, M., ed. Handbook of Noise Control. 2d ed.
    New York: McGraw-Hill Book Company, 1979. (a col-
    lection of authored papers on noise control. Be sure
    to make a card for the author and title of each article
    used.)

TD
892
H37
1979

necessary to arrange this item according to any style. You can use this card in a different paper, with a different stylebook. Note the caution: each article has a different author and must be referenced separately. The Library of Congress number appears so that listing is available if this book is ever needed again, either in this or another library using this system. Finally, the card's layout and purpose are designed to save time. The first line highlights the last name under which the card is filed. The other lines are indented to increase readability and ease in filing. Use this format for every source. If an item is missing—for example, if the book does not list an author—move on to the next item in the sequence.

Assign a number to each reference card as you write it. For example, Miller is number one because we start our research with that source, and because it is the first card written. The next source will be on card two, the next on card three, and so on. The numbers allow us to abbreviate note cards. Thus, when filling out a $4'' \times 6''$ note card with information from Miller, "Miller, 1–1" identifies the reference card, on the first line of the first card, "Miller, 1–2" on the second in the series of note cards from Miller, "Miller, 1–3" on the third, and so on (see Figure 4.10). Pencil in the number so that you may reuse the card in a different paper.

If you use this reference card format, composing the reference list at the end of your research paper or technical report is easy. Simply arrange your $3'' \times 5''$ cards in alphabetical order according to the first letter of the first item on each card. Then type and punctuate as your stylebook requires.

Note Cards. Deciding what to note requires practice, but a few suggestions may make note-taking easier. First, use $4'' \times 6''$ index cards, although some writers recommend $5'' \times 8''$ cards. We recommend the smaller card because its size forces the writer to limit the amount of information per card. Each card should carry only one idea, one fact, or one statement of opinion. The smaller card helps enforce this limitation.

Second, consider the classifications of information. A *direct quotation* is a series of words printed exactly as a person spoke or wrote them. The series should begin and end with quotation marks, and the internal punctuation must be the same as the speaker's or writer's. Place only one direct quotation on a card. On that same card, note the page of your source, and of course, the full name. A *paraphrase* is an appropriate rewording or reworking of a person's direct quotation to fit the context in which you want to use it. Quotations are unnecessary, but each

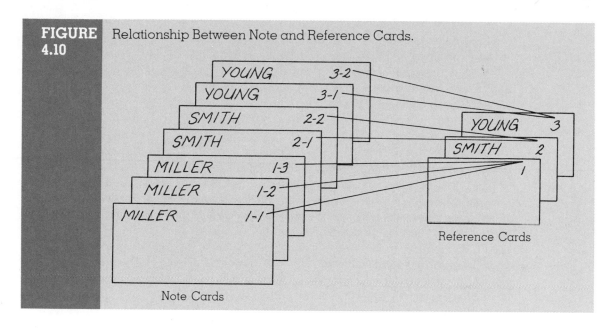

FIGURE 4.10   Relationship Between Note and Reference Cards.

note card should have only one paraphrased passage or idea. Also, note the source and the page, because you must credit the author of a paraphrased item as you would for a direct quotation. In fact, when you paraphrase, it is wise to write on your note card "Paraphrase from . . ." and then note the source. A *summary* is a statement in which you rewrite or shorten a person's statement in order to fit your context. Again, you must reference the author of the idea, citing the author by name and page number. *Dates, statistical data, figures,* and *pertinent facts* are important information that you read and write down on note cards using your own words. Nonetheless, you must credit, citing author, page, and source of any new, original, or unusual date, data, figure, or fact that your reading unearths.

If you fail to credit a source for an idea, your report may contain *plagiarism*, the use of another person's work or ideas without proper attribution. Plagiarism is illegal, and an author whose work is stolen can sue for damages. In all phases of professional and scholarly writing, plagiarism is totally unethical—even though the plagiarist may not realize the misconduct. The rule is: if you wonder whether to credit a source, be safe and do so.

As a researcher, you should also understand the difference between primary and secondary research. *Primary,* or "first-hand," *research* refers to observation or experimentation which uses the scientific process to collect original data for analysis of a given subject. Primary research is subject to tight controls and requirements that have evolved with the philosophy of science. Typically, to conduct primary research, you must write a prospectus or plan of action which your professional peers review for the project's scientific merit and, in some cases, for the morality and humanity of its methods. Once approved, the project must contain a literature review of the topic, often a statement of philosophic assumptions, a statement of the hypothesis to be tested and its alternate, a description of the manner in which the test, treatment, or process is performed, a report of *all* the results (not just the significant ones), and a discussion of those results. An example of primary research is to determine the effects of a new flu vaccine. That example should illustrate the need for exact measurement, unbiased testing, and thorough reporting of results. Primary research is first or new research, and its outcome can affect millions of people and millions of dollars.

Secondary research is typically defined as the search and presentation of already published research results. Most students and many technical writers typically conduct secondary research for several reasons. First, students broaden their experience by studying several articles of research on one topic. Second, this type of research is cost-effective. For example, business researchers often achieve better research and development results with secondary research rather than with more expensive primary research. Third, large databases, such as the data collected by the U.S. Bureau of the Census, need analysis. Human analysis has not kept pace with the vast amounts of data now collected. Thus, secondary research takes on a new significance. Practitioners in various disciplines and competent researchers must analyze and evaluate existing data in order to understand and grasp the basis for future research in government, business, and schools and other institutions.

Primary and secondary research are equally valuable disciplines. As a researcher you should know the differences since you may find yourself participating in research or authorizing and rejecting new research proposals. You should be comfortable reading and reporting the results of primary and secondary research in your field.

When you work with research materials, the notes you take must be appropriate for your audience. For example, if you are researching new developments in the pharmaceuticals industry, the types of notes you record will vary with the audience. If presenting the paper to a group of chemists, your notes will include extensive discussions of formulas and measurements of new products but no basic chemical processes. If your paper is a new product report for a marketing class or a sales team, you will limit your notes to the basic structure of the new drugs, but stress the benefits of the products as recorded in test results. Each audience needs different facts for a meaningful topic. Thoroughly analyze your audience before gathering your facts in note form. Finally, take more notes than you think you need. Extra reading will build your confidence in your knowledge of the subject and help you develop your topic for the audience. Again, it is easier to toss extra notes than to return to the microfiche reader for more materials.

INTERVIEWS

Informative interviews permit researchers to gather facts from experts working in the field. Writers can strengthen their research with reports from selected experts or from participants in the situation under study. In business, managers frequently interview to gather the facts, feelings, and experiences of the people whose tasks they must understand in order to write situation reports, case studies, and project completion reports.

Although interviews are useful in gathering new materials, they can be biased either by the interviewer's attitude, by the respondent's intelligence and state of mind, or by the interviewer's ability to record and analyze the interviewee's statements.

Interviews should be governed by your understanding of business protocol, your respect for confidentiality, and your adherence to a plan of organization.

Protocol

Request an interview either by telephone or by letter. During this initial contact, explain how you will use the interview materials, and then follow through on your promise. For example, if you intend to display pictures of the subject or of his or her work, you must have the subject sign a model release. If you intend to print 100 words or more of continuous, directly quoted text, you must obtain the subject's written permission. If possible, provide your subjects with the opportunity to review their remarks, either by mail, by phone, or in person.

Confidentiality

Respect your subject's right to privacy. If you offered confidentiality during the initial contact, keep it by assigning all the interviewees with numbers by which you identify their records. For example, if you interview selected former inmates or someone with a "social problem" in your community, you have an obligation to ensure their privacy. Do not discuss names with teachers, spouses, parents, supervisors, or anybody, including even people you respect and trust. Confidentiality means total confidentiality. You have a professional and, usually, a legal obligation to protect a subject's social reputation or ability to earn a living.

Organization

Write or type all interview questions in advance. Free-form conversation may yield useful information, but if you must interview many subjects, random conversation may bias the respondent's answers and affect your ability to interpret them. Stick to the interview questions, remain pleasant, and do not digress. Start with simple factual and personal data questions to relax your subject, then move to the more complicated items.

Caution is appropriate for students new to the research process. We do not recommend that students start their first research assignments by interviewing. You may, of course, interview an authority as one small part of your information-gathering process. The experience of a single contractor with expertise in noise control can be invaluable to your noise pollution report. But large-scale interviewing of dozens or hundreds of subjects or opinion-sampling projects is not for beginners. Even the large consulting firms handling these projects sometimes run into expensive, unexpected problems. It is very easy to write a poor instrument, but time-consuming and often expensive to write a valid, reliable instrument. If you must use interviews, take some elective courses in demography and statistics and learn to use this important but often abused process with total competence.

QUESTIONNAIRES

Questionnaires, though more formal than interviews, are not usually administered in person. Unlike interviews, questionnaires are written and often mailed to respondents for completion. In turn, the respondents return their questionnaires through the mail. Questionnaires are used extensively on-site in retail and marketing efforts. They are used to gather census data and to determine public attitudes in, for example, election polls. Because a questionnaire is usually written for mass distribution, careful testing and measurement procedures must assure that the questions really ask for the information the researcher seeks. Many excellent

FIGURE 4.11   Sample Survey Instructions and Sample Survey.

INSTRUCTIONS

This survey is to establish how much noise you feel you are typically exposed to each day. You will be asked to establish the frequencies with which you are exposed to five levels of noise.

The levels of noise are:

                soft = conversation in next room
      somewhat noisy = room full of talking people
               noisy = busy cafeteria
          very noisy = busy traffic intersection
                loud = auto horn at 20 feet

The frequencies are as follows:

               never = not at all
           sometimes = once a week
               often = once a day
          frequently = several times a day
              always = constantly

In the following example, the person has selected:

1. soft/always
2. somewhat noisy/frequently
3. noisy/often
4. very noisy/sometimes
5. loud/never

	Never		Often		Always
	1	2	3	4	5
soft 1					X
2				X	
3			X		
4		X			
loud 5	X				

**FIGURE
4.11**   Continued

SAMPLE SURVEY

I. In your home.

A. Do you ever hear your neighbors' radios, televisions, musical instruments?

	Never		Often		Always
	1	2	3	4	5
soft 1					
2					
3					
4					
loud 5					

B. Do you ever hear your neighbors' power tools, automobiles, motorcycles?

	Never		Often		Always
	1	2	3	4	5
soft 1					
2					
3					
4					
loud 5					

C. Do you ever hear street traffic noises, for example, trucks, horns, automobile traffic, auto doors, loading/unloading?

	Never		Often		Always
	1	2	3	4	5
soft 1					
2					
3					
4					
loud 5					

FIGURE 4.11   Continued

II. At your place of work.

A. Do you ever hear noises from nearby work sites, for example, machines or equipment, lathes, stampmills, conveyors, engines?

	Never		Often		Always
	1	2	3	4	5
soft 1					
2					
3					
4					
loud 5					

B. Do you hear traffic noises as you go to work, for example, trucks, autos, trains and subways?

	Never		Often		Always
	1	2	3	4	5
soft 1					
2					
3					
4					
loud 5					

C. Do you hear nearby work sites, for example, power tools, machinery, vehicles, office machines?

	Never		Often		Always
	1	2	3	4	5
soft 1					
2					
3					
4					
loud 5					

FIGURE 4.12   Typical Formats for Survey Questions.

TYPICAL QUESTIONS THAT APPEAR ON SURVEYS

Even though you may custom design a survey for your research, you will probably use either one of the format types listed here or another standard format found in survey literature. These are objective questions which can be coded for computer analysis.

True/false—or any type of question that limits the respondent to one of two choices.
   Are you employed full-time? Yes 1   No 2

Multiple choice—a question that typically asks the respondent to select one answer from among three to five choices.
   Which is your current educational status?

less than 12 years	1	master's degree	5
high school degree	2	doctorate	6
associate's degree	3	professional degree (law,	
bachelor's degree	4	dentistry, medicine)	7

Likert scale—the question that asks a respondent to evaluate a product or experience on a least-to-most basis.
   In summary, how would you rate your advisor? (circle one)
      very poor   poor   satisfactory   good   very good   excellent

Rank order—a question that asks the respondent to state how often he does something, how many times she uses a product, which activities they prefer, etc.
   Using the numbers 1 (best) through 5 (worst), name the areas of the recreation center that you enjoy most.
      _____indoor track
      _____sauna
      _____swimming pool
      _____weight room
      _____whirlpool

Combination—a question that uses two of the methods or that uses a more complex format; for example this one uses true/false and five items.

Yes	No	Is your advisor knowledgeable about:
y	n	**a.** graduation requirements?
y	n	**b.** administrative procedures?
y	n	**c.** career possibilities in your major?
y	n	**d.** other services on campus?
y	n	**e.** scheduling matters like drop/add, closed sections and other matters?

Complex format—a question that relies on its visual design for an answer, for example, a grid or a table. See our sample survey for noise pollution for an example.

statistics texts treat the subjects of text question validity and reliability. We recommend reading and following established practices.

Planning

Planning is critical to the success of a survey or questionnaire. For example, experts must first draft and test a new instrument for several types of validity. To measure its reliability, it is then pilot-tested on a small population. Then the instrument is revised, retested, and readministered. Since this process is very time-consuming, you should be certain your project assures enough lead time, maybe months, for a successful questionnaire. The careful construction of a questionnaire to sample citizens' opinions about noise in the Miskas' city could take six months to a year to plan and administer.

If you want to mail the questionnaire, the accompanying cover letter and set of instructions must also be written and printed. The production and mailing for both response and follow up will be time-consuming. Later, collation and analysis will probably necessitate computers and statistical software. Marketing research firms — the people who do the best survey work — sometimes take a full year just to ready an instrument for mass distribution. Therefore, use a production schedule much like the one in Chapter 2 to keep the production on time and the costs under control.

Writing

Thousands of tested and reliable questionnaires exist, so try using an existing survey. The fact that the instrument you use has been fully reported and analyzed in previous literature can permit you to accept your results with some degree of confidence, because you can determine from the literature the instrument's probable response rate and error margins. If you must write a new instrument, study existing ones and pattern yours on a valid document similar to yours.

Not only must you write an instrument that asks the questions you want answered, but you must also design your questions so the respondents' answers or choices can be entered on computerized scoring sheets such as Optical Scanning (OP Scan) sheets. Again, a major text or handbook in the area can assist with this procedure. More importantly, to prepare for your writing, check the institution's or firm's facilities for computerized collation. The questions must obviously be written so the answers can be entered into your unit's existing computer. These conditions restrict every instrument written, because no one has time to hand score 300 surveys, much less 3,000. You can design a custom scoring sheet, but then you will need a custom program to run the data. This is expensive and time-consuming. The real issues are time and cost: how much time and money is the information worth to you?

Avoiding Errors

Study the sample questions in Figure 4.11 so you can avoid basic design errors. Each question should relate logically to the one before it. Use objective, quantifiable answers not only for scoring purposes, but also because this procedure limits the possibility of various types of testing errors. Although procedures exist for coding free responses, they are costly and time-consuming, and trained readers must be used.

If you must compose a long survey, divide it into sections such as Part 1, Part 2, and so on. Each section should be devoted to a particular phase or subset of the survey's general topic.

To maintain your survey's coherence, write clear directions for completing the survey and show the respondent a sample question and answer. Do not expect the subject to understand your question without an example. Remember, the whole survey is new to the subject, and you do not want to bias the results by creating anxiety or by overloading the subject with too much information.

Choose the appropriate question format for the information sought. Standard surveying texts provide extensive examples of dozens of question types because different questions elicit different data. The sample noise survey (Figure 4.11) used multiple choice answers arranged on a grid, a somewhat complex format. You can use multiple choice, rank order scales, and other types of questions, but the type of question used should be appropriate for the nature and amount of the information you seek (see Figure 4.12).

Finally, package the instrument attractively. As with the final technical report, the survey's appearance catches the subject's attention first. Type size should be appropriate for the age group of the subjects. Print should be clear and legible. The layout should have enough white space to invite, rather than confuse, the subject's answer. Research indicates that within the first three seconds of opening unsolicited mail, your prospective respondent has decided either to respond or discard the questionnaire you spent so much time preparing. Remember that when you design your questionnaire.

Letter of Inquiry

Letters of inquiry are often used to establish contact with persons who live or work far from the research site. These letters should be brief, polite, and not overly demanding. Also, only use letters of inquiry if you have adequate time. When you write a letter to obtain specific information, allow a reasonable period of time, approximately 10 days, for the recipient to collect the information and respond. If the information will cost the respondent or a firm any amount of money, offer to reimburse them appropriately.

Organizational Records and Site Observations

This is a very practical type of research since the nature of the research depends upon the material available. Sometimes, students spend hours on library research because they overlook readily available information at the workplace.

Organizational records include previous reports on the topic you research. Most major industrial projects require land use or environmental impact studies, financial feasibility studies, progress reports, and project completion reports. Such internal documents are valuable sources of primary and secondary data. Because they usually contain current cost statements, your own calculations can be more timely and accurate. Major corporations have libraries, organized either by traditional methods or by computer indexes. If you work for a small firm, your "library" may be the back file in which the documents we describe will be found. Carefully study organizational records so that your report reflects your awareness of the problem's history.

Site visits can be the most important aspect of your research, especially if you work in insurance or engineering. Some simple procedures can prevent expensive, time-consuming and repeated visits to a site. Unless the site is inside a building, site visits require outdoor time. Be prepared to observe and work outdoors. Take notes at the site. Use a pad, cards, or a portable tape recorder, but record your information at the site. You may not remember all your observations upon returning to your office. You may want to design a one-page form reminiscent of the information you must collect. A form prevents missing data and helps organize your ideas for a report as you take notes.

Bring all the equipment you need to take accurate measurements and to obtain correct records. Again, a small inventory kit, even a simple carton with a list of materials attached to the top, can make it easier for you to bring all the supplies you need to the site. When you make measurements, make them twice to check for accuracy. Have an assistant work with you on measurements, so two persons can double-check for accuracy. Finally, remember the graphic component of research. Prepare a rough sketch of the site of your visit or the problem you inspect. Help the reader to visualize the site through photographs, drawings, and accurate measurements recorded at the site.

SUMMARY

In this chapter, you have seen how to limit a topic and how to check its suitability. You have also read about several basic approaches to research: personal experience, library research, informative interviews, interviews as questionnaires, letters of inquiry, organizational records, and site observations. You should be able to:

- State a proposed topic; then create logical sets and subsets of the topic.
- Determine whether or not the available research resources will permit you to complete the topic by the proposed deadline.
- Find five recent books on the topic in your local library.
- Select the correct index and locate research materials in periodicals, newspapers, vertical files, and computerized search systems.
- Fill out and start a filing system of reference cards.
- Enter useful information into a system of note cards linked with your reference cards.
- Define and locate examples of direct quotations, paraphrases, and summaries.
- Explain the causes of plagiarism and the use of citations to avoid it.
- Distinguish between primary and secondary research materials.
- Discuss factors that influence the use of interviews and questionnaires.
- Explain why a report would use office records.
- Explain what kind of information you would collect on a site visit.

ASSIGNMENTS

After examining the list of topics in Figure 4.2, select one that interests you.

1. Start the research on your topic by going through the topic limitation process, surveying your library's sources, and selecting the various types of materials you need to study.
2. Prepare your note cards with direct quotations, summaries, paraphrases, and facts. Turn in the note cards for your instructor to comment on.

3 Prepare for your instructor a set of reference cards about your topic. Use the citation format your instructor selects.

4 Explore indexes with citations about your topic. Find examples of primary and secondary research reports from two indexes. Bring these to class, and be prepared to discuss the reports.

5 Using your library's card catalog or computer, find five useful, recent books about your topic. Prepare to explain to your class why you chose these books as research sources.

6 Using your library's indexes and computer, find five useful, recent articles about your topic. Prepare to explain to your class why you chose these articles to supplement your books.

7 Use a newspaper index to find the day of the week the Tonkin Gulf Resolution was passed or the day of the week on which Winston Churchill's funeral was held. Be prepared to report to the class how much time you needed.

CHAPTER 5

Organizing Your Technical Communication

To organize your technical documents, you need to know how to compose and edit outlines. After reading this chapter, you will be able to:

- Show an outline to your associates and explain how you have organized a report or manual.
- Compose an outline that adheres to all the conventions of outline content and format.
- Compose alphanumeric or decimal outlines.
- Compose topic or sentence outlines.
- Modify a generic outline for a report you may write.
- Compose a topic or sentence outline for a manual, and then "translate" it into a table of contents.
- Edit and revise another student's outline.

CASE STUDY   **An Outline by the Deadline for Research Associates, Inc.**

The owners of Research Associates, Inc. (RAI), a relatively new technical communication firm, have become experts on noise pollution, the topic you reviewed in the library during your reading of Chapter 3. In fact, the firm hopes to encourage contractors like Steve and Elaine Miska to purchase RAI's market studies for soundproofing that complies with a new municipal ordinance.

Bill Dorsey, the writer assigned to the soundproofing project, drafted an outline for the "boilerplate," or general information report. The company hopes to redesign it for individual contractors in the Toledo, Ohio area. Because of a personnel shortage and multiple deadlines, Bill Dorsey has been reassigned to proposal writing. Someone else must finish his outline for the soundproofing report. After finishing this chapter, you will rewrite Bill Dorsey's outline to conform to all the conventions of outline format and content. His draft and first revision appear at the end of this chapter.

THE IMPORTANCE OF OUTLINES

Intermediate writers usually ask why supervisors, teachers, and editors consider outlines so important. Perhaps you have not yet learned that outlining is essential to the success of your written documents, oral communications, and group projects. In industry, you will use outlines as a management tool to plan entire manuals and reports, to organize oral technical communications, and to document your own managerial capabilities.

Outlines as Plans for Manuals and Reports

When the supervisor of a corporate publications group begins to design documentation for a new product, he/she drafts a series of outlines, one for each item in the documentation. For example, if an accounting firm plans to sell a new integrated software package for scheduling, maintaining, and financing rental properties, the supervisor of the publications group must ouline each item of documentation that will accompany the product. The list might look like this:

Installation manual   Pocket guide
Tutorial   Troubleshooting, or maintenance, or
User's manual   advanced user's manual
Reference manual

When you consider that one product may generate six publications of various lengths, the need for detailed outlines seems obvious. Only by composing an outline can the supervisor explain to the writers their tasks and deadlines for the delivery of the new product. These six publications must be uniform and consistent in many ways, including design, format, level of language usage, and vocabulary. The outline assures this continuity in a firm's publications. (For more details of the publications process, consult Chapter 17.)

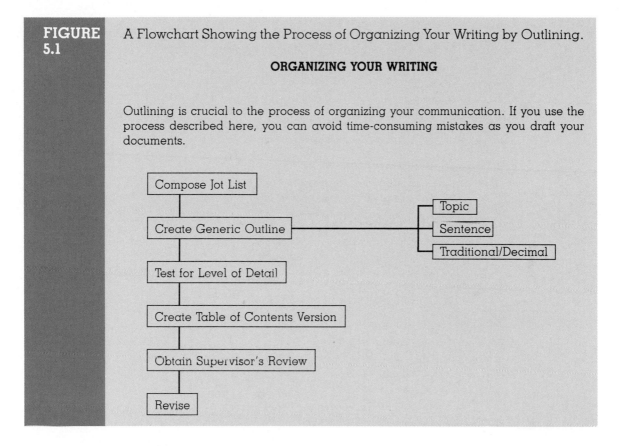

FIGURE 5.1 A Flowchart Showing the Process of Organizing Your Writing by Outlining.

ORGANIZING YOUR WRITING

Outlining is crucial to the process of organizing your communication. If you use the process described here, you can avoid time-consuming mistakes as you draft your documents.

In brief, a manager or supervisor must consider every aspect of the documentation for a new product, as the Checklist for Outlines from the Sheffield Measurement Division makes clear. It covers both the manual's content and format. The list prescribes the firm's required format for outlines, and includes, as well, a list of all the "extras," the indexes, glossaries, and examples that help to create useful, comprehensive manuals. Obviously, the outline becomes a plan, like a blueprint or a schematic, for the final product—the documentation.

Outlines as Plans for Oral Technical Communications

For more information on public speaking, consult Chapter 16. The cards Ted Wilson composes are, in fact, a "portable" outline for the remarks he will make. Ted's case is typical, because managers estimate that their technical communicators spend much of their time on oral technical communications. They research a new product through classes, interviews with the designers and engineers, and user testing. Very quickly, a technical communicator learns that outlines improve oral communications, because they permit a writer to prepare questions in advance and to group related information into major headings for temporary filing and for an eventual writing job.

FIGURE 5.2

The Checklist Used to Ensure the Completeness of the Outlines for Manuals at the Sheffield Measurement Company, Dayton, Ohio.

CHECKLIST FOR OUTLINES

Here is a checklist used for producing outlines for documentation in a measurement company. Note that this checklist covers both an outline's format and scope.

1. The manual title should be descriptive for all audiences.
2. A complete draft of the preface should include:
   a. Environment—task, performer, how often, where;
   b. Purpose—function of the manual;
   c. Scope—what is included and excluded; and
   d. Required and related references.
3. Include the Table of Contents in the outline.
4. Introduction should identify, define, and introduce the manual's subject.
5. Outline format:
   I. Section Heading
      A. ----------
         1. ----------
            a. ----------
            b. ----------
               i. ----------
               ii. ----------
         2. ----------
      B. ----------
   a. Outline should go to the third level.
   b. Each section (and major subsection) should have an introduction.
   c. Headings and subheads should be specific statements usable as text titles and subtitles.
   d. Headings should be parallel in structure and logically subordinated.
6. Indicate placement of illustrations. Include a description of the illustration.
7. Include data tables.
8. Indicate placement of examples.
9. Include placement and design of forms to be used as learning aids.
10. Include a glossary.
11. Include a section on error diagnostics.
12. Outline appendixes.
13. Include a subject index for the text.
14. Include an outline for accompanying Quick Reference summary.

Courtesy of the Sheffield Measurement Company.

Outlines as Proofs of Your Managerial Competence

Most managers and supervisors of technical publications groups whom we interview report that they spend their time primarily on outlines, meetings, and budgets. That is, as implementers of a firm's policies, managers must prepare planning documents which show how their personnel will meet company goals. Thus, supervisors and managers outline publication projects to compose budgets and explain to executives how their writers will proceed. No project moves forward without a signed, dated, and approved outline for its table of contents.

CONVENTIONS OF OUTLINING

Four conventions govern the outline.

Division

If you subdivide, you must use more than one subdivision. For example, the author of the outline for the biochemistry manual (Figure 5.5) had to correct this sort of error. In section three she had section A without section B. You have two choices when this problem occurs. You may move your subdivision "up" into its major heading, or you may write a second subdivision heading, so your heading won't stand alone.

Error	Correction
V. Nucleic acids	**V.** Nucleic acids
A. DNA isolation	**A.** DNA isolation
VI. Lab preparation	**B.** DNA observation report
	VI. Lab preparation

Correction

V. DNA isolation
VI. Lab preparation

Parallelism

Make all the headings grammatically parallel. If you start with a noun phrase, creating a topic outline, finish that way. If you start with a participial, "ing" phrase, end that way, too. Correct this problem either by editing your topic outline for parallelism or by writing a sentence outline.

Error	Correction	Correction
II. Potential hazards	**II.** Potential hazards	**II.** We classify hazards three ways.
A. Nature	**A.** Natural	**A.** Natural hazards result from the weather.
B. Technological	**B.** Technological	**B.** Technological hazards are caused by accidents with chemicals or equipment.
C. Social	**C.** Social	**C.** Social hazards result from poor crowd control.

Capitalization

Capitalize the first letter of the first word and all proper nouns. There is one exception to this convention: you may capitalize the first letter of each word if you are preparing a table of contents outline.

Error	Correction
I. Starting Up	**I.** Starting up
A. Hiring	**A.** Hiring
B. Outfitting Office	**B.** Outfitting office
C. Completing Duties	**C.** Completing duties
D. Training Personnel	**D.** Training personnel
E. Printing Materials	**E.** Printing materials
II. Communicating	**II.** Communicating

Reminder: You may leave this outline in upper and lower case if you insert page numbers to make it an outline for a table of contents.

Content

Try for an appropriate level of detail in your outline. Rather than name a general category, use its more specific label. Remember, your outline should provide "headlines" or titles for divisions of your report, proposal, or manual. Even if you add more items to the outline, the greater level of detail will improve the documentation's clarity.

Error	Correction
II. General operations	**II.** What does the reservations clerk do?
A. Organization	**A.** How do I organize the front desk?
B. Discounts	**B.** What is a discount card?
C. Miscellaneous	**C.** How do I sell discount cards?
	D. What other discounts should I know about?
	E. What is a voucher?

TYPES OF OUTLINES

Organize outlines either by the roman numeral and alphabet method or by the decimal method. You also must decide whether to compose a topic or a sentence outline.

Formats for Outlines—Alpha and Decimal

We show the same outline using two different formats. The roman numeral format appears most often in the humanities and arts for research papers, articles, or the outlines of major projects. The decimal outline is used frequently in the aviation industry, in technology, and in the sciences.

Sample Roman Numeral and Alphabetical Outline

I. Editing functions
    A. Marking text
        1. Blocking paragraph
            a) Highlighting paragraph
            b) Copying paragraph
        2. Indenting paragraph
    B. Deleting text
        1. Removing a character
        2. Removing a word
    C. Inserting new text
    D. Moving a block of text
II. Terminating operations
    A. Saving to disk
    B. Loading ''saved'' text
III. Using a sample session

Comment

Construct the more "traditional" roman numeral outline using upper-case roman numerals for first-level headings, capital letters for second-level, arabic numerals for the third, and lower-case letters for the fourth. You can extend this outline; level five would use arabic numerals with parentheses, and level six lower-case letters within parentheses. We show only four levels because we recommend a simple outline.

Sample Decimal Outline

1. Editing functions
    1.1 Marking text
        1.1.1 Blocking paragraph
            1.1.1.1 Highlighting paragraph
            1.1.1.2 Copying paragraph
        1.1.2 Indenting paragraph
    1.2 Deleting text
        1.2.1 Removing a character
        1.2.2 Removing a word
    1.3 Inserting new text
    1.4 Moving a block of text
2. Terminating operations
    2.1 Saving to disk
    2.2 Loading ''saved'' text
3. Using a sample session

Comment

The decimal system allows you to add new levels of detail simply by typing in another decimal and a number to indicate the heading's place in the order. Even though easy to use, this system too becomes cumbersome at the fourth level of heading. Therefore, try to limit your outline to three levels.

FIGURE 5.3

A Finished Topic Outline and a Revised Outline for a Supertext Word Processing Manual Showing the Level of Detail Contained in a Revised Outline.

If you have wondered how detailed to make an outline, study this draft and revision. The revision is ready to be made into a table of contents outline.

TOPIC OUTLINE FOR SECTION ONE OF A WORD PROCESSING MANUAL

I. Introduction
  A. Understanding basic features
  B. Understanding basic terminology
    1. Remembering to "Boot up"
    2. Looking at the Cursor
    3. Using "Capital" letters
II. Beginning operations
  A. Booting up
  B. Selecting operation
III. Editing functions
  A. Beginning paragraph
    1. Blocking paragraph
    2. Indenting paragraph
  B. Deleting text
    1. Removing a character
    2. Removing a word
  C. Inserting new text
  D. Moving a block of text
IV. Terminating operations
  A. Saving to disk
  B. Loading "saved" text
V. Starting a sample session
VI. Starting sample lessons for content areas
  A. Editing
  B. Moving blocks of text
  C. Inserting text
  D. Composing a letter

REVISED OUTLINE FOR SECTION ONE OF A WORD PROCESSING MANUAL

A.1 Welcome to Supertext. (Level 1 Head)
  1. Have a look at what Supertext offers. (Level 2 Head)
    1.1 Obtain more than what you see. (Level 3 Head)
    1.2 Understand your screen.
      1.2.1 Follow that flashing cursor. (Level 4 Head)
      1.2.2 Create upper and lower case letters.

**FIGURE
5.3**   Continued.

   2. Start the program—1, 2, 3.
     2.1 Insert the floppy disk.
     2.2 Give the first command.
     2.3 Select your commands from the main menu.
   3. ''Write'' a text with ease.
     3.1 Start a dynamic paragraph.
       3.1.1 ''Block'' a paragraph.
       3.1.2 Indent a paragraph
     3.2 Move the flashing cursor.
       3.2.1 Step to the left; step to the right.
       3.2.2 Step up the page; step down the page.
       3.2.3 Step more quickly—almost like flying.
     3.3 Delete small mistakes.
       3.3.1 ''Zap'' a character.
       3.3.2 ''Zap'' a word.
     3.4 Insert little changes.
       3.4.1 Get to the right spot.
       3.4.2 Insert new text.
     3.5 Move a whole block of text.
   4. Bow gracefully out of the program.
     4.1 Select a method for saving text.
     4.2 Create a filename that fits.
       4.2.1 Get back to the main menu.
       4.2.2 Create a filename that fits.
       4.2.3 Read the computer's response to omitted information.
       4.2.4 Read the computer's response to duplicated filename.
       4.2.5 ''See'' the program saved.
   5. Load it all back into the computer.
     5.1 Select a method for loading the text.
     5.2 Give the filename for your information.
A.2 Try your wings with a sample session.
   1. Start the program with a ''boot.''
   2. Write a sample paragraph like a ''pro.''
   3. Practice deleting characters by deleting ''flubs'' away.
   4. Practice inserting characters to increase your speed.
   5. Save the sample and see how easily you use Supertext.
A.3 Explore possibilities in various subject areas.
   1. Try an editing exercise for language arts.
   2. Put events in order.
   3. Make history more complete.
   4. Create a letter of inquiry.

Structures for Outlines—Topic and Sentence

Figure 5.3 shows comparison topic and sentence outlines for a word processing manual. Topic outlines permit the flexibility of a shorter document. A sentence outline, on the other hand, allows you to compose a longer statement with a greater level of detail for each item. Therefore, some writers prefer to use sentence outlines.

FIGURE 5.4   Generic Outline Formats for Various Types of Reports.

Here are some sample outline formats. You can modify these to suit your projects and reports.

FORMAT FOR THE OUTLINE OF A FEASIBILITY OR RESEARCH REPORT

Audience Analysis. The reader wants to find the sections of interest quickly without reading the whole report.

    Letter or memo of transmittal
    Title page
    Abstract or executive summary
    Table of contents
    Introduction
        Scope
        Point of view
        Restrictions and limitations
    Results
        ''Body,'' that is, technical meat of the report
        Exhibits either in the text or in the appendixes
        References within the text
    Conclusion
        Recommendations
        Bibliography or list of sources
        Attachments or appendixes

FORMAT FOR THE OUTLINE OF A PROGRESS REPORT

Audience Analysis. The reader wants to read quickly about the project. The reader wants easy-to-read information.

    Summary—previous work summarized in one or two sentences
    Work completed
    Work in progress
    Work to be started
    Special or unique concerns
    Forecast of progress for the next phase
    Exhibits, such as timelines, flowcharts, etc.

FIGURE 5.4

Continued.

FORMAT FOR THE OUTLINE OF A LAB REPORT

Audience Analysis. The reader wants to learn how the observation and test were conducted; the results and facts should be clear and concise.

    Summary
        Reason for the experiment
            or
        Summary of the reports of other scientists
    Method
        Conditions, materials, subjects, equipment, measurements
        Procedure used
    Results
    Conclusions
    Recommendations
    List of references and sources
    Appendixes and exhibits

FORMAT FOR THE OUTLINE OF A PROBLEM/SOLUTION REPORT

Audience Analysis. The reader wants to learn quickly what problem has occurred and how the writer proposes to correct it.

    Summary
        Several sentences defining the problem
        Several sentences outlining the solution
    Body
        History or review of the problem
        Discussion of the type of solution needed
        Summary of the available solutions outlining their weak and strong points
    Conclusion—detailed recommendation for a specific solution.

FORMAT FOR THE OUTLINE OF A BRIEF PROPOSAL

Audience Analysis. The reader needs to learn that the writer's firm offers the best services at the most reasonable prices. (For a short proposal, write a business letter.)

    Introduction
        Statement of shared purpose—why the writer contacts the client
        Summary of the benefits the firm offers to clients
        Summary of the proposed work
    Main Section
        Technical summary
        Management plan
        Budget and costs
    Attachments
        Routine ''boilerplates'' and assurances
        Flowcharts
        Timelines

OUTLINES FOR YOUR PAPERS AND REPORTS

Figure 5.4 shows the five types of outlines younger and adult students need to complete their various class reports and occupational writing. We have designed "generic" outlines to assist you with your writing. By "generic," we mean basic patterns for organizing your prose, patterns that you can use repeatedly to define your requirements in outlining.

Advantages of Generic Outlines

We encourage you to use and redesign our generic outlines, because standard outlines of this type offer you as a writer several advantages.

Standardization. If you are a full-time student who prepares several papers per semester, perhaps the idea of standardization or uniform appearance is unimportant. However, in industry, any routinely performed task such as writing manuals must be standardized, both for training purposes and to preserve the firm's image. For example, if you work on the design team for a new automotive training manual, your progress reports to your client should all use the same outline, perhaps the one indicated here. This repeated format and plan of organization will reassure your client that your design group can indeed deliver the promised manuals on time and in fulfillment of the design plan the client requested. In addition, standardized outlines simplify the work of writing similar reports, and save time and money.

Time Management. Drafting a correct three- or four-level outline that includes annotations for illustration, and any other notations needed for factual information is time-consuming. Therefore, you should make all your decisions about sequencing the material and your plan of logical development well in advance. The generic outlines allow you to make decisions about the audience and sequence of information when you select the outline format. You then have more time for the actual creative work, writing and editing your prose.

Completeness. The generic outlines require you to decide early in your organization process which elements your report will include, which it will omit, and which information you may still need. If you must write several documents quickly, it will be difficult to remember exactly the requirements of your firm or audience for the organization or type of information in your reports. Do you need appendixes? What about a formal introduction? Instead, do you call the introduction something other than "Introduction"? All of these and other time-consuming decisions can be made before you begin to organize your information. That way, you can concentrate on the information directly. For example, you might show our generic outline for a lab report to your chemistry instructor, modify the outline to include the instructor's suggestions, and then draft each report for the semester to fit your formula. By doing so, you avoid last-minute surprises and gaps in your information.

Tips for Improving Your Outlines

1. **Use a Jot List.** Do not try to combine your pre-writing and idea-gathering with the outlining stage. (For more details on pre-writing, see Chapters 2, 3, and 4.) If you compose your jot list on small pieces of paper or index cards, one idea per sheet, then you need only rear-

range the papers into groups of compatible ideas as you begin to outline. The time saved is certainly worth the price of the paper. When your sheets are arranged into related groups, you are ready to outline. Then, you need not struggle to recall forgotten ideas.

2. **Use a Generic Outline.** Use a general, generic outline to format your document. List all the elements you must prepare, including the introduction and conclusion — so you don't forget to write them. Then, move from the generic format to specific, factual statements. Restrict yourself to four levels of headings.

3. **Develop an Adequate Level of Detail.** Ask yourself, "Can I write my paper from this outline without notes or references within reach?" If you can, you have developed an adequate level of detail. Consider the outline in this chapter for the Supertext word processing manual. That student is ready to write, because her sentence outline lists each detail every section will present.

4. **Test for Level of Detail.** Use a simple equation to test for an adequate level of outline detail:

$$\text{subject } + \text{ point of view (approach)} = \text{complete thought}$$

The typical beginner writes an outline like any of the three needing revision shown at the end of this chapter. Among their other errors, these outlines list only their topics or subjects, and not how the writer will explain the subject or what the writer's point of view is. For example, in the Norse Paint Company's topic outline, the student has listed resins, solvents, drum stock, and pigments under the category of raw materials. What does that student intend to say about resins and the rest? Writing "Resins act as catalysts to measure contamination," reminds the experienced writer that the resins will be approached as catalysts — and nothing else. The equation applies to the sentence like this:

$$\text{subject } + \text{ point of view (approach)} = \text{complete thought}$$
$$\text{resins } + \text{ act as catalysts} = \text{to measure contamination}$$

In other words, we will not read about anything else that resins do for the paint industry.

5. **Use Your Word Processor Effectively.** Draft your outline at a computer, because the computer's technology can help refine your work. First, type in your topic outline. Then copy the outline to a new file and add point-of-view statements. Copy this second file to a third and edit your new outline to meet all the outline conventions. At this time, create either a topic or a sentence outline. Do not erase your old files until your revised outline is satisfactory, because the earlier versions may have helpful information for your later outlining process.

If you have developed your topic properly, you should now be able to copy your revised outline right into your document file. The headings in the outlines become the headings of your document. All you must do is provide specific information to develop each heading. This process can save considerable keyboarding, and it will save you the difficulty of titling each section. The outline file, with numbers typed in, can then become your table of contents as well. To use this process, study the Supertext, campground, and office management manual outlines.

Finally, use your word processor to create the formats of your outlines. Some word processors provide standard, ready-to-use outline formats. However, any word processor allows you to create a conventional format using tabs and hanging indents. The options your word processor offers can dramatically lower your error rate as you compose the outline.

OUTLINES FOR MANUALS OR LONG PROJECTS

The models in this section illustrate how writers draft and refine their outlines to create useful tools for composing a document. Before we discuss correct formats for effective outlining, we ask you to first study these models for their content.

Models of Effective Content

The biochemistry lab and Supertext word processing manuals show the process as a writer works to refine the outline's content.

Refer to Fig. 5.3, A Finished Topic Outline and a Revised Outline for a Supertext Word Processing Manual Showing the Level of Detail Contained in a Revised Outline.

FIGURE 5.5   The Draft Outline of a Biochemistry Lab Manual and the Revised Version for a Table of Contents.

DRAFT OUTLINE FOR A BIOCHEMISTRY LAB MANUAL

I. Remembering safety rules in the biochemistry lab

II. Basic experimental techniques
   A. Chromatography
   B. Radioisotope techniques
   C. Electrophoresis
   D. Centrifugation

III. Proteins
   A. Determination of a Dipeptide Sequence

IV. Carbohydrates
   A. Characterization of Simple Sugars

V. Nucleic acids
   A. DNA isolation

VI. Lab preparation

TABLE OF CONTENTS OUTLINE FOR A BIOCHEMISTRY LAB MANUAL

1. Remembering Safety Rules—Nine Steps to a Safer Laboratory
2. Understanding Five Types of Chromatography

The Biochemistry Lab Manual. The draft outline shows the writer identifying and naming topics but failing to use the topic plus point of view approach to outlining. Thus, a reader has no idea how the writer intends to develop the manual. Also, the outline shows the typical problem of using only one subheading.

The revised version shows that the writer immediately corrected the problems of inadequate subdivisions and the omitted approach by writing a grammatically parallel outline. Using participial phrases to start each topic allows the writer to describe the action or type of mental process that each section of the lab manual will describe. Words such as "logging," "understanding," and "examining" create the approach to the topic.

This outline also explains how to use your word processor efficiently to avoid duplication of effort in your writing. The author refined the topic outline, then capitalized the first letters

---

**FIGURE
5.5**   Continued.

  2.1 Purifying an Acid by Ion-Exchange Chromatography
  2.2 Logging Observations and Results
 3. Counting Radioactive Particles
  3.1 Examining an Unknown Containing ^{3}H and ^{14}C using Scintillation Counting
  3.2 Logging Observations and Results
 4. Completing Four Commonly Used Methods of Electrophoresis
  4.1 Applying Three Sample Proteins to SDS-GEl Electrophoresis
  4.2 Logging Observations and Results
 5. Completing Two Types of Centrifugation
  5.1 Separating Cell Membranes by Gradient Centrifugation
  5.2 Logging Observations and Results
 6. Recognizing Four Classes of Protein Structure
  6.1 Determining the Sequence of a Dipeptide
  6.2 Logging Observations and Results
 7. Examining Structures and Stereochemistry of Simple Sugars
  7.1 Identifying an Unknown Carbohydrate Using Polarimetry and Paper Chromatography
  7.2 Logging Observations and Results
 8. Understanding the Structure and Roles of Nucleic Acids
  8.1 Isolating Bacterial DNA
  8.2 Logging Observations and Results
 9. Preparing the Lab—Required Equipment, Materials, and Solutions
  9.1 Preparing Chromatography
  9.2 Preparing Radioisotopes
  9.3 Preparing Electrophoresis
  9.4 Preparing Centrifugation
  9.5 Preparing Proteins
  9.6 Preparing Carbohydrates
  9.7 Preparing Nucleic Acids

of the words in the headings to use them as the manual's table of contents. When the manual is finished, the author need only type in the page numbers which start the sections.

The Supertext Word Processing Manual. This student started with a good topic outline for section one of the manual. The outline needs only a minor correction to achieve grammatical parallelism in section I, B, 1 through 3.

However, consider what this author accomplished by translating the topic outline into a sentence outline. The sentences ensure that the writer has enough detail. Do you have any doubts about the information each section will present?

Because this outline is a model of content, a word about tone is useful. By "tone" we mean the attitude you will take toward your audience. You create the tone by performing a thorough audience analysis and by selecting the words and sentence style that will speak most appropriately to your audience. This manual's tone will be a bit informal. Words like "zap" and expressions like "try your wings" suggest that the writer wants to play down the difficulty of learning to use a computer. In fact, this tone reflects a comprehensive audience analysis. This manual was written for elementary school teachers who have avoided using the computers donated to their school, because they were afraid of the new technology. The author, a teacher herself, selected an appropriate tone.

Models of Effective Format

The disaster manual outline, campground manual outline, and medical office management manual outline prove their authors' abilities to use the outline format conventions to enhance, rather than to restrict, their meaning and to achieve appropriate levels of detail.

The Disaster Manual Outline. This outline presents a problem because the manual has to address many different audiences, from executives to technical workers. The manual requires direct, uncomplicated statements because, theoretically, it will be used during emergencies when the users will be under great stress, with many interruptions.

The author decided to revise the topic outline into a question-format sentence outline, because questions will hold the readers' attention. Each question speaks directly to its intended audience, so its different readers can find their sections quickly. After she created a sentence outline, the author removed the numeric notations for chapter and section numbers because the manual will be color-coded for easy reference. The deletion of the numbers creates an uncluttered table of contents that readers can grasp quickly.

The question format may seem redundant in places, but remember that this is the table of contents. The questions, as headings, will be spread throughout the book. Also, the similarity of the questions reinforces an easily readable format. Since this manual is to help university personnel recover from disasters and events with tragic consequences, the catchy, informal titles of the Supertext word processing manual would be inappropriate.

The Campground Manual Outline. The actual outline for the campground management manual was four pages long. This section from Part Two shows how you should develop an outline by "keying" your writing to your illustrations. "Keying" refers to inserting notations about your illustrations into the outline, so you know where to place the artwork and what to say about it as you draft the text. An outline that keys the artwork and text can save you time

FIGURE 5.6   A Draft Outline and Its Table of Contents Outline.

TOPIC OUTLINE

  I. Objectives of Disaster Plan
 II. Potential Hazards
    A. Natural
    B. Technological
    C. Social
III. Implementation of the Plan
    A. Hierarchy of Groups
    B. Matrix of Duties
    C. Chart of EOC
 IV. Policy Group
    A. Definition
    B. Responsibilities
  V. Communications Group
    A. Definition
    B. Responsibilities
    C. Explanation of Alert System
 VI. Action Group
    A. Definition of EOC
    B. Responsibilities as a Whole
VII. Action Checklist and Description of Duties
    A. Law Enforcement
    B. Fire/Environment
    C. Medical Services
    D. Mental Health
    E. Plant Operations
    F. Personnel
    G. Housing/Volunteers
    H. Resources
    I. Food
    J. Public Information
    K. Academic
VIII. Post-Disaster Evaluation

TABLE OF CONTENTS OUTLINE

FIGURE 5.6   Continued

What Technological Disasters Might Occur?	4
When Do Civil Disturbances Become Disasters?	4
How Is the Disaster Plan Implemented?	5
What Is the Chain of Command in Declaring a Disaster?	5
What Departments Will Respond?	6
Who Will Do What?	6
What Is the Policy Group?	7
Who Notifies Public Relations?	7
Who Works with Public Safety?	7
What Is the Communication Group?	8
How Do We Alert the University Community?	8
How Do We Alert the Municipal Community?	9
What Is the Emergency Operations Center?	10
Where Do We Report for Duty?	11
How Do We Supply Necessary Resources?	12
What Is Our Chain of Command?	12
What Are My Duties as an EOC Member?	13
What Does Law Enforcement Manage?	14
What Does Fire/Environment Manage?	16
What Does Medical Services Manage?	18
What Does Mental Health Manage?	20
What Does Plant Operations Manage?	22
What Does Personnel Manage?	24
What Do Housing and Volunteers Manage?	26
What Does Resources Manage?	28
What Does Food Manage?	30
What Does Public Information Manage?	32
What Does Academic Personnel Manage?	34
How Do I Use the Checklist for Completed Actions?	35
Can We Improve Our Response?	36
What Did We Do Well?	37
What Do We Need to Improve?	38

Courtesy of Bonnie J. Knapp.

because the outline tells you whether you have included the artwork and where the artwork belongs in the manuscript. Since this manual contained more than 50 illustrations, the sequence of the artwork was as important as the text itself. Always key your illustrations to your outline before you start to write.

This student also deleted the numeric section notations after she prepared the final revision. Thus, the detailed outline, with its figure notations listed separately, can act as her table of contents, and the absence of numbers makes the list more readable.

FIGURE 5.7   A Summary Outline for a Part of a Manual and Its Expanded Version for the Drafting of the Section.

Here is how an experienced writer "fleshes out" an outline to make it a workable document into which she can simply "pour" the text. This excerpt is from the outline for a manual on managing a major campground. Can you expand your outline to the level of detail this author did? If so, you are ready to write.

SUMMARY OUTLINE OF SECTION II FOR CAMPGROUND MANUAL

II. What are the basics of National Camping Parks (NCP)?
   A. How do I organize the front desk?
   B. What is a discount card?
   C. How do I sell a discount card?
   D. What about other special cards?
   E. What is a voucher?
   F. How do I take reservations?

DETAILED OUTLINE OF SECTION II FOR CAMPGROUND MANUAL

Give Special Card Holders Privileges	17
Fig. 1. A Discount Card	20
Sell Discount Cards: Offer Information and Applications	21
Fig. 2. A Discount Card Application	22
Give Round-the-World Card Holders Extra Time: They May Be from Another Country	23
Fig. 3. A Round-the-World Card	24
Give VIP Card Holders Complimentary Camping: They May Give the Campground Free Publicity	25
Fig. 4. A VIP Card	26
Give Rental Truck Value Card Holders Discounts When They Present Rental Truck Agreements	27
Fig. 5. A Rental Truck Value Card	28
Give Senior Citizen Golden Buckeye Card Holders a 10% Discount	29
Fig. 6. A Senior Citizen Golden Buckeye Card	30

The Medical Office Management Manual Outline. This example shows that you can use an outline to receive feedback from readers before completing a manual. The author submitted a memo which explained the manual's planning and scope. She then presented a summary outline to show the book's overall plan. Next she presented a topic outline and its revised sentence outline version, indicating that she had completed all the stages of planning for writing the actual assignment. Because the manual users have some training in health occupations and applied science, the author retained the decimal notations in the completed sentence outline.

FIGURE
5.8

The Outlining and Planning Assignment a Student Submitted for Approval Before She Wrote Her Manual.

Perhaps you have wondered how much text to write for each section of your outline. Study these documents, including the student's memo of explanation of her format, her topic list, topic outline, and table of contents outline. Do you think she had difficulty allocating her text to the topics?

To: Professor Alice Philbin
From: JoAnn Grant
Date: October 20, 1987
Topic: Proposed Medical Office Management Manual

Here are the outlines you requested for my proposed manual, *Organizing Your Medical Office*. Please keep these five points in mind as you review my outlines.

1. Since all the chapters are equally important, the units are of equal length.
2. All chapters will be short, two-page units with sample forms and charts for exhibits.
3. I have grouped the chapters into four titled sections; the first section is for the medical staff and office supervisor, while the other three sections speak directly to the entire staff.
4. The manual will have spiral binding and a wipe-off plastic cover; colored tabs will provide quick references to the sections.
5. You will receive your review copy of the manual by December 10.

Here is a summary of the manual's four sections. Please study the topic and table of contents outlines attached; I will be in to see you on November 2 to obtain your comments and suggestions.

OUTLINE SUMMARY OF *ORGANIZING YOUR MEDICAL OFFICE*

I. Hiring, Outfitting Office, Job Duties, Training, Printed Materials
II. Office Decorum, Office Space, Office Communication Network
III. Billing, Insurance, Transcription, Filing, Correspondence, Supplies
IV. Dealing With Patients, Patient Confidentiality, Phone Skills, Scheduling

TOPIC OUTLINE FOR *ORGANIZING YOUR MEDICAL OFFICE*

I. Starting up
  A. Hiring
  B. Outfitting office
  C. Job duties
    1. Office supervisor
    2. Billing clerk and receptionists
  D. Training
  E. Printed materials

| **FIGURE**
5.8 | Continued. |

II. Communication
  A. Office decorum
  B. Office space
  C. Office communication network
III. Office procedures
  A. Billing
  B. Insurance
  C. Transcription
  D. Filing
  E. Correspondence
  F. Supplies
IV. Patient care
  A. Dealing with patients
  B. Patient confidentiality
  C. Phone etiquette
  D. Scheduling

TABLE OF CONTENTS OUTLINE FOR *ORGANIZING YOUR MEDICAL OFFICE*

1. First, Hire Qualified Personnel, and Train Them Carefully.
   1.1 What Kind of Staff Do I Need?
   1.2 How Do I Hire the Right Personnel?
   1.3 What Does the Staff Need to Get Started?
   1.4 How Should Job Duties Be Assigned?
   1.5 What Training Will Produce a Well-Informed Staff?
2. The Office Communication Network Increases Business and Confidence in the Staff.
   2.1 A Good Communication Policy Is Essential.
   2.2 Office Policies Can Facilitate Communication.
   2.3 Inform Patients About Six Office Policies.
   2.4 Clear Personnel Policies Lead to Better Productivity.
   2.5 Meetings Assist in the Distribution of Information.
   2.6 Continuous Feedback and Scheduled Reviews Help in the Staff Evaluation Process.
   2.7 Handbooks and Forms Keep the Staff Informed.
   2.8 Work Space Helps with Efficiency.
3. Routine Billing Procedures Cut Down on Paperwork.
   3.1 Complete Superbills for Insurance Companies.
   3.2 Remember Five Billing and Insurance Rules.
   3.3 Use Three Standard Insurance Forms for Hard Copies.
   3.4 Transcribe Physician's Reports Each Morning.
   3.5 Complete Patients' Records While They Are in the Office.
   3.6 Write Six Types of Correspondence for Physicians.
   3.7 Maintain a Controlled Inventory.
4. Make Patient Care the Top Staff Priority.
   4.1 Use the Phone Effectively.
   4.2 Keep Patient Information Confidential.
   4.3 Use Good Scheduling Techniques to Improve Staff Performance.
   4.4 A Positive Staff Attitude Contributes to Patients' Health and Well-Being.

Each item in the outline contains a statement of the topic and an indication of its scope. To further define the outline, the author decided how much space—two pages—to assign each topic. This decision helps an author to determine the topic's scope since the space available determines how much information the manual can present. Also, because this student wrote for a mixed audience, she used color-coded tabs to identify the sections for their readers.

Outlines, then, are essential to your writing process, and not just dry exercises after you finish writing your report. Rather, an outline simplifies writing. In addition, a good outline is an efficient way for writers, supervisors, and executives to communicate expectations, problems, and progress with writing projects. Finally, and most basically, the outlining phase allows you to: plan your entire writing process; anticipate problems; budget your research time for aspects of your topic that may need further development; and tailor the presentation and format of your writing for maximum impact and usefulness.

SUMMARY

In this chapter, you learned that outlining is critical to the writing process because it determines both the document's content and format. After reading this chapter, you should be able to:

FIGURE 5.9   A Draft Outline and the Revised Version for Comparison.

DRAFT OUTLINE FOR NOISE POLLUTION REPORT

I. Introduction
II. Historical Perspective
  A. Noise Concerns
  B. Conferences
III. Ordinance
  A. Background
  B. Implications
IV. Market Discussion
  A. Residential
  B. Industrial
  C. Transportation
V. Legal Concerns
VI. Design Considerations
  A. Theory
  B. Insulation
  C. Interior Wall
  D. Buffer Walls
VII. Conclusions
VIII. Recommendations

■ Explain to an associate why an outline must precede a written draft.
■ Explain to an associate why an outline is an important tool for managing the writing and production of documents in a corporation.
■ Incorporate the four conventions of outlining in your own written work.
■ Compose an alphanumeric or a decimal outline.
■ Create a topic outline.
■ Transpose a topic outline to a sentence outline.
■ Modify a generic outline for use in your own writing.
■ Revise an outline to conform with content conventions.
■ Revise an outline to conform with formal conventions.

ASSIGNMENTS

1 Here is Bill Dorsey's outline (Figure 5.9) that we described in the Case Study section of this chapter. He has, in his revision, taken the outline from a topic outline to a partial sentence outline. Revise and edit this outline to include all the outlining conventions we have discussed.

2 Study the two versions of the outline for the Layman's Guide to Automobile Technology (Figure 5.10). Revise the sentence outline for content and format.

| **FIGURE 5.9** | Continued. |

REVISED OUTLINE

1. What this study will do.
2. Why the world is so noisy.
3. The government speaks out about noise.
4. What is County Ordinance G101.7?
    4.1 How the ordinance came about.
    4.2 What changes the ordinance will cause.
5. Soundproofing and who will need it.
    5.1 Soundproofing on the home front.
    5.2 Quieting down the factories.
    5.3 Silencing the roadways.
6. What other changes may occur because of the ordinance?
7. How to quiet a noisy world.
    7.1 What is soundproofing supposed to accomplish?
    7.2 What can insulation do?
    7.3 How can the interior wall help?
    7.4 Can buffer walls restrict noise to where it belongs?
8. Putting it all together.
9. What you can do.

**FIGURE
5.10**

An Outline That Is Revised to Obtain an Improved Level of Detail but Needs
Grammatical Parallelism.

DRAFT OUTLINE

A Layman's Guide to Automobile Technology

1.0 Anti-lock brakes
    1.1 Safety advantages
    1.2 Test results
    1.3 Availability and cost
    1.4 Components
2.0 Four-wheel steering
    2.1 Invention and design
    2.2 Components
    2.3 Advantages
        2.3.1 Safety
        2.3.2 Performance
    2.4 Availability and cost
3.0 Active suspensions
    3.1 Invention and design
    3.2 Components
    3.3 Advantages
        3.3.1 Safety
        3.3.2 Performance
    3.4 Availability and cost
4.0 Adjustable suspensions
    4.1 Invention and design
    4.2 Components
    4.3 Availability and cost
5.0 Multiple valve engines
    5.1 Method of operation
    5.2 Components
    5.3 Advantages
    5.4 Disadvantages
    5.5 Availability and cost
    5.6 Reliability and maintenance
6.0 Turbocharged engines
    6.1 Method of operation
    6.2 Components
    6.3 Advantages
    6.4 Disadvantages
    6.5 Availability and cost
    6.6 Reliability and maintenance
7.0 Supercharged engines
    7.1 Method of operation
    7.2 Components

**FIGURE
5.10**

Continued.

7.3 Advantages
7.4 Disadvantages
7.5 Availability and cost

REVISED OUTLINE

A Lay User's Guide to Automobile Technology

1.0 Do anti-lock brakes make your car safer?
    1.1 How they work.
    1.2 Why anti-lock brakes make your car safer.
    1.3 A table can show you that anti-lock brakes make your car safer.
    1.4 You don't have to wait for them.
    1.5 They are a good investment.
2.0 Why would you want to steer with all four wheels?
    2.1 The newest feature available to the new car buyer.
    2.2 It is safer with four wheels.
    2.3 Honda and Mazda are where it's at.
    2.4 It's not just a novelty.
3.0 Do you have an active suspension?
    3.1 The brains and the beauty of active suspensions.
    3.2 Why you should want a car with an active suspension.
    3.3 When you can get it.
    3.4 It's not going to be cheap.
4.0 Can you adjust your suspension?
    4.1 Hope you like Ford and Toyota.
    4.2 There is a computer in there.
    4.3 What it will cost you.
5.0 Do multiple valves bring multiple benefits?
    5.1 It makes more sense to build them this way.
    5.2 The good, but not the bad.
    5.3 Reliability you will love.
    5.4 There are more and more of them.
    5.5 The price just keeps going down.
6.0 Do turbochargers really turn exhaust to power?
    6.1 Fumes can make your car go faster.
    6.2 Power to spare.
    6.3 Why you aren't going anywhere fast.
    6.4 They are everywhere for everyone.
7.0 Are superchargers super?
    7.1 The opposite of the turbocharger.
    7.2 Power off the line.
    7.3 You won't be too impressed with the top end.
    7.4 They used to be here, and they are coming back.

3 Can you improve the Norse Paint Company's outline (Figure 5.11)? The revision has format and content problems. Try to achieve a greater level of detail in the outline.

4 Work with an associate on the radon gas manual outlines (Figure 5.12).

    4.1 Each of you should improve one of the outlines.

    4.2 Confer and compare outlines, so you can make further improvements.

    4.3 Revise your outlines.

FIGURE 5.11

An Outline Needing Further Revision.

Study these two outlines, and you will see that the topic outline reads well, but the student ran into problems as he worked toward a greater level of detail for the table of contents outline. Can you reshape or revise the outline so it is more interesting?

TOPIC OUTLINE FOR THE NORSE PAINT COMPANY'S MANUAL

I. Manufacturing overview
    A. Pre-assembly
    B. Pre-mixing
    C. Grinding
    D. Blending intermediate
    E. Filtrate and filling intermediate
    F. Blending
    G. Batch adjustments
    H. Filtration and filling
    I. Quality control recheck

II. Contamination determination
    A. Raw materials
        1. Resins
        2. Solvents
        3. Drum stock
        4. Pigments
    B. Batch blending
        1. Pastes
        2. Vehicle blended
    C. Shading and batch adjustments
        1. Quality control
        2. In-process lab
        3. Final properties
    D. Batch filling
        1. Filter type
        2. Setup
        3. Quality control check
        4. Final filling
        5. Quality control recheck

4.4 Draft a memo to your instructor explaining your analysis of the problems in the outlines.

4.5 Include in the memo your plans for the structure and appearance of the radon detection manual.

4.6 Hand the outlines and your memos to your instructor and be prepared to explain your analysis and approach to the class.

FIGURE 5.11   Continued.

FIRST DRAFT OF A TABLE OF CONTENTS OUTLINE FOR THE NORSE PAINT COMPANY'S MANUAL

1. Manufacturing: How Norse makes a quality paint
   1.1 Assembling the raw materials for paint preparation
   1.2 Pre-mixing the raw materials for paint blending
   1.3 Grinding the initial batch for paint consistency
   1.4 Blending the intermediate for batch additions
   1.5 Filtering and filling the intermediate for batch additions
   1.6 Blending the batch for paint consistency
   1.7 Making the adjustments for paint properties
   1.8 Filtering and filling the batch for shipment
   1.9 Rechecking the batch for paint quality
2. Contamination determination: How Norse ensures a pure product
   2.1 How to determine the quality of the raw materials
      2.1.1 Determining contamination in the resin
      2.1.2 Determining contamination in the solvents
      2.1.3 Determining contamination in the drumstock
      2.1.4 Determining contamination in the pigments
   2.2 How to determine the quality of the initial batch
      2.2.1 Determining contamination in the paste
      2.2.2 Determining contamination in the vehicle
   2.3 How to ensure the quality through the batch adjustments
      2.3.1 Testing the paint properties for quality control
      2.3.2 Testing the paint properties for in-process control
      2.3.3 Determining the paint properties for in-process control
   2.4 How to ensure the quality of the final product
      2.4.1 How to choose the proper filter type
      2.4.2 How to choose a proper filling operation
      2.4.3 How to determine the filling quality
      2.4.4 How to process the final filling
      2.4.5 How to test the final product quality

FIGURE 5.12

A Topic Outline and Its Revised Version in Need of Rethinking and Rewriting.

Can you identify problems in these two outlines? How would you rewrite them?

TOPIC OUTLINE FOR RADON GAS PREVENTION MANUAL

1. Introduction
   a. Radon and its sources
   b. Using the manual
2. Detection
   a. Radon detectors from private industries
   b. Avoiding radon rip-offs
   c. Government-sponsored detector companies
3. Indoor Radon Reduction Methods
   a. Overview of methods
   b. Natural and forced ventilation
   c. House depressurization
   d. Sealing radon entry routes
   e. Sealing radon sources
   f. Low indoor radon levels
4. Available assistance
   a. Government assistance
   b. Private industries

REVISED OUTLINE FOR RADON GAS PREVENTION MANUAL

1. How Avoidance of Radon Gas Could Save Your Life
   1.1 The tasteless, odorless, and colorless gas called radon
       1.1.1 Sources
       1.1.2 Locations
   1.2 Effective use of this manual
2. How to Detect Radon Gas in Your Home
   2.1 Detectors from private industry
       2.1.1 Radon rip-offs
   2.2 Government-sponsored detection
3. What are Indoor Radon Reduction Methods?
   3.1 Effective use from all methods
   3.2 Natural and forced ventilation
   3.3 Avoidance of house depressurization
   3.4 Sealing of radon entry routes
   3.5 Sealing of radon sources
   3.6 Indoor radon levels
4. Where Is There Assistance Available?
   4.1 Government assistance on methods
   4.2 Contracting with private industry

CHAPTER
6

Methods
of
Development

This chapter introduces you to the various techniques you can use to discuss your topic clearly. Traditional rhetorical methods such as definition, narration, comparison and contrast, partition and classification, and cause and effect form the technical writer's stock of appropriate techniques for the presentation of new or complex material. After reading this chapter, you will be able to:

- Explain how technical writers use definition, narration, comparison and contrast, partition and classification, and cause and effect as methods of analysis.
- Select the most appropriate methods of analysis for your topics.
- Organize your technical material by combining several methods of development, so that you explain your topic with an adequate level of detail.
- Identify the various methods of development as they appear in the technical documents you use at school or at work.
- Read another writer's work, and comment on the methods the writer can use to improve the analysis of the topic.

CASE STUDY

Sue Chin, an environmental science major at a local university, must complete a technical communication course with a grade of C or better as a requirement for a Bachelor of Science degree. With pleasure, she has noticed on her technical communication syllabus that the teacher welcomes reports within the student's field of interest.

Although Sue has compiled a strong record in her biology and chemistry courses, she does not believe she can adequately or clearly explain environmental science and its impact on society to a lay audience. She wants to write her major technical report on radon as a serious contaminant of homes. But as she goes through the pre-writing steps, she has difficulty selecting the types of details she would use to describe radon, a colorless, tasteless gas.

Sue's syllabus for the technical writing course has two class sessions planned for examining previous students' projects. She wonders if she will know better how to develop her report on radon after examining how others developed different types of abstract topics. The instructor believes many of Sue's questions will vanish after she sees the other students' reports.

How typical is Sue Chin's anxiety? In our experience, her anxiety is quite typical, because inexperienced writers often feel that they cannot control and explain their topics adequately. Often, this fear occurs because the inexperienced writer does not know consciously about the methods of developing a topic. On the other hand, experienced writers manipulate the methods of development with precision, thus saving hours of planning and pre-writing.

WHAT ARE THE METHODS OF DEVELOPMENT?

A method of development is the structure used to explain a topic. For example, you may explain how a computer's disk drive works by describing its similarity to a cassette tape player. Your method of development in this case is comparison, showing how two objects are similar. While continuing to write about the disk drive, you might try to persuade the reader to purchase a 3½-inch disk drive by describing its three best features. In doing so, you use cause and effect as your method of development. You could continue. Certainly, you can define the disk drive, and you can write a narrative of the steps used to install one. You can also describe the disk drive and explain how its parts make it operational as a mechanism, thus using partition and classification as your means of development. Most often, you will combine several methods of development to explain a topic adequately.

HOW DO THE METHODS OF DEVELOPMENT INCREASE A WRITER'S POWER?

If you learn to use the methods of development consciously, your power as a writer can increase because you can plan your documents in greater detail and save time on drafting.

FIGURE 6.1 A Review Sheet Showing the Many Uses of the Methods of Development.

APPLYING THE METHODS OF DEVELOPMENT TO TECHNICAL COMMUNICATIONS

Each method of development lends itself to a particular rhetorical purpose. Although the most experienced writers combine their methods of development to strengthen their purpose, you can analyze your topic first to see which method you should use primarily. Then you can select and combine the other approaches as needed.

Definition	Narration	Comparison/Contrast	Division/Classification	Cause and Effect
Use for	Use for	Use for	Use for	Use for
new terms for lay audience	easy reading or basic reading	moving audience from known to unknown products or ideas	intermediate or literate readers	intermediate or literate readers
glossaries	technical news stories	showing how products or concepts differ	supporting formal, informal, and extended definitions	explaining how a mechanism works
legal terms for policy statements	background news of new products or processes	explaining choices to readers	explaining the details of technical or scientific matters	explaining the steps in a process
	instructions	feasibility reports; recommendation reports	outlining solutions to scientific or technical problems	"selling" an idea or product
	procedures		explaining the scope or limitations of a report	persuading others of the accuracy of your research
			justification reports	urging others to act

Planning

Since technical communicators must outline in great detail, try to make that process as useful as possible. For example, if you write for a corporation, use your outline to obtain as many supervisory approvals as possible before drafting any documents. Experienced writers either use corporate books to select the methods of development, or they describe in the outline methods of development for each section of their work. For example, a corporate policy may require a writer to define an object first, then to list its specifications, and then to write a detailed description. If not, the writer should use the outline to obtain a supervisor's approval for the definition, specification, and description approach. This level of planning enables the writer to save time on the research and organization of the materials by defining early exactly what information is needed.

Drafting

In technical and business writing, no surprises should occur at the drafting stage. The research, approach, and outline should be so thorough that the drafting is almost a mechanical process. In other words, the creativity of finding the correct approach to a topic occurs during planning and pre-writing. A writer who can consciously select a method of development saves a company the time that an inexperienced writer spends at the computer wondering how to explain a topic. Thus as a writer's experience increases, so should the writer's speed at drafting a document.

As part of our plan of organization, we discuss five methods of development here. For a full discussion of how to write descriptions, please see Chapter 15 on descriptions, procedures, and instructions. So necessary are these three methods of development for technical writing, that they require a separate chapter.

DEFINITION

Simply, a definition tells the reader what a thing is and how an object has meaning in a particular set of cases or circumstances. Definitions may be formal or informal. You may also compose an extended definition of the formal or informal type.

Formal definitions adhere to strict principles. The object to be defined is stated first, followed by a connecting verb and the category to which it belongs. Details appear immediately after. For example, a formal definition might read, "Tardiness is lateness for work as shown by the time recorded on the employee's time card." Formal definitions adhere to a strict style because they may be "for the record" of a company's policy, an insurance policy, a renter's contract, and any of the other administrative documents that govern how people conduct their business.

Informal definitions use the same principles as formal definitions, but are written in a more relaxed style. For example, in a cookbook, a student referred to ". . . the nutritional benefits of tostaditas, the fried tortilla quarters, that make a low-calorie snack when you serve them with a vegetable dip." The tostadita is defined simply and in context with the phrase, "the fried tortilla quarters," which describes the category and details about the food. The relaxed tone is more readable and less stilted than a formal definition might be.

Extended definitions are necessary to explain your subject in great detail when a short, simple statement might generalize inappropriately. To begin an extended definition, start with either a formal or informal definition of your subject. Then you use the rhetorical strategies of description, classification, comparison and contrast, and cause and effect reasoning to delineate the topic. For example, this chapter contains extended definitions of anaphylactic shock, head lice, and airplanes. After studying these excerpts, note the different types of details the students selected for their audiences. These passages show the range of possibilities you have in selecting the descriptive matter that delineates your definition.

It would be impossible for people to communicate without clear definitions. In your personal life, the definition of certain terms can affect you dramatically. For example, if you live with your parents, how do they define "automobile privileges"? The term can mean any of these definitions: occasional use of a family car, constant use of a personal car supplied and paid for by parents, or use of a family car daily for commuting but not necessarily for social activities. Other common but important terms, such as "God," "patriotism," "wealth," "good and evil," or "relationship," are usually defined with the help of your belief system, family, education, social group, and other factors that may influence your views.

Businesses rely also on definitions for every phase of their operations. For example, a firm must have a clear definition of its policies for their employees, their dependents, and the beneficiaries of the employees' insurance. Clear definitions of jobs, titles, and responsibilities can save a firm billions of dollars as they hire, promote, terminate, and retire employees. Clear definitions can also protect a firm from claims of liability or negligence by employees and customers. Finally, definitions are critical to the success of a firm's documentation, for a firm must offer clear descriptions of its products and their uses to maintain its market positions.

How to Write Definitions

Whether a formal, an informal, or an extended definition, your definition should contain three parts: a statement of the species, the genus, and the differentia. The species is the object you define. The genus refers to the category or classification the object belongs to. The differentia tells how this object differs from the other items in this genus. This chapter's examples show how to state the species, genus, and differentia both formally and informally. This list will get you started.

Species	Genus	Differentia
An airplane is	a powered aircraft	with wings and a tail.
A head louse is	a parasite.	It is a tan, wingless, cylindrical insect.
A picocurie is	a unit of measurement	of radioactivity.
A cockatiel is	a bird	believed to be a member of the parrot or cockatoo family.
Gold is	a metal	shown as Au on the Periodic Table of Elements.
A pineapple is	a plant,	an edible member of the family of bromeliads.

The definitions in this list are formal because they follow the traditional form: object to be defined, the verb "to be," the classification, and the details. Typically you should write each formal definition as one sentence, although we show an exception with the definition of a head louse. The single sentence allows you to check your definition for the correct inclusion of the three necessary parts.

Examples of Definitions

You may also define objects and ideas informally or in an extended manner. Here are some of the examples Sue Chin's teacher showed during a class discussion of methods of development.

Example—Informal Parenthetical Definition

Heat treating, a controlled warming and strengthening of steel, improves the product life of our custom engines. Technicians heat the steel to a near-liquid stage, then quickly cool it to below 400°F. At very high temperatures, the molecules of steel form patterns that are bonded by the rapid cooling. This bonding creates a stronger microstructure than that in non-heat-treated steel.

Comment

By placing "a controlled warming and strengthening of steel" in apposition to "heat treating," the author supplies the three parts of the definition, so the word can be learned in context. This method of writing definitions enables a writer to smoothly introduce new terms without interrupting the train of thought.

Example—Informal Sentence Fragment Definition

Air changes per hour (ach)—The movement of a volume of air in a given period of time. If an office has one air exchange per hour, it means that all of the air in the office is replaced in a one-hour period. Air changes may also be expressed in cubic feet per minute.

Comment

The sentence fragment allows you to substitute a dash for the verb "to be." That dash also emphasizes the term you define and shortens your sentence structure. This method of definition is very appropriate for technical writing, especially when including the details of the differentia in full sentence commentaries. However, if you have difficulty recognizing incorrect uses of sentence fragments, use full sentence definitions.

Example—Extended Definition Using Enumeration

Anaphylactic shock is a local or generalized reaction occurring within minutes after administration of foreign protein substances such as blood products or medications such as penicillin. The generalized, systemic reaction is frequently fatal.
   Symptoms include:

☐ Apprehension and flushing (reddening) of the skin
☐ Skin itching or burning—generalized itching over the entire body indicates a general systemic reaction is developing
☐ Sneezing or coughing
☐ Hives on face or upper chest

☐ Respiratory difficulty—tightness or pain in the chest, sudden irritation of the throat (itching), followed by shortness of breath
☐ Wheezing and shortness of breath, hoarseness, respiratory stridor (noisy inhalations)
☐ Pallor (turning pale) or cyanosis (turning blue)
☐ Unconsciousness, imperceptible pulse, circulatory failure leading to coma and death

Comment

Notice how much detail this author could include by writing an extended definition. Because the audience of nurses would need to read this material quickly, the author used a list for a detailed definition. When your readers must act quickly, a list is the best choice for an extended definition.

Example — Extended Definition Using a Paragraph

The head louse is a parasite— meaning it needs a host (person) for nourishment. It is a grayish-tan, wingless, cylindrical insect. While there are over 3,000 different species of lice in the world, only three types affect humans. Head lice are the most common of the human lice species found in the United States. Treatment of head lice is difficult because patients are at constant risk of re-exposure to the parasite. If head lice are present, examine anyone who has been in contact with the infection and immediately begin prophylaxis.

Comment

Because head lice are not life-threatening, the author described them in the traditional paragraph format. This method of detailed definition is effective when you write to inform rather than to call for immediate action.

How to Use Definitions in Technical Writing

These examples have already offered some ideas about how to use definitions in technical writing. There are other major uses.

☐ In manuals and instructions, define each term that will be new to your specific audiences. If many terms need defining, create a special definition section at the start of each unit, or compose a glossary for this purpose.
☐ In technical writing for the lay audience, for example in summary reports, news stories, and informational brochures, define informally and in context with phrases in apposition.
☐ In policy statements or in documents meant for a record, for example in personnel documents, write formal definitions and compose extended definitions which further explain matters such as context, time span, or responsibilities for each term you define.

Tips for Writing Definitions

1. Use the traditional, formal species, genus, differentia format we show here. Avoid sentences that state, "Heat treating is what a company does to harden steel." This sentence skips the genus by substituting "what." Instead, write, "Heat treating is a process a company uses to harden steel." "Process," a noun, names the genus or classification heat treating belongs to.
2. Do not use a term to define itself. Do not refer to heat treating as the process of heating steel. This logic error is called "circular definition." Use other, more descriptive words such

as "warm," "bake" or "braze" in the sentence that formally defines heat treating. Then, in subsequent sentences, you may use the verb "to heat" as we did.

3. Select the appropriate genus and differentia for your audience. For example, we define a cockatiel as a bird, an appropriate definition for lay readers. However, a veterinarian or an animal care specialist may expect cockatiels to be referred to as psittacines, or small hookbills. Similarly, in writing for nurses, we refer to lice as "parasites." For a lay audience, we might call them "pests," even though that term is less accurate. Objects and ideas can fall into several classifications, so create definitions for your specific audience.

NARRATION

Although narration, the process of organizing your writing by chronology, has many uses in expository and creative writing, technical writing uses it primarily to describe a sequence of actions. Narration is also used in technical and scientific journalism to describe for the public new discoveries, techniques, or processes.

How to Write a Narrative

A narrative is the simplest writing to outline because you merely have to arrange the facts or steps in chronological order. When technical communicators prepare new manuals, they use task analysis to assure that every step is included in the correct order. (For more about task analysis, read Chapter 15 on the preparation of instructions.) Sue Chin's instructor distributed to the class this example of narrative as a method of development.

Example—A Narrative Explains a Process

To make a set of direct screen separation negatives from a transparency, the camera operator first makes a cyan separation. The pancromatic (referred to as pan) film is placed on register pins with the emulsion facing up. The transparency is placed on the pins over the pan film. The mask is placed on top of the transparency. The bump exposure is made first using the same exposure time as the main exposure and the same red filter, but with a neutral density filter added to allow only about 10% of the light to reach the film.

Next, the camera operator does the flash exposure. The operator removes the mask and transparency and places a gray contact screen emulsion down against the pan film. Throughout, the screen is placed at a 105° angle. A short exposure is made using a standard flash lamp with a very dim bulb.

The main exposure occurs next. The operator places the transparency on the contact screen and places the mask on the transparency. The main exposure is then made with the same red filter used for the bump, but without the neutral density filter. The technician must be careful not to move the contact screen between exposures (or a double set of dots may appear).

Magenta separation is fourth. Again, the flash exposure will be made first. During this exposure the technician must place the contact screen at a 45° angle. This exposure should be the same length as the cyan flash exposure. A green filter is used, but no bump exposure is needed for the magenta separation.

Step five is the yellow separation. The flash exposure is made first, and it is the same length as the cyan and magenta separations. The operator makes the exposure through the transparency and a blue filter with the screen at a 90° angle.

The technician makes the black exposure last. Usually the flash exposure will be shorter than that of the three process colors. The main exposure is made through a neutral density or a salmon filter with the screen placed at a 75° angle.

Finally, the operator can check the quality of a direct screen separation process by meeting the printing industry's standard dot percentages for offset lithography. The technician must learn to adjust the exposures so the standard dot values can be met consistently, to assure the accuracy of the color printing on good quality paper.

Comment

Note that the writer has arbitrarily divided this narrative into six steps. He could have created 12 steps by describing how the technician checks that each separation is correct. Although any process has a finite number of steps, you must decide how best to group them for your audience's needs and reading level. For example, the student who composed these instructions for basic photography students and non-majors decided that the actual process is more important than the technician's method of checking for quality. You must decide what to emphasize as you create a narrative.

How to Use a Technical Narrative

Use narration to simplify, either by emphasizing the steps in a process or by focusing on the important developments in a complicated series of events.

☐ By grouping the steps of a process into five segments of text, you enable a new user to understand, for example, how to install a disk drive, or how to insert a new memory card in a computer. Carefully select the steps you emphasize, because a lay reader may assume that those steps are the only events that can occur.

☐ By using chronology as the organizational pattern, you can simplify the events that have led to a new technology or discovery. Thus, if you want to describe an unfamiliar or even a threatening process, such as the development of AIDS, a chronological pattern enables you to emphasize the 5 or 10 main points your possibly confused or prejudiced audience must know. For this reason, much scientific and technical journalism relies on chronology.

Tips for Writing Narrative Passages

1. Complete a task analysis — even an informal one, to be sure you have included every step in the process.
2. Because a chronological plan of organization can be deceptively simple, check that the plan of presentation is sensible. Show your outline to an expert if you write for a lay audience or to a user if you write instructions.
3. Do not list each step. Group your steps into the major categories that you assign to your narrative such as preparation, steps in the process, troubleshooting, and cleaning up.
4. Avoid oversimplification, an unacceptable error. Review your narrative for factual accuracy.

COMPARISON AND CONTRAST

When comparing several objects, discuss the elements they have in common. When contrasting several objects, explain their differences.

Using comparison or contrast as the method of analysis enables you to begin with a simple object or idea that is familiar to your audience — and to progress to a more complicated or less familiar one. This technique appears often in technical manuals and business reports because it offers a straightforward plan of organization and enables you to choose appropriate figures of speech for your audience.

How to Write Comparative or Contrastive Passages

Establish the specific grounds or criteria for comparing or contrasting objects or ideas. For example, to explain how a new model of computer differs from the old, compare similar features before introducing new and different characteristics. For example, Model Two shares certain features with Model One:

Ability to use the same DOS
Ability to run the same software
Ease of installation and modification

Do not discuss all the common features. Instead, carefully select the most important bases of comparison for the users.

Since Model Two is an entirely new machine, you can choose many bases for its contrast with Model One. Again, you must select only those features that explain the product to your audience, the users. The list for Model Two's enhancements might look like this:

Increased speed
Increased memory
Improved graphics capabilities

After establishing the bases of the comparison and contrast, select a plan of organization. You may discuss all the characteristics of one machine and then those of the other. Or, you may define each basis and then discuss the machines in terms of their bases. If you compare and contrast the machines, your list might look like ours.

Model One	Model Two
Same DOS	Same DOS
Same software	Same software
Easy to install	Easy to install
Increased speed (over competition)	Increased speed (over Model One)
Increased memory "	Increased memory "
Improved graphics "	Improved graphics "

If you compare the bases as they appear in the machines, your list might look like this:

Same DOS	Increased speed
Model One	Model One
Model Two	Model Two
Same software	Increased memory
Model One	Model One
Model Two	Model Two
Easy installation	Improved graphics
Model One	Model One
Model Two	Model Two

Here are the examples of comparison and contrast that Sue Chin's instructor shared with the class.

Example — Comparison — Short Passage

Comparing an active suspension to a human being makes the concept easier to understand. Instead of the metal springs found on all conventional cars, think of an active suspension as a series of links or switches worked by muscles (hydraulic shock absorbers) controlled by a brain in response to sensors and certain rules stored in the brain. If someone sits on your shoulders, your brain tells your knees not to buckle. Active suspensions work the same way. The body of the car remains almost completely neutral while the tires and suspension control the ride and absorb the bumps. (from *A Lay User's Guide to Automotive Technology* by Joe Hartsel, p. 14.)

Comment

Note that Joe Hartsel establishes a very specific basis for comparing a car and a human being. Since this comparison is quite common, Joe avoids a trite figure of speech by focusing in on the idea of suspension.

Example — Contrast — Long Passage

Since you cannot see or smell radon, you need special equipment to detect it. The two most popular commercially available radon detectors are the charcoal canister and the alpha track detector. Both of these devices are exposed to the air in your home for a specified time and sent to a laboratory for analysis.

The charcoal canister is an activated carbon packet that is exposed in your home for a time and then sent to a lab for analysis. The test period in the home usually lasts between three to seven days. An additional two weeks is needed for the company to analyze the test, after which a written report will be issued to you. The approximate cost for the charcoal canister is in the $10 to $25 range depending on the company you choose to administer the test. The charcoal canister is a cheap, efficient method of getting an initial reading for the concerned homeowner.

The alpha track detector is a much more elaborate test than the charcoal canister. Because of its higher cost and its longer test period, the homeowner is assured of a much more accurate reading of the level of radon in the home. The minimum test period lasts between two to four weeks and the test usually costs almost double that of the charcoal canister test. Although the costs are high, a discount is usually offered for multiple purchases. Because of the extensiveness of the test, the alpha track detector is used in a follow-up fashion rather than as an initial reading. (from *The Detection and Control of Radon Gas* by Chris Bogart, pp. 3–4.)

Comment

Contrastive passages must first build from similarities. Chris Bogart points out the charcoal canister's and the alpha track detector's similarity: they both measure radon. He then describes their differences in terms of costs and uses. This type of passage, developing a whole-to-whole comparison, relies on the first type of list or outline we showed.

Example — Mixed Comparison and Contrast

Honda's four-wheel steering (4ws) design is called SAD (steer angle dependent). This means the rear wheels either phase or anti-phase depending on the magnitude of the steering input (the distance the driver turns the steering wheel in relation to the speed of the car at that given time). A computer reads the steering input and velocity of the car, calculates the magnitude, compares the magnitude to values programmed into the computer, and adjusts the rear wheels to phase or anti-phase based on the calculated data. For example, low steering inputs at normal highway speeds cause the rear wheels to phase with the front, while high steering inputs at low speeds, such as parallel parking, cause the rear wheels to anti-phase with the front wheels.

The Mazda 4ws system is much more advanced than Honda's SAD, but not as near to production. Both systems use computers to determine phase and anti-phase, but the Mazda design uses other computer software, specialized valves, and electronic switches to control the unit, while Honda's incorporates very basic mechanical parts. Mazda tries to eliminate the mechanical parts from its design to save weight and space underneath the car. Mazda will also add sensors to its design to monitor road conditions and minor steering adjustments made by the driver. Because Mazda's design is considered the most advanced, the company refuses to release more specific information concerning their design and progress. (from *A New User's Guide to Automotive Technology* by Joe Hartsel, p. 10.)

Comment

Although this passage is mostly contrastive, Joe relies on comparison to introduce the new Mazda designs in paragraph two. Much of your own comparative and contrastive writing will use this mixed method of development.

How to Use Comparison or Contrast

Use comparison or contrast extensively as you write for a lay audience because this method of analysis, more than any other, enables you to progress from a familiar object to a less familiar one.

- ☐ Use comparison to explain unfamiliar objects a reader cannot examine, such as the inside of a disk drive or the workings of a nuclear reactor.
- ☐ Use comparison to appeal to a user's brand loyalty as you prepare to describe new products; then use contrast to explain the special features of a new product.

☐ Use comparison and contrast liberally in your professional reports. This technique permits you to outline the results of a feasibility study by discussing the strong and weak points of several solutions to a problem.

Tips for Writing Comparative and Contrastive Passages

1. Restrict your bases. You cannot compare every important aspect of two or more objects, mechanisms, or ideas. Select only the most meaningful bases.
2. Restrict the number of subjects you compare or contrast. Theoretically, you can do as many as you wish. Realistically, restrict your plan of organization to no more than four subjects.
3. Use illustrations to emphasize your contrasts. Use exploded views to show the readers how specific mechanisms differ.
4. Use lists, tables, and graphs to illustrate comparisons. As you describe the similar objects, show lists of their specifications or measurements to create the correct level of detail.

DIVISION AND CLASSIFICATION

Intermediate writers sometimes confuse this method of analysis with comparison and contrast because both methods require the writer to group and to separate ideas and objects.

How to Write with Division and Classification

Division, or "partition," is the process of dividing an object or idea into its parts. *Classification* occurs as you group things into their appropriate categories and subcategories. Both of these processes allow you to show the relationship of ideas or objects to each other. Use the same outlining techniques of whole-to-whole or basis-to-basis as you would for comparative and contrastive materials.

Example — Division

Pruning shapes a plant, and it induces new growth. There are two types. Heading refers to the removal of branches back to a side branch or a lateral bud; it controls the overall size of a plant. Heading allows plants to grow more dense and compact. Thinning refers to the removal of the interior branches. It allows more light to reach more of the plant. Thinning can improve a plant's wind resistance by allowing more air to pass through the plant.

Comment

This example shows clearly that division enables the technical writer to describe the parts of a mechanism or process. Division also enables you to show how the separate parts operate together as an efficient machine.

Example — Classification Used Formally

Although all airplanes are aircraft, not all aircraft are airplanes. The classification begins with categories. Within a category, specific classes are identified. Within a class, types are identified.

Category. Aircraft that use the same method of staying aloft and similar means of propulsion are grouped into the same category. This is the broadest classification of aircraft. The FAA recognizes four categories of aircraft: lighter than air, rotorcraft, glider, and airplane.

Class. Within each category, aircraft with similar operating characteristics are grouped into a class. Within the airplane category there are four classes: single-engine land, single-engine sea, multi-engine land, multi-engine sea.

Type. When referring to a specific make and model of aircraft, you are defining its type. You may be familiar with several types: Boeing 747, McDonnell Douglas DC-10, Cessna 152.

Comment

At a glance, you see the scope that classification allows you to develop for a topic. Classification helps you to show the full dimension of a strange mechanism or idea, and how the object or idea fits into a system. Thus, the Cessna 152, a type of the class of single-engine land planes, belongs in the category, "airplanes."

Example — Division and Classification Used Informally

Phosphorus, nitrogen, and potassium are the major food sources of plants.

At the initial planting, you should apply a high phosphorus fertilizer. Phosphorus helps the roots to grow and support the entire plant. If you apply phosphorus in the fall as the foliar growing season ends, you will see a marked improvement in your plant the following spring. Because phosphorus encourages the production of fruits and flowers, you should feed all your flowering plants a phosphorus-rich fertilizer.

Nitrogen assists leafy growth. You can apply nitrogen anytime after your plant's first winter. You should, however, establish your plant's roots with phosphorus first, and then apply nitrogen. If you begin applications during mid-summer, your plant will have had time to absorb phosphorus to strengthen its roots, and its leaves will benefit from nitrogen for the remainder of the growing season.

Potassium is also essential for healthy plants. It promotes resistance to cold and disease, and overall vigor. Add a small amount of potassium at planting time and regularly throughout the season.

Nitrogen, phosphorus, and potassium, N–P–K as they are referred to on package labels, are needed in varying amounts at different times. A typical package may contain this notation: 20–10–10. This note means the formula contains 20% nitrogen, 10% phosphorus, and 10% potassium. The rest of the fertilizer is filler, generally water or sterile granules, and trace elements. Use this standard formula regularly, and ask your local garden expert for a recommendation before you experiment with other formulas.

Comment

This passage, written in a relaxed style, explains how to use three aspects of fertilizers — nitrogen, potassium, and phosphorus. This simple narrative for any new gardener shows that division and classification can simplify an unfamiliar topic for a very specific group of users. Note that the author has kept the bases simple. He discusses what each mineral can do for a plant's growth and which time of year is appropriate for each application.

How to Use Division and Classification for Technical Material

As with definition, division and classification are essential techniques for any writer preparing new product data, manuals, instructions, procedures, and descriptions.

☐ Use classification immediately after defining a new product, so the user knows how the product relates to the product line.

☐ Use division to explain the relationship of parts to the whole and to create a complete parts list for each product described.

☐ Use division and classification to explain technical problems, their categories, and the types of solutions available.

☐ In report writing, use divisions and classifications to explain the scope, limitations, and categories of the problems you define, the investigations you complete, and the solutions you propose.

Tips for Developing Passages with Division and Classification

1. As with comparison and contrast, keep the bases reader-oriented and specific. Select your bases carefully.
2. Create logical categories, and keep your subcategories equal to each other in meaning. For example, we can build a classification of the term "artist." Within the category, we can name many classes, but we choose "graphic artists," "musicians," and "writers." We skip other significant categories including "dancers," and "actors." We then compose logical lists for the categories.

Graphic artists
    Painters
        Georgia O'Keefe
        Vincent Van Gogh
    Sculptors
        Rodin
        Calder
    Printmakers
        Herbert Fink
        Salvador Dali
Musicians
    Stringed instruments
        Isaac Stern
        Itzhak Perlman
    Singers
        Kathleen Battle
        Joan Sutherland
    Percussion
        Vladimir Horowitz
        Buddy Rich

Writers
    Novelists
        Toni Morrison
        Anne Tyler
    Poets
        H.W. Longfellow
        W.B. Yeats
    Dramatists
        Harold Pinter
        Arthur Miller

For each logically defined class, we name two types. We could easily arrnage our hierarchy some other way. Instead of a system that lists Artists, Writers, Classes of Writers, and Names of Artists, we could have Artists, Writers, Periods of Writing, and Names of Writing. The requirement is that the classes correspond logically.

3. Prevent your categories from overlapping. For example, if you discuss printers as a type, define the category specifically, so you don't have to include both people who operate printing presses and the artists who design the printed artwork.

4. Use logic effectively. For example, if your categories arrange themselves naturally in a chronological order, then arrange them in that order in your manuscript.

CAUSE AND EFFECT

The cause and effect method of development is rather subtle because it requires you to move from facts to general conclusions which must result from the stated facts. Because whole books exist on the complicated subjects of causality, logic, and reasoning, we restrict our discussion to deduction and induction.

How to Write with Cause and Effect

Deduction, the simplest form of cause and effect reasoning, requires you to state your general premise and move to its specific implications. After listing the most specific case of the premise, you draw a conclusion for the reader. The reasoning might read as we describe here:

Major premise
> All undergraduates' technical reports are due on March 30.

Minor premise
> My instructor assigned a technical report.

Minor premise
> I am an undergraduate.

Conclusion
> I don't need to ask when the paper is due; it is due March 30.

Example — Deductive Reasoning

The research in this area of writing produces two arguments. Early research asserted that the right-ragged style gives the writing a more relaxed appearance that was more inviting to the reader. Researchers said that the reader's eye strayed to the wrong line less often if the text was right ragged. Justified text produced rivers of white (i.e., margins) that are distracting. The reader's eye had to do the extra work of adjusting to the spaces between the words (Goswami, 145). Publishers and printers reacted drastically to this research and many changed the entire format of all their magazines and books.

Currently, additional research in this area states that this popular belief may not be true. Stacy Keenan conducted experiments at AT&T Bell Laboratories a few years ago and found that extreme variability in line length actually slows reading because of the disrupting eye movements on return sweeps (moving from the end of one line to the beginning of the next line). When the text is justified, the reader builds an internal map of the beginnings and endings of lines. The brain then executes return sweeps automatically (Keenan, 73). The reader does not have to adjust to different line lengths and, in fact, reading may be easier. The many magazines that changed their formats may have done so in vain. (from *The Impact of Current Language Research on User Manuals*, by Kathleen English, unpublished Honors Thesis, Bowling Green State University, December 10, 1987, p. 3.)

Comment

Kathy English's passage is a classic of deductive reasoning. First, she states that she will discuss two viewpoints. Then she offers examples of each. Finally, she draws a conclusion from the facts: "The many magazines that changed their formats may have done so in vain." Notice the pleasant restraint in the presentation of her conclusion. It is rational, logical, and acceptable. This passage also exhibits the best of traditional deductive reasoning, because Kathy English relies on proofs, others' published research, as her premises. Her reference to the works of experts gives her conclusion validity.

Example — Inductive Reasoning

Recently, the general aviation industry has been plagued with a shortage of aircraft. Currently, very few single engine or light twin airplanes are being produced by manufacturers. Research shows, however, that since 1983, the actual number of hours flown in general aviation airplanes has increased from 34 million hours in 1983 to 36 million hours in 1986. This means that more people are flying their older airplanes. This new trend of "hanging on" to older airplanes has brought about the need for more aircraft refurbishing facilities. I have also discovered that few professional aviation paint shops exist in the entire Ohio, Kentucky, and Indiana area. The operation of a professional aviation paint shop could fulfill the needs of this entire area.

Comment

You can see that inductive is more subtle than deductive reasoning. This student has used a typical formula for an inductive paragraph: a thesis statement introduces the major premise for discussion, sentences two through five list minor premises (statements about the aircraft shortage), and sentence six introduces the conclusion that, "The operation of a professional aviation paint shop could fulfill the needs of this entire area."

How to Use Cause and Effect for Technical Material

As you review the examples of students' and technicians' writing in this book, note how frequently cause and effect reasoning appears. You literally cannot avoid cause and effect reasoning in your occupational writing.

☐ Use cause and effect to persuade readers of your research's validity, your efforts' worth, and the most beneficial course of action.

☐ Use cause and effect, specifically persuasion, to "sell" clients and customers your ideas, products, or services.

☐ Use inductive reasoning if you must discuss a topic in a low-key, nonthreatening, persuasive style.

☐ Use inductive reasoning if you must lead a lay audience through a discussion of a complicated topic.

□ Use deductive reasoning in formal business reports; its obvious strategy, straightforward method of organization, and direct appeal read easily.

□ Test your logic to avoid the errors listed here.

Tips for Using Cause and Effect Correctly

1. Avoid the sentence error called post hoc, ergo propter hoc. This term means "after this, therefore because of this." You cannot assume that minor premise 1 causes minor premise 2 simply because minor premise 2 is second in the plan of organization. For example, if an ethics law is passed, and in the following year ethics violations increase, can the opposing political faction assume that the ethics law caused the violation? In another case, if a shrub dies a week after it is fertilized, did the fertilizer kill the plant? A careful investigator should look for other causes. Remember, the minor premises are the factual bases for your conclusion. They do not cause each other.

2. Avoid the reasoning error called "begging the question." You make this error by assuming the conclusion you wish to prove is true. For example, do not write, "The Japanese cherry tree is the prettiest shrub you can plant, because it is more beautiful than any other ornamental tree." According to whom? What makes it prettiest? Have you shown evidence that it tested prettiest when it was shown to various focus groups?

3. Another logic error is a research error, an inadequate sample. Our cherry tree sentence shows this error because it shows no sample. Avoid the subtler forms of this error that contaminate our daily conversations. For example, you cannot write, "The new generation of IBM computers is best for our documentation system. These machines are used at the Smith Company where my friend works, and he likes them." One person is an inadequate sample. Correct this problem by listing several factual premises as we did in our examples.

4. Television networks, particularly during commericals, bombard us with faulty reasoning, as does the print media. For example, should you drink a soft drink because your favorite singer endorses it? Should you buy a car because it's linked to patriotic feelings? Should you agree with the loudest political advisor or vote for the candidate highest in the polls? You must use great caution in the composition of the persuasive passages of your technical materials. Test these passages on impartial readers and on sample product users, so they can examine your logic.

SUMMARY

In this chapter, you learned that your conscious choice and control of your methods of developing technical prose can increase your speed and efficiency as a communicator. You have read about five methods of analysis of prose, you have examined excerpts of students' technical writing, and studied suggestions for improving your own methods of presentation. After reading this chapter, you should be able to:

■ Identify and prepare formal and informal definitions organized by their species, class, and differentia.

■ Develop technical narratives with a chronological approach to tell the story of a product, process, or idea.

- ■ Use comparison and contrast as separate or combined methods of development; identify errors of comparison and contrast in the passages you read.
- ■ Apply partition and classification to technical material in product descriptions and in explanations of systems.
- ■ Use cause and effect reasoning throughout your narratives, but avoid the reasoning errors that appear frequently in popular and commercial communications.
- ■ Analyze the needs of an audience for a particular communication, so that you can select the most appropriate methods of development.
- ■ Control your writing well enough to be able to annotate your outline with your plan for its methods of development.

ASSIGNMENTS

1. By now you have selected a topic and prepared an outline for a major report. Copy your outline and mark each section with your plans for its development and presentation. Select a partner in class, or your instructor may assign one. Confer about your mutual outlines and plans for their development. Obtain your partner's advice. Now rewrite your outline so it is neat and readable. Your instructor may ask you to submit this outline for a conference or a grade or both.

2. Turn to our chapter on short reports. Select a report and "annotate" it. Specifically, mark with a pencil the methods of development used in the report. Prepare to speak with the class for about three to five minutes about the report you reviewed. How would you improve its analytic power?

3. Repeat this assignment with a selection from our chapter on long reports.

4. Examine several periodicals that are important in your present or future occupation. Find an article or several passages you think are exceptionally well developed. Copy those passages, annotate them, and prepare to explain to the class why you think these passages are well written.

5. Choose a topic of science or technology about which you know very little. For example, you may know nothing of radon contamination, Sue Chin's topic. In your library, examine articles on this topic for the lay audience. Copy and bring to class the three articles that interest you most. Be prepared to explain why these articles are effective, what methods of development the authors used for the lay reader, and what types of questions about your topic these articles answer.

CHAPTER 7

Effective Revision and Editing

This chapter introduces the techniques of revising and editing technical documents. After reading this chapter, you will be able to:

- Read a document and prepare a set of suggestions for its revision.
- Identify simple ways to improve your writing with strong verbs, specific nouns, and powerful sentences.
- Use or design a checksheet that will help to improve your documents.
- Understand the uses and limitations of readability formulas.
- Decide how extensive the revisions and editing must be in order to prepare a well-written manuscript.
- Prepare a revised document to an individual's or a firm's editorial specifications.

CASE STUDY   Revising and Rewriting for a Better Grade

Laura Heberlein, a junior majoring in marketing, is nervous about her required technical writing course. She avoided this course until junior year, and she now fears a bad grade.

Assigned to write a set of instructions for a simple process, Laura Heberlein wrote instructions for planting a lilac bush because she works at a local nursery, where she acquired some valuable experience in retailing. The instructor returned the instructions with a "no grade" notation, meaning the document needed major revision before it would be suitable for a grade. Although the instructor commended Laura for her attractive format and her understanding of the process's needed steps, he listed various errors in sentence structure, spelling, punctuation, pronoun reference, subject-verb agreement, and other, more general types of problems.

The instructor stated that, while he wanted to grade Laura's assignment, what she had presented was a "discussion" draft of her ideas, not a finished draft for review and evaluation. Because Laura seems to have major problems with her writing, the instructor suggested she collaborate with a classmate to improve her grade. Laura's first draft is at the end of this chapter in the assignments section. Your instructor will ask you to read and comment on the draft, to make the necessary revisions, and to present the improved document for an evaluation and possibly a grade.

Laura Heberlein's case is somewhat typical. A student with a successful record in her major enrolls in a writing course and learns that substantial improvements are necessary to achieve a better-than-average grade. After dealing with the initial discouragement, the student's next reaction is, "How do I fix this paper?" Often the student feels that the instructor has marked every word for special attention and that the paper will never be good enough.

Actually, a student such as Laura learns that revision is a job like any other. As with any other large project, writing and revision can be accomplished through a clearly defined, well-organized set of steps. In some fashion, all writers pass through the stages of revising and editing this chapter discusses. Before we proceed, here are some operational definitions of the terms this chapter uses:

Editing—This term refers to the marking of a draft that has already undergone one or several revisions. Editing includes minor revisions such as the transposition of certain words and phrases, word substitutions or changes, and marking for both usage and printing. Think of editing as very detailed cosmetic surgery.

Reviewing—A process in which you read the work of other writers in order to comment on the improvements, changes, and deletions that can make the document easier to read or use. Think of reviewing as consultation with a doctor before surgery.

Revision—We use this term to refer to all the rewriting, changing, deleting, and general alterations you make on a draft of a document. Think of revision as major surgery.

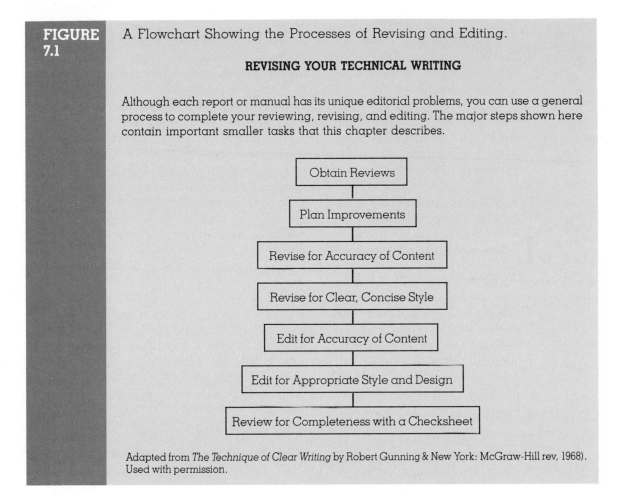

FIGURE
7.1
A Flowchart Showing the Processes of Revising and Editing.

REVISING YOUR TECHNICAL WRITING

Although each report or manual has its unique editorial problems, you can use a general process to complete your reviewing, revising, and editing. The major steps shown here contain important smaller tasks that this chapter describes.

Obtain Reviews

Plan Improvements

Revise for Accuracy of Content

Revise for Clear, Concise Style

Edit for Accuracy of Content

Edit for Appropriate Style and Design

Review for Completeness with a Checksheet

Adapted from *The Technique of Clear Writing* by Robert Gunning & New York: McGraw-Hill rev, 1968). Used with permission.

Typically in industry, your work may be reviewed, revised, and edited several times before it is distributed to customers. First you need to understand how to read a document for reviews.

READING A DOCUMENT FOR REVISION AND EDITING—FOUR STEPS

We encourage students to work together as they review, revise, and edit their assignments. This practice can help you prepare in several ways. First, you can practice working with others as you accept their suggestions for improvements and as you offer your own constructive criticism. This collaboration can help you prepare for a career in industry, where your supervisors, customers, and associates will review your documents. Second, collaboration tends to produce good writing because you can get suggestions from people who are still "fresh" as they approach your topic. As with any complex task, you achieve better results if you define your reviewing as a series of tasks. Consider these four steps:

Step One—Read for Overall Comprehension.

At this point, read the document to understand what the author wants to say. Do not start marking errors. Instead, read steadily and thoroughly, to learn about the document's audience, objectives, and overall content. Refrain from judgments. Concentrate on learning about the author's material.

To ensure your accuracy and concentration, when you have finished reading the document, write one sentence on a sheet of $8\frac{1}{2}''\times11''$ paper that summarizes the content. For example, of a recent process analysis paper, a student wrote, "Jim (the author) has written five pages describing what a carousel slide projector is and how to load slides into the tray of the mechanism." Such a sentence can help you to focus on the author's objective.

Use your sheet with the summary sentence on it to complete the review steps. In fact, avoid writing on the author's draft. In multiple reviews, a "dirty" draft becomes confusing, even causing the reviewers' comments to be ignored. Also, a heavily marked draft can cause the writer to feel as Laura did when she received her ungraded lilac bush instructions. You will elicit cooperation by keeping your comments off the manuscript and on a separate sheet unless you are told to mark up the copy.

Step Two—List All the Positive Features for Development.

Quickly make a list of the items done well, the document's strengths that the author can develop to improve the copy. For example, on the instructions assignment we mentioned, the reviewer commended the simple format, well-chosen illustrations, good punctuation, and simple direction. Your list not only salves the author's ego, it also identifies the author's best efforts, so he or she can repeat them in the revisions and in future documents. The list also provides a starting point for the conversation you will have with the author about the document's revisions.

Step Three—List All Areas Needing Improvement.

The phrasing of this step is no accident. You should avoid thinking in terms of the document's "weaknesses" because defining what is negative or missing takes a great deal of time and explanation. Concentrate instead on the best method for revising the document. For example, a weakness occurred in the slide projector directions, because they were too general. In a positive manner, the reviewer stated, "Jim, you have two examples in this entire text. Could you insert two examples for each step?" Although this suggestion involves rewriting for the author, the editor has identified specific improvements Jim can make. He can avoid the frustration of guessing what is "too general" because the reviewer has requested a specific improvement.

Step Four—Write Instructions for Improvement.

For this step, use an enumerative style and write in the active voice, so your suggestions are easily read. List the four or five major improvements the author must make. For example, on Jim's instructions, the editor wrote in the imperative mood.

1. Proofread for the word "visual" throughout, and fix its spelling.
2. Write a warning about the light bulb. Tell the user not to touch a hot bulb. Also run fan to cool the mechanism.

3. Double-check Figure 3—you have the cord cover on the left bottom, and it is on the right bottom.
4. Boldface the troubleshooting steps. Sequence is okay, but I need to see them more quickly.

When you have finished the four steps, sign and date your review, so the author can keep an accurate file of reviews and reviewers.

How to Discuss the Review

It is better to discuss the document than simply to return a sheet of comments to the author. Our students have their conferences in class; you might also schedule a conference for the library or some other location the author selects. Place your comments next to the document and use the comments—not your personal reactions to the document—as your starting point. Tell the author about the items you have listed on your comment sheet, and ask the author to comment on your review. In short, the conference should be a mutual sharing of ideas about the document. At the conference's end, present your comments and the manuscript to the author.

A Word About Time

We have found that the review sessions are most productive when the reviewer practices time management. For example, in our classes, the students complete the four steps in 20 minutes, and then have 10-minute conferences. If you read with attention, one thorough reading should allow you to find the areas needing the greatest revision. Therefore, you should make an appraisal when you receive a document for review. For example, block out 30 minutes for your reading and written comments about a medium-length document. Allow 50 minutes for a document of 10 to 20 pages. Read steadily, avoid interruptions, and try not to reread. Remember that your overall impression, which will form as you continue to read, is as important for the revision and editing process as is your attention to the details of correct usage.

REVISING YOUR DOCUMENTS

Revising and editing are very specific language skills, which require a feeling for appropriate language. Because an effective editor can alter a document in many ways, we prefer to start with the most obvious improvements you can make. After mastering these techniques, you can explore the many nuances of professional editing, a highly creative and rewarding field.

Most stylists concentrate on three ways to improve your writing.

1. Use strong verbs.
2. Select specific nouns.
3. Develop powerful sentences.

Although numerous other techniques can also improve your style, these three steps alone can significantly strengthen your documents. Let's examine each item in detail. (We also urge you to consult your Handbook section for the fine points of correct usage.)

Use Strong Verbs

Use active instead of passive voice verbs. Research tests and users' responses indicate that readers remember text in the active voice better. Review these examples of the passive voice transposed to the active voice.

Passive Voice

A technique which has been used to increase horsepower from a given displacement in an internal combustion engine is turbocharging.

Active Voice

Turbocharging **increases** horsepower from a given displacement in an internal combustion engine.

Comment

The use of the active voice eliminates eight extra words in this sentence and increases its readability.

Passive Voice

For 20 years, turbocharging has been applied to gasoline engines, and in the last decade the process has been commonly applied to production automobile engines.

Active Voice

For 20 years, engineers **have applied** turbocharging to gasoline engines, and in the last decade **engineers have applied** the process to production automobile engines.

Comment

Active voice introduces a specific subject, "engineers," into the sentence, making the level of detail more specific.

Passive Voice

Bearing failure and heat migration to the compression turbine can be prevented by oil-cooled bearings and connecting shafts between the exhaust turbine and the compressor.

Active Voice

The oil-cooled bearings and connecting shafts **prevent** bearing failure and heat migration to the compression turbine.

Comment

Active voice clearly shows the "doer" and "receiver" of the verb's action, because the subject appears before and the object after the verb.

We understand that certain forms of writing require the passive voice to create the level of objectivity that particular types of subjects require. For example, most medical research writing and applied medical writing, such as postoperative reports for insurance forms, is done in the passive voice because the tradition of science requires that the presence of the "doer" of the action be kept in the background. Similarly, the authors of reports in biology, chemistry, and related fields should use the passive voice because the style books of these disciplines require the absence of the experimenter's views and personality from the text. These disciplines adhere to strict, traditional forms tied to the history and philosophy of science. The passive voice has its place.

However, in all writing for a lay audience, use the active voice frequently. This is such an important editorial task that many computer editing programs now count the number of passive-voice verbs in the text. Many programs print the sentences with the passive voice so the author can easily revise them. But you do not need an expensive software package. Here is what our students do:

1. Use a pencil to circle all the passive-voice verbs in your draft.
2. Isolate the wordiest, most troublesome sentences by underlining them.
3. Transpose these sentences to the active voice, checking that you have used specific nouns as subjects and for descriptive, active-voice verbs.

To assist with this task, make a "verb help list" before you start to compose. For example, if you decide to write about turbochargers and superchargers, write down all the actions these processes involve. Here are some verbs we thought of:

Verb Help List for Turbo and Superchargers

apply	defeat	produce
attach	descend	promote
blast	dissipate	pull
blow	increase	run
compress	operate	use

After making the list, place it where you can see it. The list allows you to find descriptive, active-voice verbs as you draft your documents. You may add to the list as you write. When you rewrite, the list assists in the transposition of passive voice to active by eliminating that feeling that you are helpless to "think up" descriptive, active-voice verbs.

Select Specific Nouns

Early drafts may seem dull because they lack a specific level of detail about the subject. Although several techniques of style for creating an appropriate level of detail are available, you should first learn to use descriptive, factual nouns as subjects and direct objects. Create further details by choosing appropriate adjectives to qualify and emphasize the nouns. Consider these examples from the essay on turbocharging:

General

An ordinary gasoline or diesel motor uses the pistons to draw the air or air-fuel mixture into the cylinders for power.

Specific

A normally aspirated internal combustion engine uses the vacuum of the descending pistons to draw the air or air-fuel mixture into the cylinders at the ambient air pressure for compression and combustion.

Comment

The specific sentence substitutes the correct, technical nouns and adjectives for the general words, thus improving the factual accuracy of the sentence.

General

The first fan acts as a drive fan which is attached to the second fan, which pushes air into the engine.

Specific

The exhaust gas drives the first turbine, connected by a short shaft to a second turbine, the high-speed air compressor.

Comment

We revised this sentence for verbs and nouns. The verb "drives" now substitutes for the less specific "acts." The level of detail improves greatly with the insertion of the correct technical noun, "turbine," and the adjectives and noun, "high-speed air compressor."

General

Surfaces that receive increased pressures due to increased horsepower must be improved to lengthen wearability.

Specific

The special valves, pistons, and bearings with closer tolerances control the raised cylinder pressures that produce the increased horsepower.

Comment

For "surfaces," we have substituted the names of the specific parts receiving the wear. We have inserted "the" to make the sentence easier to read. For "receive" we have substituted the more emphatic "control." Again, specific nouns arranged in an obvious subject-verb-object order improve the sentence's clarity and level of detail.

Develop Powerful Sentences

Two techniques, coordination and subordination, create emphasis and balance within a sentence. When you use the technique of coordination, you link two or more ideas in one sentence with a conjunction such as "and," "but," "or," "for," "nor," "yet," or "so"—the coordinating conjunctions. Subordination, on the other hand, allows you to group or classify ideas as dependent upon one another. Subordinating pronouns and adjectives such as "that," "which," or "who" and subordinating conjunctions and adverbs such as "while," "when," "because," and "where" help you show how one idea relates to another by time, place, cause, effect, or purpose.

For review, here are sentences of each type about the same topic.

Coordination

You hear the pitched whine of turbocharging in the familiar whistle of the diesel locomotive, and you can sometimes hear the same whistle in the smaller performance cars.

Note how the use of "and" makes the two clauses independent and equal in importance within the sentence. The "and" creates a balanced style.

Subordination

You hear the pitched whine of turbocharging in the familiar whistle of the diesel locomotive because the manufacturers need not control the noise volume on trains as they must for automobiles.

Note how the "because" establishes the relationship of causality between the main clause about the pitched whine and the subordinate, explanatory material about the manufacturers.

For more information on the uses of coordination and subordination and for lists of conjunctions and adverbs that will help create emphasis in your sentences, check any grammar handbook. These two tools will substantially emphasize the meaning of your technical writing. Here are the examples of the improvements we made in sections of the turbocharging description.

Lack of Emphasis

Engineers have used turbocharging for over 50 years. It is a simple process. The turbocharging compresses the air or air-fuel mixture before the engine requires it. The compression causes the engine to use a carefully limited quantity of fuel. More power is produced, though.

Emphasis through Subordination

Engineers have used turbocharging for over 50 years. A simple process, turbocharging compresses the air or air-fuel mixture before the engine requires it. The compression causes the engine to use a carefully limited quantity of fuel, while producing more power than a normally aspirated engine.

Comment

Here you see subordination used simply to create variety and to eliminate a pattern of "babyish"-sounding sentences. Sentences two and three are combined in the revision with sentence two in apposition to the noun "turbocharging."

Unclear Order of Ideas

The two critical problems are heat and rpm. The turbocharger runs at high (20,000-100,000) rpm. Therefore, bearings and balance are critical. Second, the exhaust gases are at a high temperature, the higher the better, i.e., more energy, and heat damages bearings and other parts. Further, the compression of air produces heat in the air-fuel charge, decreasing density and defeating the purpose of turbocharging.

Emphasis on Cause and Effect Through Subordination

The two critical, related problems with turbochargers are heat and rpm. Because the turbocharger runs at high speeds (20,000 to 100,000 rpm), the bearings and balance must meet very close tolerances. At the same time, the bearings must withstand the high temperatures produced by the exhaust gases. Furthermore, the heat produced by compressing the air complicates the problem of destructive temperatures.

Unclear Order of Importance

The new buyers of performance cars often confuse supercharging and turbocharging, because supercharging draws power directly from the engine's horsepower. Turbocharging, on the other hand, uses the heat energy of the exhaust gases that have already left the engine.

Coordination to Create Balance

The new buyers of performance cars often confuse supercharging and turbocharging, the two very different methods of increasing an engine's power. Supercharging draws power directly from the engine's horsepower, but turbocharging uses the heat energy of the exhaust gases that have already left the engine.

Comment

In the first version, the writer has attempted to create a nonexistent cause and effect relationship. Since turbocharging and supercharging are equal as ideas in this discussion, they should be linked in a comparative or contrastive presentation through the use of coordination. We have revised the sentence with the coordinating conjunction "but" to create balance as we emphasize the difference between the two processes.

Readability Formulas—Wise Revision or Unnecessary Restriction?

During the 1950s Robert Gunning, Rudolf Flesch, and others who train writers began to advocate the use of various readability indexes. These measurement tools allow the writer to assign a school grade level to work as an indicator of the ease or difficulty readers may experience with a document. For example, many newspapers measure at the fourth- to sixth-grade reading level; many military manuals measure at the sixth-grade level; and a magazine for an educated or expert audience, such as *Scientific American*, measures at above the twelfth-grade reading level.

These formulas have caused much discussion among writing teachers. Many experts feel the indexes encourage the nonartistic, noncreative, "all business" approach to writing. It is true that the readability indexes focus on measurement, not on the invention of new ideas or specific figures of speech. Thus, writers should use the indexes with caution.

Nonetheless, the quantitative approach is a reality for the entire communications profession. Newspaper and magazine publishers, industry publications groups, and individual technical writers have adopted these indexes as the preferred means of standardization. Used judiciously, the indexes can encourage the even style, simple vocabulary (usually one- and

---

FIGURE 7.2   Fog Index.

1. Pick a sample of writing about 100 to 125 words long. Calculate the average number of words in each sentence. Treat independent clauses as separate sentences. "I came, I saw, I conquered" would count as three sentences.
2. Count the number of words of three syllables or more. Do not count capitalized words; combinations of short easy words like "bookkeeper;" or verbs made into three syllables by adding "-ed" or "-es," like "created" or "trespasses." Divide the count of long words by the length of the passage to get the percentage.
3. Add the two factors above (average sentence length and percentage of long words). Multiply by 0.4. Ignore the digits after the decimal point. This gives you the Fog Index. It corresponds roughly with the number of years of schooling a person would require to read a passage with ease and understanding.

Adapted from *The Technique of Clear Writing* by Robert Gunning (New York: McGraw-Hill rev. 1968). Used with permission. "Fog Index" is a service mark of Gunning-Mueller Clear Writing Institute, Inc., Santa Barbara, California.

two-syllable words), and short, information-filled sentences (nine to eleven words) that are so badly needed in product documentation and training materials. So current is this practice in industry that most of the editing programs for mainframe computers have readability indexes built in as the standard measures of a text's level of difficulty.

We show Robert Gunning's Fog Index and a short form of the formula he created to determine how clear or how "foggy" your writing is. Experiment with this index as you revise your materials. Although the editing software for a mainframe computer can perform the analysis for you, you can also complete an analysis with the formula and a pencil and paper.

After determining the grade level of your text, you can perform two tasks during your revision that may "standardize" your work a bit, and result in a lower-grade reading level.

1. Circle all your passive voice verbs. As you transpose them to active voice, can you eliminate the excess words as we did in some of our examples? Eliminating these words will lower the grade reading level of your text.
2. Circle all the words that have more than two syllables. Can you substitute descriptive one- and two-syllable words? If so, your grade reading level will drop even further.

EDITING YOUR DOCUMENTS

Editing can be an individual act, a collaborative act between a writer and an editor, or a systematic approach to a product, the document. Since we devote several chapters to the systematic approach, Chapters 16, 17, and 18 contain information about editing and production in corporations. Our concern here is with editing methods for individuals and collaborators. In preparing technical documents, the author, acting either as his or her own editor or with another professional editor, should concentrate on the document's content, including the method of analysis, the technical accuracy, and the format (the overall appearance).

Each editorial office tends to develop its own checksheets and style books describing its policies for preparing copy. To approach the topic of editing in a specific way, we discuss the most common errors we see, the content requirements of well-edited documents, and the format editing needed.

Ten Common Errors

This list shows the errors our students make most often. Note that these items have little to do with the document's content. Instead, the items identify the obvious errors that quickly distract a reader and undermine the writer's credibility. Let's consider the items as a group.

Our handbook section discusses many of this list's errors and offers examples of the corrections to make in your copy. Lately three errors have increased greatly.

First, even writers on the intermediate level seem to have trouble with subject and verb agreement. Be aware of this problem as you review and proofread a document. Consult the handbook for the correct forms for agreement.

Second, watch for overuse of the passive voice. Some writers seem unaware that the two voices, active and passive, represent different ways of presenting a subject. After identifying

FIGURE 7.3	The Ten Most Common Student Errors.

COMMON ERRORS

Here are 10 errors students make most often. Edit your work to eliminate these problems.

1. Misspelling throughout. Check names, headings, titles, and captions.
2. Subject-verb agreement. Correct for a plural subject and singular verb as in "there's three of them." Although you hear this phrasing in conversations and on television, it is nonstandard in writing. Singular subjects and plural verbs are, of course, equally nonstandard, though admittedly less frequent.
3. Incorrect use of the apostrophe or no apostrophe at all.
4. Failure to use quotation marks correctly. Do not substitute apostrophes for quotation marks.
5. Incorrect uses of hyphens and dashes.
6. Inconsistent verb tense. Try to write in one consistent tense. Do not mix future or past tenses unless you use a chronological organizational pattern, or unless the logic of the sentence demands such a shift.
7. Overuse of passive voice. Your text should contain at least 70% active voice and 30% passive voice. When in doubt, use more active voice. Circle all your passive voice verbs and transpose them to active voice.
8. Inconsistent person. Try to write an entire document in one person—first, second, or third. If you must change person, do so in a new section of the document.
9. Spacing after the period. Use two spaces.
10. De-gender your documents. Review their content and usage for equality in the representations of men and women.

all the passive voice verbs in a document, decide if the manuscript's content really calls for the contemplative point of view that the passive voice creates.

Third, watch for sexism. Often the same author who uses "he and she" throughout will also refer to a "craftsman" or a "chairman." These terms also reflect sexism and should be changed to "skilled worker" or "technician" and "chair," respectively. Watch for sexism of form and content.

Editing for Content

We include two student-designed checksheets to help with the processes of revising and editing. Read these lists as we discuss the topics of content and format.

Content includes everything from a check of the factual matter to an examination of how the copy looks on the page. You should review in five areas:

Facts. Can the author prove his or her point with adequate facts? For example, if an author claims that certain chemicals in the water cause childhood leukemia, does the author use reputable demographics and studies that support the claim?

FIGURE 7.4	A Typical Checksheet for Rewriting and Editing.

CHECKSHEET FOR REWRITING AND EDITING

Essential Elements	Date and Initial
Title page	
Table of contents	
Letter of transmittal	
Body/text type	
Annexes/index/glossary	
Omitted material	
Missing pages	
Missing paragraphs	
Missing sentences	
Missing words	
Audience analysis	
Appropriate technical language	
Audience environment	
Defined target audience	
Verified facts	
Plagiarism	
Technical accuracy	
Graphics	
Accurate table and figure references	
Proper use of tables	
Proper use of diagrams	
Mechanics	
Spelling	
Punctuation	
Sentence structure	
Consistency	
Overall appearance	
Headings—spelling, parallelism	
References—all included in text	
References—list complete at end	
Annexes—all included in correct format	
Graphics—all artwork keyed to text	

| **FIGURE 7.5** | A Typical Checksheet for Revising and Editing a Technical Text. |

EDITOR'S CHECKSHEET

Front matter and headings

_____Headings logically correct?
_____Headings parallel?
_____Headings capitalized and punctuated correctly?
_____Headings listed in correct order?
_____Short titles correctly used?
Comments _____
_____

Visuals

_____Caption accurate and descriptive?
_____Figure or table title and number correct in sequence?
_____Illustration referred to correctly in text?
_____Conventional symbols and abbreviations used?
_____Correct proportions in graphs and diagrams?
_____Tables complete and in correct format?
Comments _____
_____

Text—usage

_____Passages broken into readable "blocks"?
_____Copy checked for spelling and punctuation?
_____Sentences checked for jargon?
_____Sentences translated to active voice?
_____Passages checked for consistent use of person?
Comments _____
_____

Text—technical matter

_____Major topics listed in the introduction?
_____New terms and their abbreviations introduced correctly?
_____Latest form of standard abbreviations used?
_____Appropriate level of technical detail for audience?
_____Visuals and captions readable for audience level?
_____Numeric text checked for factual accuracy?
_____Technical, symbolic, and numeric text checked for errors?
Comments _____
_____

Logic. Are the reasons, facts, comparisons, contrasts, descriptions, and definitions all correctly used and related to each other? Here, watch for level of detail. Does the author offer enough details to make his or her claims believable? Too many details, on the other hand, may result in verbose or tedious text.

Organization. Of course, organization and logic are related. But the plan of organization in technical writing should be especially clear. Make the organization obvious with sequence markers, connecting and subordinating conjunctions, and format markers such as underlining, enumerating, and other methods of "chunking" the text.

Tone. Did the author complete an audience analysis? Does the treatment of information reflect that analysis? For example, are the sentences simple enough for a lay audience or dense even for a group of experts? Is the tone consistent throughout the manuscript?

Usage. How does the author use the language? Note that one checksheet lists five items for correct language usage. Under this heading, the editor should check for consistent usage as well, that is, the repeated use of the same forms. For example, numbers and measurements should be presented consistently.

Editing for Format

By *format* we refer to anything affecting the publication's appearance. In test situations, document users more frequently notice errors of format rather than of content. Thus in editing, always consider a document's appearance. Format includes these categories:

The Front of the Document. Often readers find errors in the spelling of an institution's name on the title page, omitted headings from the table of contents, or missing copyright material. Does the front of your report, proposal, or multivolume study include everything you need?

Chapter Titles, Headings, and Other Labels. Examine all the titles in the document. For example, do the chapter titles agree with the table of contents titles? Are all titles correctly spelled? Are they formatted according to the practices of your handbook?

Visuals. Check your thumbnail sketches and finished artwork. Have you labeled everything? Are labels grammatically parallel and correctly spelled? Does each visual have a caption, a one-sentence explanation of its purpose? Are the proper nouns used in the drawings spelled correctly?

Pages. Does each page have everything it needs, including (if they are part of the format) headers, footers, numbers, correct text, and correct references to the visuals and appendixes?

Appendixes. These sections at the end of the body of the text are also called "annexes." Have you included all the attachments? Do they refer back to the text easily? Are they correctly numbered? Do their labels allow for easy use?

Note that for personal editing of college papers, we stress editing that treats the text itself and the types of errors that word processing and typing introduce. We also concentrate on very general editing problems. Our checksheets identify the specific categories you should

FIGURE 7.6	A Checklist for Reading and Reviewing a Text Containing Multiple Equations, Data, or Numeric Information.

CHECKLIST FOR NUMERIC TEXT

Facts—Analysis and Development

> **Verification**—Has the author used the correct technique for analyzing the information?
> **Substantiation**—Has the author cited the appropriate sources for a given mathematical procedure or technique for analysis?
> **Originality**—Has the author explained why the text uses a new method of analysis or presentation of the numbers?
> **Condensation**—Has the author explained in at least minimal detail what a given score or result means? Can the reader generalize from this result?

Symbols—Appearance and Logic

> **Correct Type**—Has the author inserted the proper symbols for the mathematical equations or logical statements? Do you need to order special symbols or type fonts to print the text correctly?
> **Correct Format**—Has the author used a standard format for the presentation of numbers? For example, does the line graph used in the text represent the correct proportions discussed in the text? Has the author chosen an appropriate symbol system, for example, one developed by the American Standards Association, to present the numbers or data?
> **Correct Context**—Has the author supplied enough information to explain what the numbers mean?

check. Copy the checksheet you like best and use it to complete the projects for this class. As you proofread for each item, check it off on the list. Then sign and date the list at the bottom. These records, you will see, become quite important when you edit work for others.

Editing Numeric Text

We include here a special checksheet for "numeric" text, that is, any technical narrative dense with equations, symbols, and numerals. Since such text may contain more errors than nonnumeric text, edit it with special attention. Check the text in the seven areas our checksheet outlines and defines.

What should you do if the numerals or equations seem intimidating? Have an expert review the content for its technical accuracy. Ask the expert to explain the most frequently used terms, so you can acquire a working understanding of them. Obtain that particular discipline's stylebook and proofread the text according to that discipline's requirements. Also, consult our lists of technical and statistical abbreviations in the handbook.

Proofreading

Students most often lose points on their grades because of their errors in proofreading—not in content or format. These "little" errors are most bothersome for professional writers as well. Assume neither that you can proofread your document only once, nor that your software spelling checker can identify every superficial error in a document. Using the professional proofreading symbols in the handbook, proofread at least three times. Read for a specific objective each time as our list indicates:

First Reading	Spelling, punctuation, the 10 errors
Second Reading	Factual accuracy, logic, style, ease of reading
Third Reading	Overall appearance of the document, inclusion and accuracy of any last-minute changes

Use other "tricks" professional proofreaders have developed. First, with a piece of paper or a ruler, cover the lines you are not reading so you can concentrate on the words and letters at hand. Second, read the copy aloud in order to "hear" any errors your eyes may miss. Finally, after proofing the entire document, proof it from back to front. Seeing the document this way enables you to catch errors you might have missed by following the manuscript's front-to-back logic.

Three Typical Revisions—How the Students Improved Their Writing

Here are three typical assignments that students have revised and edited. Think of them as from major surgery—total revision—to minor surgery—minor editing. Read the discussion draft, the revision, and our comments on the author's progress with this progression in mind:

Example 1—topic, the vise—substantive rewriting needed.

Example 2—topic, the proposal letter—fairly good usage; rewriting needed for ease of reading and emphasis.

Example 3—topic, the disaster manual letter—very good draft; only minor editing needed.

Why the Vise Essay Needed Major Surgery—Complete Revision

These students formatted their discussion draft well because they used blocks of text and an appropriate list. We suggested they retain the format. The headings seemed acceptable, too.

Nearly all verbs in the discussion draft were in the passive voice. Since the students wanted to explain a bit about the history and use of vises, they used the passive voice inappropriately, causing the colorless, "actionless" sentences of the first draft. Before revising, the students listed the verbs that describe the vise. They incorporated verbs such as "fasten," "clamp," "use," and "secure" in the revision, not only to develop the active voice but also to increase the description's technical accuracy.

The section headed "The Purpose of the Metalworking Vise" was most revised. Passive voice made this section stilted and seemingly irrelevant to the reader. Because the subjects and direct objects were obscure, the organizational pattern seemed unclear. The active voice

FIGURE
7.7
A Discussion or Review Draft of a Student's Description of a Vise.

THE BEGINNER'S GUIDE TO THE VISE

A Brief History of the Vise

The Greeks and Romans were the first people to use the holdfast, the predecessor to the vise. The holdfast was a heavy crude metal bar with a hook at the top; similar to a walking cane. The bar was placed through a hole in a workbench so that the end of the hook could rest on the object desired to be held. the holdfast was only effective on objects that lay flat on the surface of the workbench.

The vise developed during the 17th century out of a need to hold objects that could not lay flat on a workbench. The first workable vise was simply a hook-shaped piece of wood nailed to the workbench. Through the top of the hook was a screw that could be adjusted to hold oddly shaped objects.

Later, the vise was modified further to specifically hold objects off the surface of the workbench. This required that the holding device be in an upright position. Thus, it became the modern-day vise. Now it is much easier to hold those materials that need filing, bending, or chiselling.

The Purpose of the Metal Working Vise

One of the most common types of vises is a metal working vise. This is a necessity in any home workshop. The metal working vise consists of two jaws which can be closed or drawn together by a screw, a lever, or a cam. this tool will hold most any type of material to be worked upon. When it is used for holding the item during hand operations such as filing, hammering, or sawing, the vise may be permanently bolted to a strong and sturdy workbench.

For specialized jobs, a vise with a swiveling base is very handy. This allows the object to be swung into the best position from which it can be worked upon. A vise with a solid base is recommended for heavy, rough work involving hammering or chiselling. If the work consists of delicate jobs (furniture or finished jobs), the jaw plates may either be changed or extra "padding" may be used. This padding may be rubber, leather, or fiber. Covers can also be placed over the plates.

The plates of the jaws can be changed, depending on the jobs performed. Soft plates, such as rubber pading, can be used to protect the work. Lead plates are good for finished metal products. Since lead is malleable, it can be formed to fit a

FIGURE 7.7   Continued.

job. Brass plates are good for those jobs with rough finished surfaces. For wood-working jobs, fiber plates (leather or cloth) may be desired.

Instead of changing the plates to meet the job, the mob may be wrapped with different types of padding. The padding may also be placed between the jaws and the work.

Parts of the Vise

Jaws—moveable and fixed; drawn together to hold the object securely.
Plates—interchangeable parts to fit the needs of the material being used.
Anvil—the surface on which materials can be hammered.
Slide—tube to which the moveable jaw is connected.
Swivel base—allows the vise to rotate.
Swivel locks—locks swivel base into the desired position.
Screw—draws the moveable jaw toward the fixed jaw.
Handle—rotates the screw.
Pipe holding jaws—securing pipes or other cylinder-like objects.

Safety Tips When Using the Vise

Use only a strong and sturdy vise.
Make sure that the interchangeable plates on the jaws are secured to avoid slipping or movement of the object while working.
Place the object securely in the center of the vise.
Keep the jaws closed when not using the vise.
Place the handle in an upright or vertical position to avoid personal injury or damage to other objects.
Never use an extra lever or hammer to tighten the handle down further. Use only hand strength.
Place long objects as low as possible in the vise. If necessary, use a block of wood to support the workable item.
Avoid heavy hammer blows to the object. This may cause serious stresses on the slide, preventing the vise from turning. for this reason, it is recommended that a solid-based vise by used.
Keep the vise clean.
Oil the screw occasionally.

FIGURE 7.8	The Rewritten and Edited Version of a Student's Description of a Vise.

A BEGINNER'S GUIDE TO THE VISE

A Brief History of Vises

The Greeks and Romans are known to be the first people to use the holdfast, a predecessor to the vise. The holdfast consists of a heavy, crude, metal bar with a hook at the top, similar to a walking cane. The user drops the bar through a hole in the workbench causing the weight of the holdfast and the hook to secure an object to the bench. An effective holdfast works only on those objects that lie flat on the bench. However, the object must weigh less than the holdfast.

Skilled workers developed the vise during the Seventeenth Century because they needed to hold objects that could not lie flat on a workbench. Nails secure a hook-shaped piece of wood to the side of the workbench. A vertical screw through the top of the hook allows for adjusting the vise. The user has to refrain from tightening the screw too much to avoid cracking the wood.

The use of hardened steel makes the modern-day vise stronger and more durable than its predecessor. Manufacturers mold vises of high-quality steel or even cast iron. Vises with jaws in the upright position raise objects off the surface of the workbench, providing object mobility. This makes work involving filing, hammering, or sawing much easier.

The Purpose of the Metalworking Vise

You can use a metalworking vise in any home workshop. (Simply go to your local hardware store to find the appropriate type.) Make sure to bolt the vise permanently to a strong and sturdy workbench. When you fasten and clamp metal objects, you realize the necessity of the metalworking vise. The vise consists of two jaws which, when pressed together by a screw, a lever, or a cam, secure the objects. Vises restrict the movement and ensure the stability of various objects while you work.

You need to use a vise for hand operations, such as filing, hammering, or sawing. The most convenient vises have a swivel base. You may position the objects in many ways, allowing for increased accessibility. This will simplify your work. Use a vise with a solid base for heavy, rough work that involves hammering or chiseling.

When you work on delicate objects, you should change the jaw plates to match the job. For example, rubber plates protect virtually any type of material. You can use lead plates when you work with finished goods, because you can

**FIGURE
7.8**   Continued.

pound the lead to fit the job. Use brass plates for those jobs with rough finished
surfaces, and for woodworking surfaces, fiber plates work well. To save time, you
can simply wrap the plates in different types of padding such as leather strips.
Padding on the jaw plates may consist of rubber, leather, or fiber materials. You
may also use covers that fit securely over the plates. Ask the clerk at your local
hardware store for more details.

Parts of the Vise

Jaws—moveable and fixed; drawn together to hold the object securely.
Plates—interchangeable parts to fit the needs of the material being used.
Anvil—the surface on which materials can be hammered.
Slide—tube to which the moveable jaw is connected.
Swivel base—allows the vise to rotate.
Swivel locks—locks swivel base into the desired position.
Screw—draws the moveable jaw toward the fixed jaw.
Handle—rotates the screw.
Pipe holding jaws—securing pipes or other cylinder-like objects.

Safety Tips When Using the Vise

Use only a strong and sturdy vise.
Make sure that the interchangeable plates on the jaws are secured to avoid slipping
    or movement of the object while you are working.
Place the object securely in the center of the vise.
Close the jaws when you are not using the vise.
Place the handle in an upright or vertical position to avoid personal injury or damage
    to other objects.
Never use an extra lever or hammer to tighten the handle down further. Use only hand
    strength.
Place long objects as low as possible in the vise. If necessary, use a block of wood to
    support the workable item.
Avoid heavy hammer blows to the object. This may cause serious stresses on the slide,
    preventing the vise from turning. For this reason, it is recommended that a solid-
    based vise be used.
Keep the vise clean.
Oil the screw occasionally.

helped the students reveal the pattern of organization in the revised version. They also incorporated the second person point of view to personalize the essay for a user.

Can you identify the other improvements the students made?

What the Nutrition Manual Proposal Needed—Major Surgery

Although the discussion draft of the nutrition manual proposal is much more readable than the vise essay, it needs improvement. We commended the author's excellent phrasing of the project's objectives, with her use of action statements starting with verbs. We also encouraged the author to retain the overall logical plan of organization.

But this document, like the vise manuscript, also needed major changes. The author used the passive voice too often, omitted the articles "a" and "the," and failed to clarify the project's plan of organization.

Note that the revised document uses "chunking" to direct your eye. The enumerated, grammatically parallel action statements for the project objectives remain, but a clearly enumerated plan of work has replaced the old narrative. Labels and italics draw the eye to read about the steps. This author has achieved an effective technical format.

During the rewriting, she also inserted "the," "an," and "a" as needed, to smooth her style. She incorporated the pronoun "you" as well, to include the reader in the message. She transposed many of the passive voice verbs in the Plan of Work section into descriptive action statements in the active voice, thus duplicating the style of the Project Objectives section and obtaining a rhetorical balance. This author really improved her persuasive style through revision. Can you find any other strengths to imitate in this proposal?

Why the Disaster Manual Letter Needed Cosmetic Surgery—Editing, Not Rewriting

At first glance, the disaster manual letter may prompt you to ask, "Why change it? It looks pretty good!" It does. Good material such as the logical message of this letter really allows you to improve the message. This document needed good, cosmetic editing, so its meaning could emerge with precision.

The author, a fairly strong intermediate writer, still had a certain stiffness of style caused by the absence of articles, transitional phrases, and conjunctions between ideas. This author needed to study punctuation a bit, too; she failed to use the comma to achieve the emphasis that the letter needs.

The edited version contains carefully inserted commas to create the pauses that emphasize the message. Note, too, that with the change of a few words, the edited version emphasizes Mr. Bess's ability to serve the community rather than discusses his multiple roles as an administrator. That single argument strengthens the letter's message. Finally, we inserted articles throughout to improve the document's readability.

You may ask, "Why revise so much?" Perhaps you think that large-scale revision, major surgery that catches all the text's big problems, is enough. Editing to improve the document's nuances may seem too nitpicking. The answer is that the best edited document often gets results. Two of this chapter's examples, the nutrition proposal and the disaster manual proposal, netted substantial consulting contracts for these students.

Now what do you think?

FIGURE 7.9   Unedited Draft of the Narrative Section of a Short Proposal for a Nutrition Manual.

PROPOSAL

I propose to write an informative nutrition manual designed to provide nutrition information and motivate your patients to change bad eating habits and to reinforce good eating habits.

PROJECT OBJECTIVES

1. Acquaint patients with the nutritional problems caused by alcoholism.
2. Help patients to appreciate the role of good nutrition in restoring health.
3. Help patients to recognize how good eating habits and a nutrient dense diet may help to reduce alcohol craving and thus promote sobriety.
4. Provide patients with basic nutrition information necessary to choose a well balanced diet.
5. Motivate patients to improve eating habits and increase the nutrient density of their diets.

PLAN OF WORK

Needs Assessment of your rehabilitation program will be my first step in planning for writing of this manual. Due to my past summers work including nutrition education with residents in an alcohol rehabilitation program and to extensive review of the research literature written on nutrition and alcoholism, I have a well developed concept of the general nutritional needs of recovering alcoholics. I would like to visit your organization, and with your assistance apply my past experience to determine your specific needs.

A topic audience matrix will be designed, and a structured outline will be written. Modules will be created after getting specific input from you, your staff and your patients. A storyboard will be utilized to further design and revise the manual. After the first draft is written, I would like to meet with you again for an evaluation to further revise, debug, and rewrite the manual. After the manual meets your specifications and my standards it will be printed according to your request for materials, cover, and printing specifications.

After the manual has been used for three weeks, with your assistance, the help of your staff, and a written evaluatiron from your patients, I would like to meet with you to evaluate the effectiveness of the manual, and its contribution to the nutrition education and total rehabilitation of your residents.

FIGURE 7.10   Rewritten and Edited Draft of the Narrative Section of a Short Proposal for a Nutrition Manual.

PROPOSAL

I propose to write an informative nutrition manual designed to provide nutrition information and motivate your patients to change their bad eating habits and to reinforce their good eating habits.

PROJECT OBJECTIVES

The project has five objectives.

1. Acquaint patients with the nutritional problems caused by alcoholism.
2. Help patients to appreciate the role of good nutrition in restoring health.
3. Help patients to recognize that good eating habits and a nutrient-dense diet may help to reduce alcohol craving and to promote sobriety.
4. Provide patients with basic nutrition information necessary to choose a well-balanced diet.
5. Motivate patients to improve eating habits and increase the nutrition density of their diets.

PLAN OF WORK

I will begin the planning for the manual with a needs assessment of your rehabilitation program. The needs assessment will include a review of the research literature on nutrition and alcoholism and a visit to your hospital to determine the specific needs of your program's recovering alcoholics.

I will prepare your manual using a five-step process.

Step One. Complete a topic-audience matrix showing the concepts your audience needs to understand.

Step Two. Write a structured outline that shows you not only which topics the manual covers but also how the manual presents them.

Step Three. Draft the modules, or units of instruction about nutrition, the manual will contain.

Step Four. Assemble a storyboard that shows the manual's instructional sequence for your approval.

Step Five. Prepare a draft of the entire manual for your review.

After I complete the first draft, I would like to meet with you again so that we can evaluate, revise, debug, and rewrite the manual. When the manual meets your specifications and my standards, I can contract for the printing and production as you request. After you have used the manual for three weeks, I plan to evaluate both its effectiveness and its contribution to the nutrition education of your residents.

**FIGURE
7.11**

A Letter Needing Editing, Not Rewriting.

BONNIE J. KNAPP
121 Palmer Street
Bowling Green, OH 43402

June 1, 1988

William Bess, Director
Department of Public Safety
Bowling Green State University
Bowling Green, OH 43403

Dear Mr. Bess:

After reviewing data that Leah Ness-Prescott compiled, as well as the previous disaster
manual, I certainly agree with you that your department needs an organized, workable
document which will give you and your staff proper directives should a disaster occur forcing partial or complete closing of the university.

The lack of such a manual could have grave consequences should a disaster occur. Organization is essential during such a crisis as certain duties must be discharged in a minimal
amount of time with greatest possible effectiveness. An organized procedures manual will
allow the Emergency Operations Center to execute recovery techniques quickly and effectively; as a result, personnel will be able to focus energies on practical aspects of public safety
and protection.

You will find my expertise as a technical writer/editor invaluable in coordinating this project. I am accustomed to working quickly, but efficiently, under pressure of deadlines;
organizational skills will provide the basis for this project.

Ms. Ness-Prescott's preliminary work is notable and will simplify my job. I propose to take
information you have given me, extract workable documents and create new ones when
necessary, and put all of this into terms that people in charge during an emergency can
understand easily. Probably my most important objective is to organize and index information, so that staff can use it easily as a resource when time is limited. Not only will your
staff benefit from such a useful manual, but those people affected by the disaster will also
benefit, as recovery will occur more quickly.

With this manual, your job as head of Campus Security will run more smoothly during the
course of a disaster. Furthermore, you will have a concise, working document for the
members of the campus community to review, as well as others who may call you in their
efforts to create such a manual. I look forward to working with you on this project.

Sincerely,

Bonnie J. Knapp

| FIGURE 7.12 | An Edited Letter. |

BONNIE J. KNAPP
121 Palmer Street
Bowling Green, OH 43402

June 1, 1988

William Bess, Director
Department of Public Safety
Bowling Green State University
Bowling Green, OH 43403

Dear Mr. Bess:

After reviewing the data that Leah Ness-Prescott compiled, as well as the previous disaster manual, I certainly agree with you that your department needs an organized, workable document. Such a manual will give your staff and you the proper directives, should a disaster occur forcing a partial or complete closing of the university.

The absence of such a manual could have grave consequences should a disaster occur. Organization is essential during such a crisis because certain duties must be discharged in a minimal amount of time with the greatest possible effectiveness. An organized procedures manual will allow the Emergency Operations Center to execute the recovery techniques quickly and effectively; as a result, personnel will be able to focus their energies on the practical aspects of public safety and protection.

You will find my expertise as a technical writer and editor most useful in coordinating this project. I am accustomed to working quickly, but efficiently, under the pressure of deadlines, and my organizational skills will provide the basis for this project.

I propose to take the information you have given me, extract the workable documents and create new ones when necessary, and put all of this information into terms that people can understand easily. Probably my most important objective is to organize and index the information, so your staff can use it easily as a resource when time is limited. Not only will your staff benefit from such a useful manual, but those people affected by the disaster will also benefit, as their recoveries will occur more quickly.

With this manual, your job as the Director of Campus Security will run more smoothly during the course of a disaster. Furthermore, you will have a concise, working document for the members of the campus community to review, so they can understand their roles in coping with a crisis. I look forward to working with you on this project.

Sincerely,

Bonnie J. Knapp

SUMMARY

In this chapter, you learned that reviews, revision and editing improve a technical document's readability. You have studied a four-step process for reviewing a manuscript. You have learned that to revise effectively you should concentrate on active voice verbs, descriptive nouns, and emphatic sentences. You have also reviewed two editing checksheets that are helpful when you examine a manuscript's content and form. After reading this chapter, you will be able to:

- Differentiate among editing, reviewing, and revising as tasks in producing documentation.
- Read a document to identify its content, strong points, and areas for development.
- Prepare a set of instructions for an author to use in revising and editing a document.
- Conduct a short conference with an author to explain how to improve a manuscript.
- Use the skill of transposition to replace passive with active voice verbs as you revise.
- Substitute descriptive, technically accurate nouns and adjectives for general, colorless ones.
- Use the technique of subordination to create causality and emphasis as you rewrite documents.
- Use the technique of coordination to create equality of ideas as you rewrite balance sentences.
- Identify your objectives for a document as an editor.
- Use or modify a checksheet to complete an editing task.
- Identify 10 common errors and edit them out of your documents.

ASSIGNMENTS

1 Here is Laura Heberlein's draft of the instructions for the planting of a lilac bush. Examine the draft and prepare to collaborate with her on the revisions, so she can obtain a better grade. Your tasks are enumerated in the items below.

2 Think about how Laura should illustrate these instructions for the inexperienced user. Visit a local nursery and check a public library for examples of planting instructions. Select some illustrations that you find effective. Sketch the rough illustrations you would like, and bring your sketches to class.

3 After examining Figure 7.5, Editor's Checksheet, modify it to meet your needs for your major technical writing project. Feel free to insert new sections, delete others, and rewrite.

4 Rewrite Figure 7.4, Checksheet for Rewriting and Editing, so you can use it to revise, edit, and proofread the instructions you will write during this semester. Edit the checksheet freely, but remember to write a special section pertaining to instructions and procedures.

5 In your college library, find articles about other readability indexes. Find at least two other ways of measuring the difficulty of a text. Bring your articles to class, and be prepared to explain how the indexes you found differ from Gunning's Fog Index.

FIGURE
7.13
A Working Draft of a Student's Instructions—Ready for Rewriting and Editing.

INSTRUCTION MANUAL FOR THE PLANTING AND CARING OF THE LILAC BUSH

Congratulations, you have just purchased your own healthy Lilac Bush. As a homeowner, you should be interested in growing healthy and attractive bushes on your property. Your Lilac Bush is a local bush, and with the proper care it will grow to be a beautiful and healthy attraction to your landscape. these set of instructions are for you and will assist in your planting and caring for your Lilac Bush. If these instructirons are followed, your Lilac Bush will grow healthy and look healthy.

Table of Contents

Equipment and Materials

Shovel	peat moss	top soil
fertilizer	water	root fertilizer
stake		

When to plant your Lilac Bush

Let us begin by informing you when the Lilac bush should be planted. For best results, your Lilac bush should be planted in the early spring. This time is best because the Lilac Bush follows certain care instructions for each season, and the Spring will assure the plant healthy and progressive seasons.

Where to plant your Lilac Bush

The next question you may ask is where to plant your Lilac Bush. For best results in the growth of your bush, the Lilac Bush should be planted in a semi-shaded area. The spot should offer equal amounts of sunlight and shade. If the bush is planted in direct sunlight or total shade then their may be little success in your Lilac Bush growth. Also follow these planting ingredients for your particular type of soil in your home area.

Sandy Soil—⅓ peat moss, ⅓ topsoil, ⅓ ground dirt

**FIGURE
7.13**

Continued.

Clay Soil—⅓ peat moss, ⅓ sand, ⅓ ground dirt

Average Soil—½ peat moss, ½ ground dirt

How to plant your Lilac bush

Step 1. Dig a hole that is 1½ times the size of the pot which your bush comes in. It should be dug this size in height and width.

Step 2. Apply a base of ground dirt at the bottom of the hole to raise your bush above ground level. This will allow the bush to settle to ground level within the first year. Apply enough dirt so the bush is 1½ to 2 inches above ground level.

Step 3. Mix together your dirt ingredients according to your type of area soil.

Step 4. Remove your bush from the pot, and place it in the dug hole. Take your ground ingredients and cover your bush until it is steady. Remember that the bush should be above ground level to allow for settling purposes.

Step 5. If your bush is heavy and will not stand straight, then use a stake and tie the bush to the stake to allow straight growth.

Step 6. After spreading your ground ingredients, water and feed your bush with root and feed fertilizer. The root fertilizer will aid your bush in adapting to the shock of a new environment. The feed fertilizer will give your bush food for better growth and survival; follow directions on fertilizer for correct amount.

Watering Process of your Lilac Bush

The watering amoung depents on the area in which you plant your bush. YOu should water the Lilac Bush when you notice the ground around it is dry. Follow these instructions of the watering process depending on your area.

If you plant your Lilac Bush in a high level of ground then you should water it often because a high level ground becomes dry before a low level ground.

If you plant your Lilac Bush in a low level area, then you will water it less because a lower level area tends to be moist already.

The Seasonal Upkeep of your Lilac Bush

Spring   **A.** Plant your bush in early Spring.
         B. Water and fertilize your bush on a regular basis.
         C. Fertilizer should contain 10% Nitrogen, 10% Phosphorous, 10% Potassium, and 70% filler.
         D. You should watch for disease during this time.

**FIGURE
7.13**

Continued.

Summer **A.** Prune after flowers fall in mid to late summer.
   B. Water and fertilize your bush on a regular basis.
   C. Fertilizer should contain 10% Nitrogen, 5% Phosphorus, 5% Potassium, and 80% filler
   D. Watch for insects at this time.

Fall   **A.** Water and fertilize your bush on a regular base.
   B. Fertilizer should contain 5% Nitrogen, 5% Phosporous, 10% Potassium, and 80% filler.
   C. Watch for insects at this time

Winter **A.** If you plant your bush in an open area where there may be excessive wind then cover it with a burlap bag, and remove it in early Spring (do not use a plastic bag because it will cause excessive moisture inside the bag.)
   B. Water if you find the area around your bush is dry at times, but in the winter the ground is already moist enough.

The growth of your Lilac Bush

Your Lilac Bush will bud (form flowers) in early to mid summer. It will not grow very much in height because of the initial shock it grows through during the first year of its new environment. Expect your bush to grow approximately 6-12 inches the first year.

The pruning of your Lilac Bush

You may prune your Lilac Bush because of odd branches or to shape. When you prune your bush you should follow the following instructions.

   A. Cut only after the flowers had fallen in the mid to late summer.
   B. Cut the branch ¼ to ½ inches below the bud.

spraying for diseases and insects

Diseases   A powdery mildew may form on your Lilac Bush because of the excessive moisture it has received. To eliminate this disease, spray your bush with a spray that contains Sulfur. The Sulfur will burn the mildew off the leaves, and your bush will look healthy again.

Insects   There are many different insects that can appear on your Lilac Bush. The following are some insects and what they will attack on your bush
     Red spider—branches
     Leaf Miner—leaves
     Japanese Beetle—leaves and flowers
     Black Weevil—roots.
    A spray with Sisthemic in it is recommended in order to kill these insects.

PART TWO

Application

CHAPTER

8

Graphs and Tables

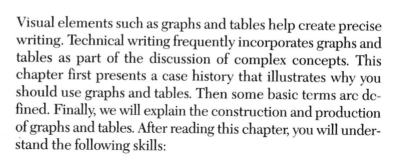

Visual elements such as graphs and tables help create precise writing. Technical writing frequently incorporates graphs and tables as part of the discussion of complex concepts. This chapter first presents a case history that illustrates why you should use graphs and tables. Then some basic terms are defined. Finally, we will explain the construction and production of graphs and tables. After reading this chapter, you will understand the following skills:

- Reading and designing pie graphs, bar graphs, and line graphs.
- Interpreting and constructing formal and informal tables.

CASE STUDY

Pat Pryer, a college senior majoring in computer science, has written a software package designed to schedule intramural and intercollegiate athletic events. The software package enables coaches to schedule athletic events by manipulating rosters electronically, a virtually foolproof process.

Since most athletic coaches and their assistants are not computer experts, Pat's manual must be "user friendly." Pat has written and tested the documentation, but the users have complained that it is difficult to follow page after page of printed commands. All the reviewers have suggested that her manual include visual elements.

Pat wants to improve her software package so she can sell it to a major software distributor. She can write documentation, but she does not know how visual elements will improve her software package.

Pat Pryer, along with any technical writing student, must learn about the basic construction and production of graphs and tables. Pat will soon learn that graphs in her sales literature will help convince the reader that her software package is cost-effective. Pat will also learn to construct tables that list all her program's major commands and that allow potential program users to determine instantly if her software package will run on their computers.

WHY GRAPHS AND TABLES?

Graphs and tables are symbolic visual aids that represent quantities. *Graphs*, referred to in text as figures, present numerical data in a visual form. *Tables*, referred to in text as tables, present data in columns and rows. Both graphs and tables clarify your text and aid quick reader comprehension by allowing your reader to see and to understand. Use graphs and tables to shorten text, clarify information, emphasize data, establish relationships, and summarize material.

PRODUCTION: TRADITIONAL OR COMPUTER-ASSISTED DESIGN?

Today the production of documents changes constantly because of the advent of computer-assisted design software packages and desktop publishing techniques. As a technical communicator you may work for a company that uses either traditional production styles or the latest computer technology.

We have designed this chapter to help you understand the basic principles involved in designing and constructing graphs and tables. Whether producing the visuals by hand or computer, you must understand the basic principles in order to select the visuals that best communicate your data to your reader. We encourage you to become well acquainted with any computer-assisted design programs available to you, but we also stress the importance of the information in this chapter.

DEFINITIONS

Here are some operational definitions of terms that will help you in your introduction to the use of graphs and tables.

Visual — any graphic representation of an idea or object, a basic term to describe all the illustrations in a text.

Figure — any illustration; traditionally the term includes graphs but excludes tables.

Constant — a number or value that does not change.

Variable — a number or value that does change or that is measured repeatedly.

Graph — a diagram that shows how the number value of a variable changes.

Abscissa — the horizontal line of a bar graph or a line graph used to record constant values and represented by the letter X.

Ordinate — the vertical line of a bar graph or a line graph used to record variables and represented by the letter Y.

Table — a format for arranging data in columns (vertically) and rows (horizontally), showing statistics on facts in a compact space.

GRAPHS

We will discuss the use and design of these visuals along with some cautions about their possible shortcomings.

Pie Graphs

Also known as circle graphs, pie graphs are the simplest graphs available to the technical writer. A pie graph is a circle divided into wedges. Each wedge is labeled to indicate percentages of the total amount (100 percent). When used correctly, pie graphs enhance the text and emphasize important relationships among facts.

Use. Use pie graphs for a group audience, particularly a group that may have varying levels of interest or background in the subject. Use them for quick readability and comprehension. Pie graphs give an audience the "big picture," that is, basic groupings, relationships, or classifications of data. Use pie graphs to show how the whole consists of parts or how one segment affects the whole.

Design. To design a pie graph, draw a 360° circle with a compass. Since 3.6° equal 1 percent, a 72° wedge in your pie graph represents 20 percent. Each wedge should equal its percentage in degrees. To find the degrees, multiply the percentage of your wedge by 3.6. Use a protractor to measure the number of degrees your wedges need within the pie graph. All the wedges of the circle must add up to 360°, that is, 100 percent. Don't try to "eyeball" how large a wedge should be to represent a component part of your whole circle. Differences that look reasonable on a piece of paper may look ludicrous if projected onto a large screen. To properly design a pie graph, be sure to follow these instructions:

Pie Graph Showing Basic Construction Formula.

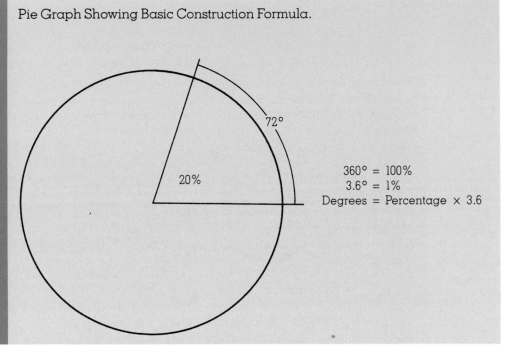

$$360° = 100\%$$
$$3.6° = 1\%$$
$$\text{Degrees} = \text{Percentage} \times 3.6$$

Properly Designed Pie Graph.

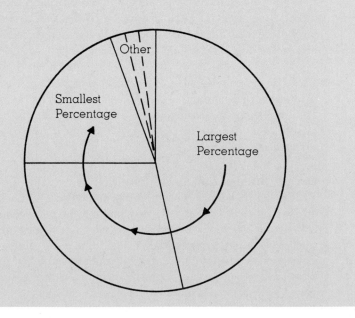

1. Place the first radial line at twelve o' clock on your pie graph. Placing the line at that point is a good starting point, and it is a good reference point for your reader.

2. Starting at twelve o' clock, arrange the wedges clockwise in descending order with the largest wedge first. If the logic of your discussion prevents the arrangement in descending size order, be sure that a system of order, such as alphabetical or chronological order, is visible for your reader.

3. Use at least three wedges in the graph (only two wedges results in a silly-looking graph), and no more than five (six or more wedges can be confusing). If you need more than five wedges, consider another type of graph or a new system of classification.

4. Make no wedge smaller than 2 percent (7.2°). If you have more than two wedges this small, try grouping them together into one larger wedge labeled "other."

5. Label all sections horizontally. Use grammatical parallelism for your labels to create continuity. Include the percentage of each wedge in its label. Try to label within the wedge. If the wedge is too small for a label, place your label outside the graph with a straight line drawn to the wedge.

FIGURE 8.3   Pie Graph Showing Federal Government Dollar.

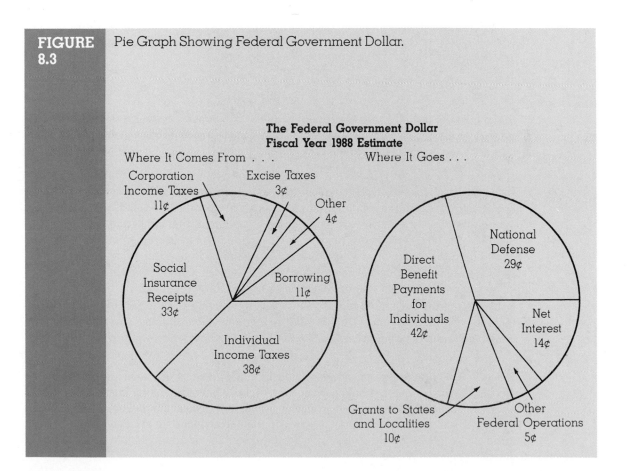

Cautions. When using and designing a pie graph, consider this information. The purpose of the pie graph is to communicate general information to your reader. If the data you are presenting contain very specific statistics and information, consider another kind of graph. In designing your pie graph, be careful if you are detaching wedges from the circle to create emphasis. By detaching more than two wedges from a pie graph, you lose the circle's unifying effect.

In addition, be wary of using color for your pie graph. Colors can enhance the comprehension of the pie graph, but they can also confuse the reader. Make certain that your colors neither distort the data by creating too much emphasis nor confuse your reader to the point where your graph "lies." When used together in a pie graph, "cool" colors, like greens and blues, make wedges appear less significant. "Warm" colors, like yellows and oranges, will make wedges appear more significant. For example, do not color a 20% wedge dark green when it must appear next to a yellow 50% wedge. The "cool" green wedge will look smaller than it should when contrasted with the "warm" yellow wedge. Instead, color the largest wedge yellow, and the next largest wedge orange unless, of course, you have a good rhetorical reason to emphasize the smallest wedge. You should use colors for your pie graph ethically and with good rhetorical reasoning.

If your pie graph is in black and white, follow the same basic rules that apply to colors. For black and white graphs, tints and crosshatches will delineate the wedges. To retain the graph's readability, never shade it more than 18% gray.

Bar Graphs

Bar graphs, also called *histograms* when they show frequencies in statistics, are usually considered for direct comparisons. A bar graph shows the relationship of a variable to a constant, a series, or an interval of measurement.

Use. Use bar graphs with all kinds of audiences to show trends rather than individual facts. Specifically, bar graphs most often show measurements of one item at different times or intervals (one student's yearly math score for five years), measurement of different items during the same time (five students' math scores for one year), or measurements of different parts of one item that make up a whole (one student's scores in math, reading, science, social studies, and physical education for one year). Use simple bar graphs to explain simple relationships; use segmented bar graphs to explain highly technical relationships. Use pictographs, a variation of the bar graph, to show the actual item being counted or compared. In pictographs, the bars are made up of picture symbols such as male and female figures. Newspapers and popular magazines like *Time* or *Newsweek* favor pictographs.

Design. The design of a bar graph should be direct, logical, and consistent. For quick readability, the bar graph should have no more than 10 bars. Variables and constants should be labeled in as simple and direct a language as possible. The amount of data on a page and the page's trim size affect the layout of your bar graph. It is important to be consistent. Try not to violate the reader's natural sense of logic. Use the following information to design your bar graphs:

**FIGURE
8.4**

Bar Graph Showing Enrollment in Institutions of Higher Education.

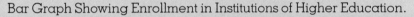

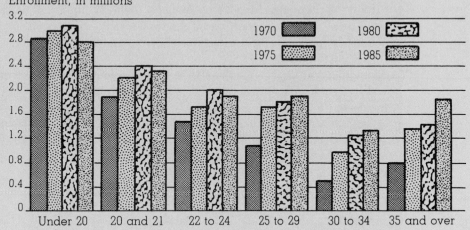

Source: U.S. Department of Education, Center for Education Statistics, ''Fall Enrollment in Colleges and Universities'' surveys; and U.S. Department of Commerce, Bureau of the Census, *Current Population Reports*, ''Social and Economic Characteristics of Students,'' various years.

**FIGURE
8.5**

Bar Graph Showing Trends in Bachelor's Degrees.

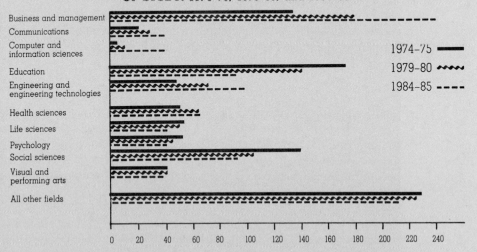

Source: U.S. Department of Education, Center for Education Statistics, ''Degrees and Other Formal Awards Conferred'' surveys.
Digest of Educational Statistics 1987—Public Domain.

1. Decide whether your bars will run vertically (see Figure 8.4) or horizontally (see Figure 8.5) on the graph although, of course, you should be consistent and logical in your choice. For instance, altitudes should probably be shown vertically, whereas distances traveled would be best shown horizontally. Technical artists suggest that you arrange the bars vertically to illustrate the amount of a variable at different intervals of time, and horizontally to illustrate different measurements of several variables at the same time.
2. Your scale for measuring the bars must be consistent (Figure 8.5 shows the number of degrees in thousands).
3. For similar reasons your bars should be the same width. Technical artists suggest that a width of ½ inch works best, but the height and space available often determine bar width in graphs.
4. If possible, arrange your bars in descending order, with the longest bar at the left and the shortest at the right (see Figure 8.4). This arrangement emphasizes the first and last bars, as well as the rest of the bars' relationships. If the data prevent this arrangement, make sure that the logic of your arrangement is obvious.
5. If possible, separate your bars, but maintain equal distance between bars. If bars must be joined, make certain the illustration's labeling is still readable.
6. Do not place labels on the bars. Instead, place your labels to the left of the bars or beneath them, away from the graph's axis (see Figure 8.5).
7. If you color your bar graph, place the darkest value on the left and use lighter shades as you move to the right. Be certain that color values between bars change by at least 20 percent to maintain obvious distinctions between the bars and to ensure clear printing.

Cautions.   Remember, if the bar graph lacks proportion, the data may appear distorted or exaggerated. Therefore, be certain the bars are wider than the spaces between them. Too much space between the bars may confuse the reader. When using many bars, eliminate the spaces between them. Be certain the bars have the proper length and width to present the data accurately. Tall, slim bars tend to exaggerate the units of measurement represented, whereas fat bars tend to minimize the units of measurement. When in doubt, use our suggested ½-inch bar width.

Line Graphs

In the line graph the constant is plotted horizontally, that is, along the X axis, and the variables are plotted vertically along the Y axis. Line graphs are best used to show the relationship between X and Y as a set of discrete measurements. The line graph can show a series of exact measurements and a trend created by these measurements. Thus, it is appropriate for presenting data that must be recorded as a series of facts, for example, amounts spent monthly or dollars spent per person.

Use.   Use line graphs for all audiences when you want to show a trend rather than a comparison. Line graphs may be the most frequently used graphs because they allow the user to show discrete data as part of a trend or to strikingly illustrate a prediction. Use the line graph when you want to emphasize data without distorting their effect. The line graph can replace pages of explanation, allowing the comprehension of complex ideas in just a few moments.

Design. Since the line graph represents a ratio relationship between X and Y, the visual representation should suggest and enhance that ratio. The following information will help you design your line graph.

1. Create your axes (X and Y) logically. Draw your horizontal axis in equal intervals from left to right, starting with the lowest number on the left. Draw your vertical axis with zero at the bottom and the highest number at the top. Remember, your lowest numbers will be where X and Y meet (see Figure 8.6 and Figure 8.7). You must keep a constant ratio on these two lines so the lines give a logical picture of the data. If you do not keep a constant ratio or if your drawing is either too wide or too tall, your data will appear distorted because

| FIGURE 8.6 | Line Graph Showing Projected Demand for Teachers. |

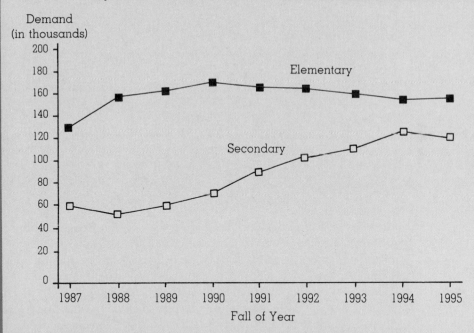

TRENDS IN PROJECTED DEMAND FOR NEW HIRING OF TEACHERS, BY LEVEL

- The projected demand for new hiring of elementary school teachers is expected to increase in the late 1980's and then decline in the first half of the 1990's.
- The projected demand for new hiring of secondary teachers is expected to increase in the first half of the 1990's.

Source: Center for Education Statistics, unpublished forecasts.

Digest of Educational Statistics—Public Domain.

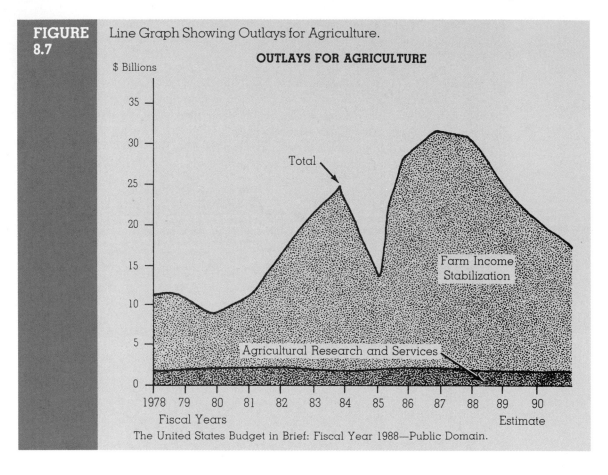

**FIGURE
8.7**   Line Graph Showing Outlays for Agriculture.

OUTLAYS FOR AGRICULTURE

$ Billions

Total

Farm Income
Stabilization

Agricultural Research and Services

1978 79   80   81   82   83   84   85   86   87   88   89   90

Fiscal Years   Estimate

The United States Budget in Brief: Fiscal Year 1988—Public Domain.

the lines will appear either too steep or too flat. For example, if the *Y* line measures months, all twelve months should be listed and given equal spacing. If the *X* line lists units of production, and you use intervals of 100, then the line must move from 0 to 100, to 200, to 300, etc. No gaps should occur in the labeling of your line.

2. Whenever possible, show zero as the starting point of both *X* and *Y*. If space problems require you to start with a higher number, do so clearly with a white space and a marker showing where you begin to start the intervals. This technique is called a "suppressed zero."

3. Label your intervals clearly, preferably horizontally and next to the intervals on the *X* and *Y* axes. Use a legend (a key identifying the points and lines) to label the points you plot and to identify the lines you draw to show trends. Identify all facts for the reader; place your key for identification on the lower right of the graph so that any citations or source notes can be placed at the lower left.

4. If your graph is making multiple comparisons and requires more than one trend line, use contrasting lines (see Figure 8.6). For a color graph, use contrasting colors; for a black and white graph, a heavy line, a broken line, and a dotted line.

5. You may want to use a grid to give your data more accuracy and to enhance the readability of your graph. Use care that inserting a grid does not clutter your graph's design.

Cautions. Because the line graph compresses so much data, you may be tempted to insert too many multiple lines and show relationships among too many variables on one graph. Don't. Most graphic artists advise that a graph contain no more than three lines. Too many lines confuse the reader. If all the comparisons are arguably necessary, plan a series of line graphs, using each graph to compare two or three variables.

TABLES

Our discussion of graphics includes tables. A table arranges large amounts of data in columns and rows that allow the reader to study them all together. In this section, we define tables, explain the construction of a correct table, and discuss some table design conventions. A well-designed table can be more accurate than a graph, and also present much more information. Tables are classified as formal and informal.

Formal Tables

Formal tables may be necessary for handling very complex information in scientific writing, statistical and numeric reports, economic and business articles and reports, and engineering studies. Formal tables condense data in a quickly-understood form and place them in a limited space. A well-designed formal table can replace pages of expository text.

Use. Use the formal table for expert audiences. Use a formal table or a series of tables when you must present all the data relevant to your topic or when you must present large amounts of data (see Table 8.1).

Design. Many conventions exist for constructing formal tables. Professional associations in their publications manuals will specify their conventions. For example, if very specific types of tables must be prepared to present statistical results, consult a special text such as the American Psychological Association's *Publication Manual* to arrange your data correctly.

This section explains the basic design of a formal table. We have constructed an example table that might appear in a college annual report. We have labeled the example table with numbers in parentheses, and we will discuss the parts using the parenthetical numbers:

1. **Table number.** Printed in capital letters and centered horizontally, the table number should indicate the sequence of the table in its chapter or segment of your work. Number your tables by chapters in works that are divided into chapters. Use decimal points and consecutive arabic numerals to indicate sequence within the chapter, as in TABLE 8.1, TABLE 8.2, TABLE 8.X. Another way to identify tables is with hyphenated numbers, as in TABLE 8–1, TABLE 8–2, TABLE 8–X. Sometimes the table number is printed in boldface type. Do not letterspace the table number; space only between the word TABLE and the number.

TABLE 8.1   Table Showing Outlays by Agency.

TABLE 5
OUTLAYS BY AGENCY, 1986–92

(In billions of dollars)

	1986 actual	Estimate					
		1987	1988	1989	1990	1991	1992
Legislative branch...	1.7	2.1	2.2	2.2	2.2	2.2	2.3
The Judiciary...........	1.1	1.2	1.4	1.5	1.5	1.5	1.6
Executive Office of the President........	0.1	0.1	0.1	0.1	0.1	0.1	0.1
Funds appropriated to the President....	11.4	11.8	11.2	13.3	12.7	12.3	11.7
Agriculture..............	58.7	55.1	50.7	46.5	42.9	39.5	37.6
Commerce..............	2.1	2.4	2.3	2.4	3.3	2.0	1.9
Defense—Military[1]...	265.6	274.2	289.3	303.7	321.0	340.0	361.0
Defense—Civil.........	20.3	20.9	22.1	23.4	24.5	25.5	26.5
Education................	17.7	16.8	14.7	14.4	13.3	12.8	12.4
Energy....................	11.0	10.6	10.2	10.9	11.8	12.5	13.5
Health and Human Services, except Social Security.....	143.3	145.3	146.8	156.3	165.2	176.4	185.9
Health and Human Services, Social Security..............	190.7	202.9	214.5	227.0	241.4	256.2	270.6
Housing and Urban Development........	14.1	14.6	13.9	14.3	14.0	13.6	13.7
Interior...................	4.8	5.2	4.4	4.7	4.6	4.4	4.3
Justice....................	3.8	4.8	5.8	5.6	5.4	5.5	5.6
Labor.....................	24.1	24.5	25.4	25.4	25.9	26.2	26.8
State.....................	2.9	3.3	3.6	3.7	3.9	4.0	3.9

2. **Title.** The title must describe the table's contents briefly yet accurately. You should not use cute, eyecatching, or unconventional titles on formal tables. When space allows, use complete words in your title; if you must abbreviate, be consistent. Print your title in capital letters, and center each line of the title horizontally.

3. **Box heading.** This caption identifies the information the columns contain. Center the box heading over the columns it defines rather than over the entire table. Our box heading, Qualifications and Specialization (see Table 8.2), identifies the three columns below it, not the column at the far left. Be specific with your box heading.

4. **Subhead.** Subheads are labels for the individual columns. They must be grammatically or numerically parallel. Usually the subhead is centered horizontally over its column, or it is aligned on the left with the information in its column. Subheads should be in capital and lower-case letters.

| TABLE 8.1 | Continued. |

	1986 actual	Estimate					
		1987	1988	1989	1990	1991	1992
Transportation.........	27.4	26.2	24.6	25.9	25.5	25.0	25.4
Treasury................	179.2	180.2	187.3	197.0	200.7	202.9	196.0
Enviromental Protection Agency.........	4.9	4.6	4.6	4.7	4.7	4.6	4.5
General Services Administration......	0.2	-0.1	-0.4	-0.3	-0.2	-0.3	-0.7
National Aeronautics and Space Administration......	7.4	7.9	9.5	11.1	11.1	11.0	11.0
Office of Personnel Management........	24.0	27.7	26.8	28.4	29.9	31.5	33.1
Small Business Administration......	0.6	0.1	-0.3	-0.1	0.4	0.4	0.4
Veterans Administration......	26.5	26.8	27.0	27.6	27.9	28.5	28.9
Other independent agencies..............	11.4	17.9	11.5	9.2	10.5	11.9	11.8
Allowances[2]............	—	—	-0.8	0.5	2.8	4.9	7.1
Undistributed offsetting receipts................	-65.0	-71.8	-84.2	-90.2	-99.1	-110.8	-118.0
Interest................	(-32.0)	(34.7)	(-38.8)	(-44.4)	(-50.6)	(-56.8)	(-62.4)
Other...................	(-33.0)	(-37.1)	(-45.4)	(-45.8)	(-48.5)	(-54.0)	(-55.6)
Total outlays......	989.8	1,015.6	1,024.3	1,069.0	1,107.8	1,144.4	1,178.9
Off-budget......	16.7	19.5	39.7	51.7	64.3	74.4	81.4

[1]Includes allowances for civilian and military pay raises for the Department of Defense.
[2]Includes allowances for civilian agency pay raises and military pay raises for the Coast Guard.

Source: The United States Budget in Brief Fiscal Year 1988.

5. **Stub heading.** The stub heading identifies the far left column as a series of objects, persons, or events about which the table presents information. Because the stub heading tells the reader how to interpret the labels in the far left column, it should be accurate and specific. Print the stub heading in capital letters.

6. **Line heading.** Line headings identify classifications under the stub heading. In our example, Table 8.2, the line heading, Professor, is a classification under the stub heading, FULL-TIME FACULTY. Since each line heading describes only one classification, it must be accurate and specific. If the formal table does not have classifications under the stub heading, you will not need to include line headings.

7. **Subheading.** Subheadings identify each of the objects, persons, or events about which your table presents information. In Table 8.2, the subheadings are the names of the faculty members. Subheadings should be grammatically or numerically consistent.

(1) **TABLE 8.2**

(2) **CHARACTERISTICS OF ACADEMIC STAFF:**
COLLEGE OF EDUCATION AND ALLIED PROFESSIONS.*

	QUALIFICATION AND SPECIALIZATION (3)		
(5) **FULL-TIME FACULTY**	**Degree** (4)	**Institution**	**Teaching Specialty**
(6) Professor			
(7) Brown, John F.	Ph.D.	Michigan State	Human Development
Smith, Sarah L.	Ed.D.	Ohio State	Higher Education
Associate Professor			
Jones, Paul. G.	Ph.D.	Boston University	Administrative Theory (8)
Carter, Sam T.	Ph.D.	Colorado State	Educational Media
Assistant Professor			
Cole, Nancy T.	Ed.D.	Purdue University	Adult Learning
Black, Jane R.	Ed.D.	Teacher's College	Human Development
(9) **EDUCATIONAL CURRICULUM AND INSTRUCTION**			
Professor			
Garcia, Joseph H.	Ed.D.	Ball State	Science/Math Education
White, David L.	Ph.D.	Kent State	Social Studies
Wilson, Susan K.	Ph.D.	Michigan State	Library Science
Associate Professor			
Taylor, Wilson S.	Ph.D.	Akron University	Language Arts (8)
Walker, Jennifer L.	Ed.D.	Ohio State	Reading
(9) **HOME ECONOMICS**			
Professor			
Gates, William F.	Ph.D.	Florida State	Foods and Nutrition
Hartman, Janice K.	Ph.D.	Miami University	Family Development
Associate Professor			
Becker, Donna E.	Ph.D.	Florida State	Interior Design
Woods, Ralph W.	Ph.D.	Ohio State	Textiles/Clothing (8)
Assistant Professor			
Miller, John P.	M.S.	Purdue University	Consumer Studies
Reed, Edward R.	M.S.	Michigan State	Human Development

(10) *"Allied Professions" are areas related to education.

(11) *Source:* Philbin and students, 1987.

8. **Body.** This section should contain all the data that illustrate your discussion of your topic. Consistency is very important when reporting your data. If your data are reported in words, be consistent with your form, punctuation, and abbreviations. If your data are numbers, they should all be expressed in the same units. Whole numbers should be aligned by the right column; decimal numbers should be aligned by their decimal points.

9. **Spanner heads.** Spanner heads are the table's major divisions. Our spanner heads show the major divisions of the college of education and allied professions. When you consider using spanner heads to show the divisions of one formal table, also consider whether a series of tables would serve your data better.

10. **Footnotes.** Use footnotes to explain any parts of the table that may be unclear. In our sample we used a footnote to explain allied professions.

11. **Source line.** The source line credits the person or institution that provided the information used to create the table or the table itself. The source line should appear at the lower left of the table and should contain the proper noun names of the institution or text of origin, any authors' or illustrators' names, and the date of composition.

Other. As an author you must decide whether to integrate tables into your text or to insert them after your text as a series of appendixes. For a lay audience or for easy readability, include the tables in the text. When a table is to appear in the text, prepare the reader by discussing the table in the text several sentences just before the table itself. Refer to the table by its name, "TABLE 8.X." For an expert audience, place your series of tables in an appendix. Remember, to experts, your document itself may be less important than its tables. Experts want to study all the facts. In the appendixes, give them tables of the data and provide summary tables as well.

When designing your appendix, arrange the tables in logical order rather than by random choices. In the text refer to your tables by name, and include at the first mention of each table a note in parentheses "(see Appendix X)" and the page number if your appendixes are paginated.

Cautions. Keep your tables clean and readable. Beginning writers often try to fit too much data and too many subheadings into formal tables. To avoid this problem, we urge you to use only first- (column and line headings) and second-level (column and line subheadings) headings in your tables. It is also important that your language and type styles are consistent.

Be conservative about using colors in tables. In formal tables, grey screens may create contrasts, emphasize strategic data, and thus improve the readability of multiple columns. Formal tables should blend with the colors of other artwork, but they should never be colored for the sake of color.

A caution about layout may be useful. As with other graphic elements, design the table so that it fits vertically on the page. If the number of columns prohibits vertical layout, arrange the table with the title facing the left margin of your document. Then arrange the sequence of the tables so that all the quarter-turned tables appear consecutively.

Informal Tables

The informal table is designed to appear directly in your text. It can have an informal title, but it need not have a table number.

Use. Use informal tables for lay audiences and for children. Informal tables are often used by publications with large circulations, such as newspapers.

Design. The informal table should fit on one-half page or less of the document. If the table is larger than three-quarters of a page, redesign it as a formal table or another type of illustration. We have designed a sample informal table using again the idea that this table, a freshman profile for a particular year, may appear in a college annual report. We will use our sample to describe the design of an informal table.

1. Use simple complete terms to label your columns. Note how many terms are written out so our reader will not be put off by needlessly technical terms or abbreviations.
2. Use the fewest number of labels and subdivisions that will inform your reader about the table and still keep the information simple and direct. We have used only three headings, and avoided subdividing the columns.
3. Use lines, tab settings on your typewriter or word processor, and spaces to create your columns; do not box or place double lines in any section of the table.
4. Align the columns at the left-hand margin or by your tab settings. If your table includes decimal numbers, align the columns by the decimal point.

Cautions. Avoid lengthy explanations in footnotes. Although such notes may be customary and useful, the purpose of an informal table is to present the information simply and clearly. Details should appear in the text immediately before or after the table.

Avoid mixing units of measurement in the table. For example, do not mix feet and inches or hours and minutes. Use as few units of measure as the logic of the table will permit.

TABLE 8.3 Sample Informal Table.

Freshman Class Profile: Fall, 1988

Chosen College	Total Enrolled	ACT Composite Score
Arts and Sciences	751	21.9
Business Administration	823	21.3
Education and Allied Professions	551	20.8
Health and Human Services	286	20.4
Musical Arts	181	22.2
Pre-Major Advising	417	21.2
Technology	121	21.0

SUMMARY

In this chapter you have learned that graphs and tables are used primarily for illustrating quantitative information. You have seen how to select graphs and tables for appropriate illustrations of your technical writing. Specifically, you should be able to do the following:

- Select an appropriate pie, bar, or line graph for a collection of data.
- Draw a correct pie graph.
- Draw a correct bar graph.
- Plot a correct line graph.
- Look at a pie, bar, or line graph and state whether or not it is designed correctly.
- Construct a formal table.
- Construct an informal table.
- Know whether a formal table or an informal table is designed correctly.

ASSIGNMENTS

1. Start a file of graphs and tables that you like and consider well done. Use them as models for your own graphs and tables.
2. Find out what software packages are available for your word processing system. What graphs and tables can you produce with these packages? If the packages are available to you, produce some sample graphs and tables.
3. Evaluate the graphs and/or tables in a technical document. What types of visuals are used? Do they have a purpose? Can you see what you need to see? Choose one graph or table and explain how you would improve the visual.
4. Draw a pie graph to represent the student activities budget at your school, or another budget that you know well. Can you represent the same data in a bar graph? Which is the better visual? Why?
5. Construct a formal table, similar to our sample (Table 8.2), using the faculty of your school or the faculty of a college at your university.

CHAPTER
9

Drawings and Photographs

Drawings and photographs are pictorial visuals that represent an object's shape or physical appearance or show a pattern of organization. Technical communicators use drawings and photographs as they would graphs and tables (Chapter 8), to help the reader see and understand. Photographs and drawings shorten text, clarify information, emphasize data, establish relationships, and summarize material. In this chapter we present a case history that illustrates the need for drawings and photographs. We also describe the construction and production of drawings and photographs. After reading this chapter, you will understand the following skills:

■ Selecting, constructing, and producing drawings.
■ Using and producing photographs.

John Sanchez has arranged to complete an internship with a small, local manufacturer of reasonably priced, commercial office chairs. John is a senior technical communication major with a background in marketing. His major project for his internship is to construct and produce all the documentation for a new secretarial chair the company is adding to its line.

Because John's company is relatively small, he is responsible for all phases of documentation. He needs to produce the sales literature that the company's catalog will include and the parts and assembly instructions that will be shipped with the chairs. John is confident that he can write the documentation, but less confident about producing the drawings and photographs his documents will require.

John's documentation will rely heavily on drawings and photographs. He will need photographs of the new secretarial chair for his sales literature and line drawings for his parts list and assembly instructions. John could include a callout to show his reader all the parts of the chair; he could also include an exploded view and a segmented drawing to help his reader assemble the chair.

As a technical communicator, John knows that the success of his documentation depends on his drawings and photographs. Keeping his company's policies and budget in mind, John needs to decide what production methods to use. He can investigate compatible computer-assisted design packages and contact professional illustrators and photographers. He must be knowledgeable about the design of drawings and photographs in order to communicate with these professionals and to make his decisions about hiring them.

DRAWINGS

Pictorial visuals are produced in a number of ways. As a technical communicator, you need a basic knowledge of the construction and production of drawings. A technical communicator may construct and produce drawings, using either traditional methods or graphics software packages. Professional illustrators and printers, either hired or employed by the writer's company, may actually construct and produce the drawings. No matter what situation exists, the technical writer must have knowledge about drawings.

Flowcharts

The flowchart is a drawing that presents the overview of a process. It uses repeated geometric forms such as circles, ovals, or rectangles to define parts and lines connecting the forms to show a process or a hierarchy. Within this classification, we can list organization and milestone charts, hierarchical task analysis charts, and various kinds of block diagrams. For details about the numerous conventions of designing flowcharts, consult the publications of the American National Standards Institute (ANSI), in particular, *U.S.A. Standard Flowchart Symbols and Their Usage in Information Processing* by ANSI.

Use. Use flowcharts for all audiences to show how a system operates. Flowcharts illustrate how actions can be divided into understandable parts. For example, flowcharts show how an institution is structured (these kinds of flowcharts are called "organizational charts"), when a sequence of deadlines is to be met or what steps compose a particular process.

Design. Standardization is the key to successful flowcharts. Be consistent in your use of logical order and standard symbols.

FIGURE 9.1   A Sample Hospital Organizational Chart.

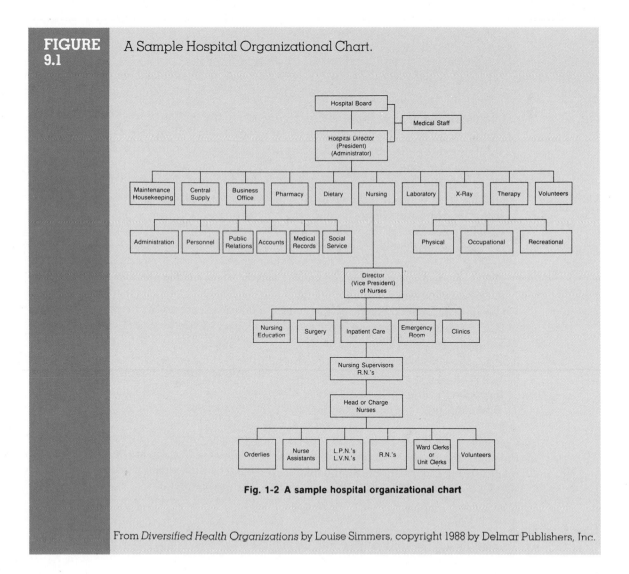

Fig. 1-2 A sample hospital organizational chart

From *Diversified Health Organizations* by Louise Simmers, copyright 1988 by Delmar Publishers, Inc.

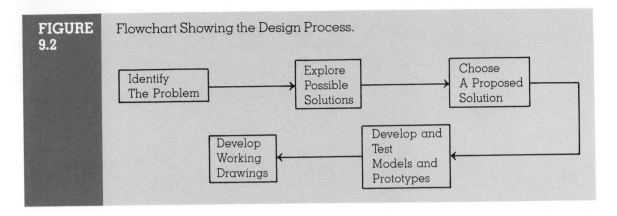

FIGURE 9.2   Flowchart Showing the Design Process.

1. Flowcharts should move from the top of the page to the bottom or from left to right. An organizational chart moves from the top of the page to the bottom reflecting the organizational structure of the institution. A milestone chart moves from the left to the right reflecting a period of time.
2. Since the ANSI has standard symbols for different functions, use these symbols whenever possible. The rectangle is the standard symbol for units in an organization. Be consistent; always use the same symbol for the same function.
3. Label the flowchart with a figure number as in Figure 9.1. Then assign it a descriptive title. Each element of the chart should be clearly and consistently labeled. In the organizational chart, the function comes first, and the name of the employee comes second. In the milestone chart, only functions are named, and above each circle indicating a function, the date for completion is printed.
4. Use lines to connect sequential functions on a flowchart; use arrows to connect functions that do not appear to be part of the natural sequence, such as feedback loops. Use arrows to represent flow in all pictorial items.
5. Place the flowchart within the text that describes it.
6. If several shapes of symbols appear in your flowchart, provide the reader with a key or legend in the lower left-hand corner of the chart.
7. Whenever possible, design your flowchart to fit on one page.

Cautions. Flowcharts are suddenly very popular, but don't overuse them. Keep the design of the chart simple, either vertical or horizontal. If the process that you are describing has too many feedback loops or returns, you may dazzle—and confuse—your readers. Consider instead a series of short, simple charts, one for each group of functions.

Line Drawings

Technically, any illustration made with the application of lines on paper is a line drawing, but we use this term to refer to illustrations created by illustrators or machines (other than cameras) to depict objects, persons, and processes. As a student, learn which type of line drawings apply to your field so that you can practice producing and using them. Our section on line drawings focuses on the materials and ideas you should supply for an illustrator, because you will probably hire a trained illustrator or work with an illustrator inside the same company. We discuss line drawings in capsule form to provide you with some general guidelines.

Use. Use line drawings for all audiences. For example, lay audiences would use line drawings for assembly instructions, and electrical engineers, as an expert audience, would use line drawings of electrical circuits. Consult our list of terms below for fine points concerning the uses and details of line drawings. The vocabulary used to define hand or machine illustration called "artwork" may vary from one region or country to another, but these terms get you started. For each term, we have listed some specific uses. Using your knowledge of your particular field, begin to think about other effective uses of each technique.

Callout (see Figure 9.3) The callout is a drawing with names, letters, or words inserted to identify the parts of a mechanism. It is used to show the reader all the parts of the object, how the parts look, or where the parts belong in the mechanism.

Cutaway (see Figure 9.4) The cutaway, as its name suggests, is a drawing that shows not only the external view of an object but an internal view as well. It is used to show the reader how the internal pieces fit within the external shell of the mechanism. Our example shows the internal parts of a drum controller.

Exploded view (see Figure 9.5) The exploded view is a drawing that shows the parts of a mechanism in proper sequence, but the drawing is separated to show fit, shape, and relationship. It is used to list or describe all the parts of mechanism when the reader must be aware of all the parts, how they fit together, and what process they complete. Our example shows the parts of an electromagnetic compressor clutch.

Perspective drawing (see Figure 9.6) The perspective drawing is a picture of a mechanism on a plane surface illustrating how the mechanism would appear to the viewer. It is used to illustrate proportions or to show the reader the relative sizes of objects in a collection or to show proportions.

Segmented drawing (see Figure 9.7) The segmented drawing shows all of a mechanism's separate parts but in logical order. It is used to show the reader how the parts look, how many parts she should have to perform a task, or how to purchase replacement parts. Our example is from the assembly instructions packaged with a model rocket.

Most scientific and technical disciplines may have their own uses for the types of line drawings defined here. For example, perspective drawings appear in architecture as elevations of a building and as renderings for customers.

Design. In order to estimate production costs and to schedule production time, line drawings are usually considered as simple, medium, or complex in scope. The production costs of line drawings may vary by geographic region. Obtain estimates from several technical illustrators in your area before picking one illustrator. Most skilled illustrators will estimate their costs in terms of hours, so you can also use their estimates to schedule production time for your line drawings. Experience quickly teaches that the requirements of line drawings may demand a great deal of the illustrator's time and other expenses. We have composed an informal table to assist you with your own estimates as you design your line drawings. Large companies generally have "in-house" illustrators, but even these companies are concerned about the costs and scheduling of illustrations.

Our guidelines for designing effective line drawings are based upon the classifications in the table.

TABLE 9.1	Classification of Line Drawings.			
		Simple	**Medium**	**Complex**
Work hours of illustrator		0–4	2–8[1]	6–X
Number of sketches		1–4	3–6[2]	6–X
Level of subjects		informal	advanced	expert
Number of processes		2	3–6	7–X
Handlettering		none	simple	decorative
Freehand sketches		none	1–2	3–X
Special type		none	1–4	5–X
Sizes		standard	squeezed	oversized
Design		standard	revised	original
Quality of roughs		clear	average	poor

[1]Ranges of the illustrator's hours overlap because it is possible to spend 3 hours on a simple or medium drawing.
[2]Ranges overlap because more than the number of sketches affects their complexity.

FIGURE 9.3	Callout.

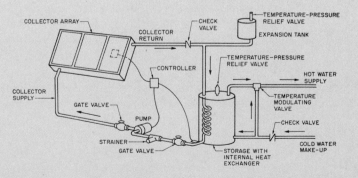

From *Refrigeration and Air-Conditioning Technology* by William Whitman and William Johnson, Copyright 1987 by Delmar Publishers, Inc.

**FIGURE
9.4**

Cutaway.

Courtesy of Cutler-Hammer Products.

**FIGURE
9.5**

Exploded View.

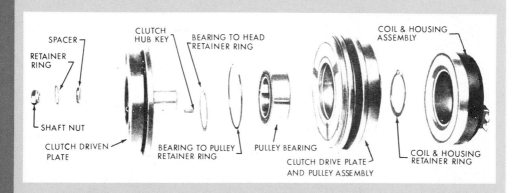

Courtesy of AC-Delco, General Motors Corporation.

FIGURE 9.6   Perspective Drawing.

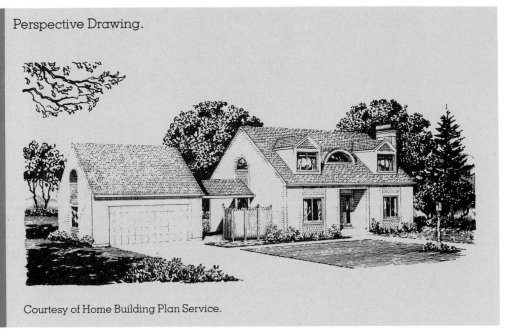

Courtesy of Home Building Plan Service.

 FIGURE 9.7   Segmented Drawing of Astro™ Model Rocket.

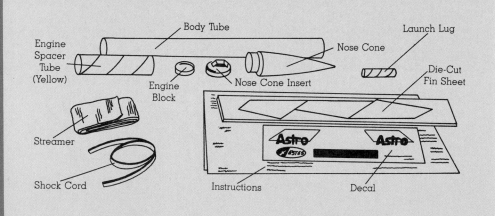

Courtesy of Estes Industries.

Work Hours of Illustrator. This is your most expensive item. The higher the skill level of the illustrator, the faster the illustrator will work and the more time-saving strategies he or she will have mastered, but the higher the skill level, the higher the fee. To cut time and cost, design your line drawing to show only the details that the reader needs to see. The more you add to the drawing, the longer the illustrator must work.

Number of Sketches. Beginners learn with some surprise that an illustrator has to sketch the drawing several times and in several ways to complete a rendering. Make certain you have the illustrator's estimate of the number of sketches needed to complete the drawings.

Level of Subjects. Are you illustrating nuclear fusion for easy readability within a widely circulated publication, or are you illustrating for experts, nuclear physicists? The audience's interest will affect the level of detail in your drawing, but your overall concern is for the ideas you are illustrating.

Number of Processes. Many times, finished line drawings require a mixture of other illustrative techniques. For example, cutaway views or exploded drawings may use airbrushing to create interest. Each improvement to the drawing costs more. Learn the illustrator's terms and be certain you understand what the illustrator must do to create your finished drawing.

Handlettering. Depending on your company's computer-aided design capabilities, the illustrator may need to handletter logos, trademarks, or scientific or language problems. Plan drawings that use offset type letters and labels so the illustrator must only place, not draw, the words in the proper place.

Freehand Sketches. To save time on repeated requests for illustrations and to maintain a consistent image over a "run" of illustrations, most corporate and many freelance illustrators keep "mechanicals" or camera-ready designs of your objects, logos, and trademarks in their files. New sketches for each line drawing run up the illustrator's time. For the same reason you should simplify all equations so the illustrator need only paste up a few mechanicals.

Special Type. Keep your design simple and label your work with either your company's chosen typeface, a standard bookface, or a standard display face. If special fonts or unique symbols must be ordered from subcontractors, the illustrator may attach a surcharge, and your drawing may become needlessly complicated and expensive.

Sizes. Standard sizes exist for your format so your document can fit the printer's specifications and the designer's requirements. Choose line drawings that will fit your plan. The drawing must be easy to see, but it should not absorb too much useful print space. Do not clutter a small drawing with extra labels, but if you have a whole page for your drawing, label every relevant part.

Design. Define your design purpose as you would your text's objective. While the drawings should be uniform in style and appearance, they should also reflect your care in choosing an appropriate design. Do not provide a full perspective drawing of a mechanism when a segmented view will allow you to label the parts more clearly and save space. Use a cutaway only when you need to show the relationship between internal and external parts.

Quality of Rough. Give the illustrator a clear, clean sketch or picture of the drawing you need. Show, rather than tell, the illustrator. A sloppy, unclear rough forces you and the illustrator to spend extra time on the project, and it may prevent the illustrator from understanding the concept to be illustrated.

Cautions. Customarily, technical writers prepare tables, and they often prepare graphs, but line drawings require both the training and the talent of professional illustrators. Usually the most cost-effective approach to designing line drawings is to work with a professional illustrator. Unless you have formal training or natural talent, do not attempt your own line drawings.

Note as well that line drawings sometimes require human figures or parts of human figures. Check for and eliminate sexism and racism from all drawings that include figures or parts of figures. Whenever feasible, hands, feet, or parts of the human body should usually appear sexually neutral. For American audiences, women and girls should be shown as equal participants in the action with men and boys, and men and boys should be shown as equal recipients of the action with women and girls. For example, in instructions for the use of a word processor, show a line drawing of two figures, a man and a woman, side by side at two computers. Do not show the woman typing while the man stands next to the computer, appearing to supervise her. Line drawings for the lay audience of the United States should portray a racial mix. Show black workers with whites, Hispanics, Orientals, and other appropriate figures to keep the drawing representative of all possible ethnic users. Show both sexes and all races as executives, managers, supervisors, workers, and co-workers. Test your drawings with members of ethnic American groups to detect any cultural or sexual bias.

PHOTOGRAPHS

Carefully selected photographs can offer a cost-effective and contemporary solution to the problem of illustrating technical exposition. We limit this discussion to black and white photographs, because production expenses may prohibit the use of color.

Use. Use photographs for all audiences when you want to introduce the audience to experiences that they will probably never have. For example, the user of a home computer may never have seen the printed circuit boards for electronic equipment. Including a photograph of the circuit boards will help the reader understand your text. Use photographs to take the reader to places he or she would not normally visit. For architectural studies, show the reader the facades of little-known buildings (see Figure 9.8); for text explaining industrial application of horizontal machining centers, show the reader a horizontal machining center operating in an industrial setting (see Figure 9.9).

Design. Since a photograph can show every detail, use it to your advantage. Keep the composition of the photograph simple and focus on one object as the center of your study. The

FIGURE 9.8   Photograph—Building Facade.

Courtesy of California Redwood Association.

FIGURE 9.9   Photograph of a Typical Horizontal Machining Center.

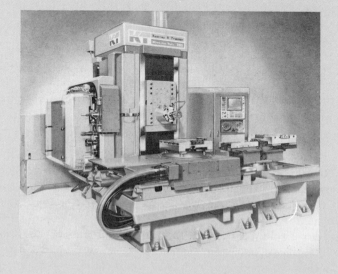

Courtesy of Kearney-Trecker Corporation, Milwaukee, Wisconsin.

FIGURE 9.10   Photograph of an Engine Dynamometer.

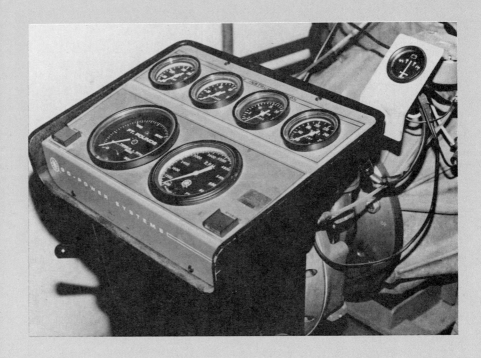

From *Motor Automotive Mechanics* by Anthony Schwaller, copyright 1988 by Delmar Publishers Inc.

subject should fill the frame (see Figure 9.10), allowing no distracting objects to clutter it. To produce the photographs that you require, we recommend hiring a technical photographer whenever possible. The training these experts provide will improve your document's visual impact. Whether hiring a photographer or taking the photographs yourself, evaluate your negatives according to several principles of well-designed photographs.

Clarity The photograph should have sharp lines and a clear focus.

Depth of field The photograph should emphasize what you want: near and far objects clear, near objects clear and far objects blurred, or near objects blurred and far objects clear.

Balance The photograph should place the object of emphasis in the foreground, centered horizontally (see Figure 9.10). Action photographs should conform to the basic composition rules of photography.

Contrast The photograph should contain a full range of white, gray, and black, not a "muddy" collection of grays (see Figure 9.10).

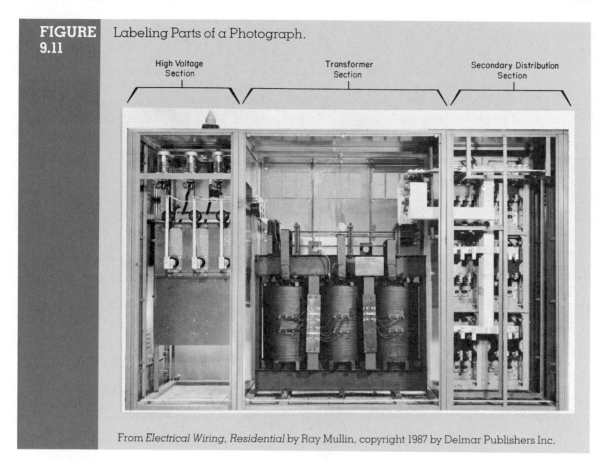

FIGURE 9.11   Labeling Parts of a Photograph.

High Voltage Section   Transformer Section   Secondary Distribution Section

From *Electrical Wiring, Residential* by Ray Mullin, copyright 1987 by Delmar Publishers Inc.

Remember to integrate your photographs correctly into your text. Label the parts of the photograph with straight black lines that lead to the border; then place identification lines beneath or alongside the photograph (see Figure 9.11). Give the photograph a figure number, but not a formal title. Instead, place a caption or short description beneath the photograph. Place the credit for any source in the lower left.

Cautions. Since the camera catches every detail, photographs may become dated much more quickly than line drawings. If your document must circulate five years or more before revision, rely on line drawings so the illustrator can emphasize only the objects you select. If you use models in your photographs, be certain they dress conservatively. Then they do not distract from the products or processes the photographs illustrate and their clothing styles do not date your photographs. Edit any photograph as you would a line drawing to delete racial, sexual, and cultural biases offensive to readers and users. Carefully plan your photographs so you can avoid excessive cropping or enlarging. These techniques can distort the images and reduce your photographs' clarity.

SUMMARY

This chapter explained how to select drawings and photographs for appropriate illustration of your technical writing. You have learned that flowcharts can represent systems, processes, and hierarchies; that various kinds of line drawings create emphasis as they show objects; and that photographs capture detail. Specifically, you should be able to do the following:

■ Construct flowcharts that can represent systems, processes, or hierarchies.
■ Decide whether a line drawing or a photograph is more appropriate for a visual presentation of a particular object or process.
■ Choose an appropriate line drawing—callout, cutaway, exploded view, perspective drawing, or segmented drawing—to illustrate a particular object or process.
■ State the level of difficulty—simple, medium, or complex—for a line drawing.
■ Select an appropriate photograph from an assortment of shots taken of the same object or process.

ASSIGNMENTS

1 Start a file of line drawings that fit into the categories listed in the chapter. Include in your file both traditional drawings and computer-assisted design drawings. Use them as models for future projects.

2 Construct an organizational flowchart for the administration of your college or university.

3 Go to your library and read about the uses of color in visuals. What are some of the uses of color, aside from its use in color photographs?

4 Evaluate the drawings or photographs used in the assembly instructions for an object (toys, bikes, and models make good examples). What types of illustrations are used? Do they have a purpose? Is the assembly process easy to follow? Choose one illustration and explain how you would improve it.

5 Find out if any computer-assisted design packages are available at your school. If the packages are available to you, produce some sample visuals.

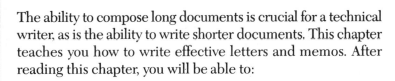

CHAPTER
10

Correspondence

The ability to compose long documents is crucial for a technical writer, as is the ability to write shorter documents. This chapter teaches you how to write effective letters and memos. After reading this chapter, you will be able to:

- Prepare your correspondence using any of four possible formats.
- Compose effective letters of request.
- Draft confirming letters to create a written record.
- Write effective letters to create a written record.
- Prepare appropriate letters of recommendation about your associates.
- Develop letters which deliver bad news directly but without cruelty.
- Draft basic form letters for your routine correspondence.

CASE STUDY   Correspondence Builds a Business

Joe Rachiele owns a small computer sales and repair business. Specifically, he sells Apple and IBM-compatible desktop computers to businesses and families. A Viet Nam veteran who trained in electronics during his military service, he has also earned a baccalaureate degree in electronics. Because of his training in electronics, Joe's business has expanded into repair and servicing computers. He has found that his repair business generates repeat sales of desktop computers and their accessories.

Business has been kind to Joe. His firm, started with a GI loan, has expanded. He now supervises six employees, and has started to sell large orders using direct marketing. The problem, however, is paperwork. Nothing in Joe's education prepared him to place orders using letters of confirmation or to use routine correspondence to track his orders. He also has difficulty collecting refunds from suppliers for defective merchandise, because he occasionally fails to explain in adequate detail the breakdowns or defects that occur in his new stock under warranty. Because of the technical nature of descriptions of computer products, Joe, the resident computer expert, drafts his own correspondence instead of delegating the task to an employee.

Recently, Joe failed to deliver some repairs on time, because either he did not order the correct parts, or he received unordered items which he neglected to replace. Joe did not expect a lack of correspondence skills to threaten his business. Now he must develop letters for requests, confirmations, and the other communication routines of a growing technical business, or risk the loss of accounts to his better-organized competitors.

PURPOSES OF CORRESPONDENCE

You might think that a business owner such as Joe Rachiele would have no need of correspondence. After all, can't he telecommunicate with his suppliers via a modem? Can't he purchase a software package that generates orders and form letters? Yes, technology enables Joe to communicate by modem and to manage some of his responsibilities using computer-generated messages and forms. Also, electronic mail offers exciting possibilities for an owner-operator such as Joe.

Nonetheless, many communications must be drafted individually, so that specific problems are thoroughly presented and explained. Thus, although letters cost more in personnel hours than telecommunication or computer communication, correspondence remains a necessary tool of the successful manager. Effective managers must delegate time, either daily or weekly, to deal with the various orders, queries, and random messages scattered on their desks or workbenches. In this chapter, we show you the types of letters we helped Joe develop, but first you may want to understand a bit about correspondence as a means of communication.

Consider the letter in Figure 10.1, a typical letter to obtain action that Joe had to compose and send to a supplier. Joe's letter shows that any student who expects to work in the business world must develop the appropriate correspondence skills. Initially, the letter acts as a request, that is, a call for an adjustment, the delivery of the badly needed manuals. Note

**FIGURE
10.1**   A Typical Letter to Obtain Action.

RIVERSIDE COMPUTER COMPANY

208 Georgia Avenue
Bowling Green, Ohio 43402
Phone: (419) 354-2219

January 4, 1989

Ms. Sally Johnson
Sooner State Electronics
P.O. Box 2000
Moose City, OK 73468

Dear Ms. Johnson:

I wrote to you in December to explain that we have had trouble with the Multi-I/O cards you sold us for our IBM-compatible computers. We cannot make the clocks and calendars of the machines run, despite your guarantee of 100% IBM-compatibility.

You stated by phone on December 10 that Sooner State Electronics had documentation in production that would offer clear instructions, so that our customers could make their clocks and calendars work with the DOS 9.12, which they now use. You said the manuals would be in the mail to Riverside Computer by January 1, but we have not received them.

We need your manuals because our customers purchased their computers in good faith, and they expect them to run. Because of the problem with the Multi-I/O cards, I have had to give refunds on several machines, and I have lost sales to my competitors, who can make the clocks and calendars run using DOS commands. As you know, I cannot return the cards to you and ask for a refund because the cards are in machines I have sold to customers. A recall of your product would damage our reputations and our customers' confidence in their computers.

Would you assist us in the following ways:
1. Find out when the manuals will be shipped.
2. Call me this week with a firm date for shipment.
3. Mail the manuals directly to my customers who have their machines equipped with your cards. I can provide gummed labels to make the mailing job easier.

With your attention to a speedy shipment of the manuals, we can keep our customers happy. Since I find the Multi-I/O card that you sell to be quite reliable and affordable, I would like the opportunity to continue to order your products, assuming you can guarantee the standard of quality that my customers require.

Very truly yours,

Joseph Rachiele

that Joe refers to his request in December that Sally Johnson supply the manuals. The letter also confirms that Joe did indeed explain to his supplier, Sally Johnson, his need for the manuals in order to operate her products (the Multi-I/O cards) properly. The letter also serves as a record of the actions Joe has taken, so that he and Sally will have a mutual record for their use. If they fail to agree on a settlement, a court will have a statement that records Joe's efforts to obtain an appropriate remedy within a reasonable time.

Joe's letter also serves as a complaint, that is, a statement of record that the Multi-I/O cards have not lived up to the supplier's promise, and that Sally must remedy the situation by delivering the manuals. Because the letter exists as a record, Sally must now address Joe's complaint, or face expensive legal actions. However, no actions can occur unless Joe proves that he attempted to seek "reasonable" remedies. This letter shows Joe's reasonableness because he twice — once by telephone and once by letter — notified his supplier about the defective cards and his need for manuals to correct the problem. As a complaint, the letter contains a recommendation, that Sally Johnson deliver the manuals directly to Joe's customers or face extra expenses and inconveniences, to say nothing of the damage done to a good relationship with Joe Rachiele, a repeat buyer. Finally, the letter delivers bad news because Joe informs Sally that the problem with the card continues, will not disappear, and must be corrected.

In one example, Joe's letter shows that business correspondence related to technology exists for any or several of the following purposes.

- ☐ State a request.
- ☐ Confirm an informal or spoken agreement, usually a previously made agreement.
- ☐ Create a record of an agreement for the protection of all the persons involved.
- ☐ File a record of a complaint when informal, oral discussions have failed to obtain action.
- ☐ State a recommendation for a course of action to solve problems (or to recommend an individual for a position).
- ☐ Convey news, especially bad news — unpleasant messages that the recipient will have to understand, because action may be required.

This chapter does not include application letters because we view such letters as part of your resume preparation. Also, we do not comment on "good news" letters since, unlike "bad news" letters, these create neither difficulty nor necessity for action. Finally, we do not discuss sales or direct marketing letters here since the business communications and marketing literature already contain a considerable collection of such materials. Our concern centers on the communication needs of a person such as Joe Rachiele — an individual who must describe technology repeatedly in his letters of record as a function of both his occupation and his rare technical skill.

FORMATS OF LETTERS AND MEMOS

Before we discuss the tone and attitude letters should assume, you need to understand that the format of the message helps to establish its tone. Although terminology for the formats of technical business letters may differ from one source to another, we refer to *four forms*:

☐ Full block.
☐ Modified block (also known as semi-block).
☐ Alternate form for block.
☐ Memo form.

To illustrate the different formats, we show the same message in four different formats. Some of the differences, such as vertical spacing, are rather subtle. Note, always check and follow the format that is acceptable at your place of business. In fact, you may find that formats differ greatly even within divisions of your workplace. For example, in the Bowling Green State University Department of English, we use a modified block format, but the College of Technology uses a full block format. Since this sort of discrepancy exists in many institutions and businesses, always check format requirements.

Full Block Format

Full block format uses vertical, not horizontal, spacing to create the divisions within the letter. The most modern of the formats, its clean direct look lends itself well to technical communication. Its simple vertical format tends to reduce the error rate since you simply do not have to be concerned about horizontal spacing to indicate new paragraphs. Figure 10.2 lists and identifies its seven components.

1. Heading
2. Address
3. Salutation
4. Body of the letter
5. Complimentary close
6. Signature
7. Identification line

The *heading* generally includes the date the letter is written. If you do not use letterhead stationery, include your address over the date.

The *address* above the salutation, or *inside address*, is the letter recipient's full address. Complete, this address should occupy no more than four lines and match exactly the address on the envelope. This address should also contain the correct United States Post Office state abbreviation and Zip Code.

For the *salutation*, the line that begins "Dear . . .," use either *Mr.* for men or *Ms.* for women. Avoid *Mrs.* or *Miss* unless the recipient has informed you of her preferred form. Instead of using "Dear Sir" or "To Whom It May Concern" as forms of the salutation, obtain a name to whom you can address your letter. Even "Dear Editor" or "Dear Salesperson" is preferrable to the vague salutation. The salutation should end with a colon.

The *body* of the letter contains the text of your message. You will find detailed suggestions about the content, attitude, and style of your letter in this chapter's section on types of letters and special messages. Two format details are useful: first, limit your message to one page so that it receives the reader's attention; second, use conscious "chunking" techniques such as lists, tabs, and short paragraphs to hold the reader's attention. Finally, doublespace and create new paragraphs at the left margin.

**FIGURE
10.2**

Full Block Format for a Letter.

895 North Main Street
Bowling Green, OH 43402

February 16, 1988

Mr. John Smith
Customer Relations
Ohio Telephone Company
133 Buckeye Street
Columbus, OH 43217

Dear Mr. Smith:

I received a collection notice from Ohio Telephone on February 13, 1988. The letter states that I owe a past due balance from the June 16 to July 16 billing period in 1987. The letter also states that my service will be disconnected unless I act immediately; however, my service was already disconnected on August 15, 1987.

I called and spoke with one of your representatives about this matter back in August when I received the unitemized bill. He took my phone number and assured me he would call back the next day after he reviewed my account. He failed to reach me, and unfortunately, I failed to get his name. I called again only to hear a recorded message stating that your representatives were busy. After trying for a week to speak with a representative, I wrote to the billing center (see attachment). I received no reply until February 13, 1988 (five months later), when I received the collection notice.

I am willing and able to pay the bill, but I must have an itemized bill so I can split the long distance calls with a roommate. I would like this matter settled by March 15, 1988, so would you please send me an itemized bill promptly.

Sincerely yours,

Angela Knauss
(419) 352-5555

cc: Dix and Dix
Encl: copy of Feb. 13, 1988 collection notice

The *complimentary* close should appear two lines below the last line of text, at the left margin. Capitalize only the first letter of the first word and always end the line with a comma.

Your *signature* should be in blue or black ink. Make your signature easy and large enough to read. Use simplified penmanship so the reader can understand the correct spelling of your name. Allow four vertical spaces for your signature.

Then type the *identification line*. This line contains your name and, if you have one, your title. Do not insert your nickname or your abbreviated name. Assuming that your correspondence may act as a letter of record, sign and type a legally correct name and title. You may also type a title or position title below your identification line. For example, a college professor's identification line reads, "I.M. Learned, Ph.D." Just one line below this appears the position title, "Associate Professor," for example, to show profession titles, institution positions, or academic ranks.

Two vertical spaces below the identification line, type notices about *enclosures* or *distribution of the letter* on single lines. For example, in Figure 10.2, Angela Knauss types "cc: Dix and Dix," the names of her local attorneys, so Mr. Smith knows they will receive a copy of the letter and its attachments. She also types "Encl: copy of Feb. 13, 1988 collection notice," a complete description of the attachment she includes with this letter.

This line serves two purposes. First, it creates a record that Ms. Knauss has sent a copy of the troublesome collection notice to Mr. Smith's attention. Now he cannot claim to have lost or never received the bill. Second, the line tells Mr. Smith about each item Ms. Knauss's correspondence contains. If, for example, her attachments are lost during the opening or filing process, Mr. Smith can inform Ms. Knauss of his need for a copy that has been lost through handling. As a matter of courtesy, you should always inform the receiver of the distribution list and the list of attachments for that letter.

An exception to this courtesy is the case of *blind copies*. The term "blind copy" ("bcc") refers to a copy you send to someone who may need to have your letter as a record for future action or to complete a file. If your letter treats a sensitive or private issue, such as a personnel matter, you may not want the receiver to know you are sending the letter through the necessary channels of readers that your company's policy or the law may require. In such a case, you omit a distribution line from the original letter and instead type "bcc:," followed by a list of their names on the copies you send to the concerned readers. The term "bcc" will inform them that the receiver does not know you have sent them copies. Your abbreviation notifies the receivers of the "blind copies" that they have a responsibility to treat the sensitive information you convey in the letter according to the company's guidelines.

Modified Block Form

Figure 10.3 shows that modified block format differs from full block in the placement of the heading, date, complimentary close, signature, and identification lines. These components should all start about halfway across the page, and be aligned vertically. The sample also shows that you begin the paragraphs of a modified block letter with a one-tab or five- to seven-space indentation. You should retain the vertical doublespace between paragraphs.

Although business seems to prefer full block format, modified block is quite acceptable. Many consider the modified block's appearance more balanced and traditional. Consider modified block format when corresponding with somewhat traditional institutions or businesses. Full block, with its formatting and editing ease, certainly lends itself to form letters

**FIGURE
10.3**   Modified Block Format for a Letter.

895 North Main Street
Bowling Green, OH 43402

February 16, 1988

Mr. John Smith
Customer Relations
Ohio Telephone Company
133 Buckeye Street
Columbus, OH 43217

Dear Mr. Smith:

I received a collection notice from Ohio Telephone on February 13, 1988. The letter states that I owe a past due balance from the June 16 to July 16 billing period in 1987. The letter also states that my service will be disconnected unless I act immediately; however, my service was already disconnected on August 15, 1987.

I called and spoke with one of your representatives about this matter back in August when I received the unitemized bill. He took my phone number and assured me he would call back the next day after he reviewed my account. He failed to reach me, and unfortunately, I failed to get his name. I called again only to hear a recorded message stating that your representatives were busy. After trying for a week to speak with a representative, I wrote to the billing center (see attachment). I received no reply until February 13, 1988 (five months later), when I received the collection notice.

I am willing and able to pay the bill, but I must have an itemized bill so I can split the long distance calls with a roommate. I would like this matter settled by March 15, 1988, so would you please send me an itemized bill promptly.

Sincerely yours,

Angela Knauss
(419) 352-5555

cc: Dix and Dix
Encl: copy of Feb. 13, 1988 collection notice

and correspondence for placing and confirming technical or scientific orders. For example, many universities require modified block format for their official correspondence. Because of the emphases on both technical research and on older scholarly traditions in universities, the modified block format seems a logical choice. Note, however, that Joe Rachiele, the owner/operator of a high-technology business, chooses full block for his orders, his confirmations, and other related correspondence.

Alternate Block Form

Note that Figure 10.4 looks like Figure 10.2, with one change: a subject line. Many companies now use this form because the subject line helps speed the letter's distribution. The subject line tells the receiver that the letter calls for decisions, conveys information and results, or records events for storage. The subject line can save time for the staff who open and process new correspondence and for the final receiver. Note that we show the alternate form with a full block format only. The businesslike attitude of a subject line does not lend itself to a modified block format.

Memo Form

Typically, use letters to communicate from your workplace to another company, and memos to communicate within a business, organization, or institution. Lately, however, we have seen increasing flexible use of memos. They are effective in formal, external communications as well as in informal, internal communications. Figures 10.5 and 10.6 illustrate how Angela Knauss's letter looks in the two acceptable memo formats.

Both figures contain a formatted "chunk" or set of lines, at the top of the page, separated from the memo's message. These lines identify the recipient (and his or her position if you send a memo from outside the firm), the sender and your position or firm (note that Angela Knauss's phone number identifies her as a telephone customer), the full date, and an accurate description of the subject in one line if possible. The sender's signature or initials should appear to the right of his or her name, not at the bottom of the message.

We present two forms of the memo so that you have a range of acceptable formats to select from. Some firms provide memo pads that you simply fill out to their specifications. Some companies require the use of the word "Memorandum" at the top of the memo; others omit that word. But you must use the "chunked" arrangement of *to, from, date,* and *subject* consistently to create an acceptable memo.

Figure 10.5 shows a format many workplaces use. Figure 10.6 shows a prepared format from a popular word processing program. When the format appears on the screen, you simply type in the identifying information and message. Many word processors now offer formats for letters and memos. You can save time and produce attractive documents using them. Just be sure that the formats your word processing software offers conform to the current, accepted correspondence protocols outlined here.

Before briefly discussing continuation pages of letters and the use or omission of letterhead, we remind you that, in business and technical communication, the current practice accepts quite a degree of variation in the appearance of correspondence. For example, some companies require a ragged right margin, while others use a justified right. Some word processors produce letters in modified block style with their headings at the left margin instead of centered on the page. The inconsistencies can be maddening if you try to find one correct

FIGURE 10.4   Alternate (Full) Block Format for a Letter.

895 North Main Street
Bowling Green, OH 43402

February 16, 1988

Mr. John Smith
Customer Relations
Ohio Telephone Company
133 Buckeye Street
Columbus, OH 43217

SUBJECT: Your collection notice of February 13, 1988

Dear Mr. Smith:

I received a collection notice from Ohio Telephone on February 13, 1988. The letter states that I owe a past due balance from the June 16 to July 16 billing period in 1987. The letter also states that my service will be disconnected unless I act immediately; however, my service was already disconnected on August 15, 1987.

I called and spoke with one of your representatives about this matter back in August when I received the unitemized bill. He took my phone number and assured me he would call back the next day after he reviewed my account. He failed to reach me, and unfortunately, I failed to get his name. I called again only to hear a recorded message stating that your representatives were busy. After trying for a week to speak with a representative, I wrote to the billing center (see attachment). I received no reply until February 13, 1988 (five months later), when I received the collection notice.

I am willing and able to pay the bill, but I must have an itemized bill so I can split the long distance calls with a roommate. I would like this matter settled by March 15, 1988, so would you please send me an itemized bill promptly.

Sincerely yours,

Angela Knauss
(419) 352-5555

cc: Dix and Dix
Encl: copy of Feb. 13, 1988 collection notice

**FIGURE
10.5**

Memo Format.

TO: John Smith, Manager, Customer Relations

FROM: Angela Knauss, (419) 352-5555

DATE: February 16, 1988

SUBJECT: Your collection notice of February 13, 1988

I received a collection notice from Ohio Telephone on February 13, 1988. The letter states that I owe a past due balance from the June 16 to July 16 billing period in 1987. The letter also states that my service will be disconnected unless I act immediately; however, my service was already disconnected on August 15, 1987.

I called and spoke with one of your representatives about this matter back in August when I received the unitemized bill. He took my phone number and assured me he would call back the next day after he reviewed my account. He failed to reach me, and unfortunately, I failed to get his name. I called again only to hear a recorded message stating that your representatives were busy. After trying for a week to speak with a representative, I wrote to the billing center (see attachment). I received no reply until February 13, 1988 (five months later), when I received the collection notice.

I am willing and able to pay the bill, but I must have an itemized bill so I can split the long distance calls with a roommate. I would like this matter settled by March 15, 1988, so would you please send me an itemized bill promptly.

cc: Dix and Dix
Encl: copy of Feb. 13, 1988 collection notice

method. Check your work by comparing it to our examples since we have drawn on our experience with the most obvious current business practices. Check your work against similar examples from your correspondence files for a detailed collection of acceptable and useful formats. Because the English language, business practices, and formats change, use the best of currently available formats such as full, modified, or alternate block.

Continuation Pages

For second and following pages, use plain stationery of the same type as your letterhead. Create a very simple heading for subsequent pages, such as this one for Angela Knauss:

Mr. John Smith   -2-   February 16, 1988

**FIGURE
10.6**
Memo Format from a Word Processing Software Package.

MEMORANDUM

TO: John Smith, Manager, Customer Relations
FROM: Angela Knauss, (419) 352-5555
DATE: February 16, 1988
SUBJECT: Your collection notice of February 13, 1988

I received a collection notice from Ohio Telephone on February 13, 1988. The letter states that I owe a past due balance from the June 16 to July 16 billing period in 1987. The letter also states that my service will be disconnected unless I act immediately; however, my service was already disconnected on August 15, 1987.

I called and spoke with one of your representatives about this matter back in August when I received the unitemized bill. He took my phone number and assured me he would call back the next day after he reviewed my account. He failed to reach me, and unfortunately, I failed to get his name. I called again only to hear a recorded message stating that your representatives were busy. After trying for a week to speak with a representative, I wrote to the billing center (see attachment). I received no reply until February 13, 1988 (five months later), when I received the collection notice.

I am willing and able to pay the bill, but I must have an itemized bill so I can split the long distance calls with a roommate. I would like this matter settled by March 15, 1988, so would you please send me an itemized bill promptly.

cc: Dix and Dix
Encl: copy of Feb. 13, 1988 collection notice

Triple space after the heading, and start your text. Then use a similar heading for all the pages that follow.

Arrange your text so that no single line or last line of a paragraph—much less a single word or short phrase—appears alone as the first line of the next page. Begin your new page with at least two, preferably three, lines of the paragraph from the previous page. If possible, begin the new page with a new paragraph—to improve your text's readability. Here, too, your word processor may help. Most programs enable you to build in and repeat page headings. We recommend checking your word processing software package's features, since it often performs such tedious tasks more efficiently than writers.

Letterhead

Figures 10.1 and 10.11 show how to arrange your heading to accommodate a letterhead. Because the letterhead reports the sender's address, you need type only the dateline at least two vertical spaces below the letterhead's last line or illustration. Align your dateline with the letterhead. For example, because Joe Rachiele's letterhead is centered, so is his dateline. If your letterhead appears on the upper left side of the page, place the dateline two spaces below it on the left. In the absence of letterhead, type into the heading a complete address that the receiver can use to contact you. Angela Knauss, a student who writes without a letterhead, gives the complete address to which John Smith can send an itemized bill.

USE LETTERS FOR FIVE TYPES OF MESSAGES

Our experience with Joe Rachiele, as well as with large companies and students, suggests that letters are needed for five types of activities in which records can become easily confused because of changes of plans, stressful news, or the passage of time. The letter creates a record of your decisions and requests, so learn to develop letters and memos with these five types of messages: requests, confirmations, complaints, recommendations, and bad news. Since Joe Rachiele's business letters use these types of messages, study Figures 10.7 through 10.11 as we describe the stylistic features of Joe's prose.

Requests

Joe Rachiele's letter of request, Figure 10.7, uses full block format, the easiest to produce. Joe's letter accomplishes its purpose. When you read it, is there any doubt about what he wants? His document's several strengths define an effective request.

Brevity. Can you find any extra words to delete? We can't. The letter's brief style both states its purpose and outlines the details in paragraph one. Furthermore, the letter combines a cover letter and purchase order into one effective document, reducing the extensive (and expensive) paperwork plaguing Joe. Note also that Joe uses simple language. One- and two-syllable words keep the complicated technical letter readable.

Details. This letter's level of detail is excellent because it lists the number of products, the manufacturer (so there can be no confusion among similar items), a brief description of each item, and finally the advertised price. Thus, a packer can check Joe's shipment for accuracy according to four different standards. Joe also gives his information source (*Computer Seller* magazine) so the people at RGB Electronics can check for any miscommunications.

Repetition. Wisely, Joe's letter repeats his shipping address: it is on the letterhead and also at the close. Although such redundancy does not create the best expository prose, it reminds the shipper of the most important item—to ship the equipment to Joe Rachiele's office.

Visual Cues. Your eye travels easily over this letter because Joe groups his information—almost as though he were completing a form. Since recent research has shown that instruction users better understand and remember "chunked" blocks of text, Joe's form-like text is appropriate. Consider how he creates visual cues for the staff at RGB: the first sentence reads in one line, the column headings are underlined for emphasis, the manufacturers' names are

**FIGURE
10.7**

A Typical Letter Placing an Order.

RIVERSIDE COMPUTER COMPANY

208 Georgia Avenue
Bowling Green, Ohio 43402
Phone: (419) 354-2219

January 25, 1988

Mr. Larson Chips, Sales
RGB Electronics
1234 Memory Lane
Silicon Valley, CA 96807-1228

Dear Mr. Chips:

Please ship the items listed here with my monthly UPS delivery:

No.	Mfgr.	Description	Cost
1	ATLAS	JCN-YU Computer w/ 640K	500.00
1	PICE	Serial/Parallel card w/ clock and calendar	100.00
1	OCHE	EGA card w/ software	150.00
1	MLOV	Modem, 300-1200 bps, Hayes compatible, Bell 300, internal, speaker, w/ software	100.00
2	KOLACH	Disk Drives, 363k, DS, w/ cables, @ 100.00	200.00
1	PRST	20 MB Fixed Disk, Peagate compatible	250.00
1	NOHE	Floppy Disk Drive Card-Dual Drive	50.00
1	ZOUBI	Hard Disk Controller, Everex compatible, w/ cable and software	200.00
1	KNEDLIK	EGA Monitor 650 x 350, 32 colors	500.00
		total	2050.00

The items listed here are described in your advertisement in the Computer Seller magazine, March 1987, page 38, as the RGB Electronics ZOOM 3000 + PRO PAK computer, at the price of $2050.00 plus $20.00 shipping and handling charges.

I have enclosed a certified check for $2070.00.

Please ship the entire order to:
    Riverside Computer
    208 Georgia Ave.
    Bowling Green, OH 43402
    Ohio Vendor (Tax) #135790

Very truly yours,

Joseph Rachiele
General Manager

Encl: Certified check for $2,070.00

abbreviated in capital letters to stand apart from the text, the descriptions are written in grammatically parallel fragments to create a sense of logic and continuity, and the shipping address is indented to read as a block.

Joe Rachiele's letter of request illustrates several advantages of well-prepared correspondence. After stating its purpose in line one, the letter immediately introduces a useful level of detail. Its sparse one-page format and style allow easy duplication and mailing. This last point is worth careful consideration. Joe can store this letter on his hard disk or floppy disk systems and modify it anytime to place another order. Depending upon his skill and software capabilities, he can either create a file with this format and enter his specific message for each order, as we did with Figure 10.6, or write a short program, a "macro," and call it to his computer's memory by pressing the key he assigns to the program. Theoretically, Joe can store a collection of such form letters in his keyboard. He may find, as others do, that revising an old letter is faster and easier than starting each letter anew.

Confirmations

Figure 10.8 shows a typical letter that seeks confirmation of a business transaction. Joe Rachiele has placed his order using a bank check, as the advertisement in the *Computer Seller* magazine requested. Now he must find out why RGB Electronics has neither sent a letter of confirmation, sent any of the shipment, nor returned the money.

A word about business routines may explain the importance of Joe's second letter. It is courteous to allow the representatives of companies about 10 days to address a routine inquiry or order. This is time enough for them to send a confirming order or to contact the customer by telephone if filling and shipping are delayed. Joe sent his request for confirmation because RGB Electronics has had his money for 10 days while he has received nothing. Why did Joe select this particular method of composition for his letter? The answers are the speed of response, the requirements of decision-making, and the importance of starting a record.

Speed of Response. Joe wants a timely response, so he drafts a document the receiver can answer on Joe's letter, copy for the files, and mail to Joe. Joe's audience analysis reveals that the people in the sales and inventory departments of the electronics shipper are probably not intellectuals. They prefer doing jobs rather than writing about their work or to their customers. Joe's document leaves space for a handwritten reply, or a member of the RBG Electronics staff can quickly draft a response to any of his three questions. Note that Joe does not ask, "What happened to my order?" This reasonable question is too general. Instead, he asks for one of three possible explanations. Because his letter is easier to answer, it is more likely to receive a response. Many routine query letters use an "answer form" such as Joe's.

Requirements of Decisionmaking. Joe asks the people at RGB Electronics only for information he really needs to know in order to run his business properly. Will he have the machine? Part of it? When? What about his money? He can't have checks floating unattended. Joe must know what to tell his customers, how far in advance to order his electronics equipment to maintain a tight inventory control, and whether he should find some alternate wholesalers. This information is obviously important to Joe's business. Joe uses his letter effectively to obtain specific information instead of useless, general guesswork.

**FIGURE
10.8**
A Typical Letter Seeking a Confirmation of a Transaction.

RIVERSIDE COMPUTER COMPANY

208 Georgia Avenue
Bowling Green, Ohio 43402
Phone: (419) 354-2219

February 10, 1988

Mr. Larson Chips, Sales
RGB Electronics
1234 Memory Lane
Silicon Valley, CA 96807-1228

Dear Mr. Chips:

I write to find out what has happened to my January 25 order, and I have
attached a copy of that order for you to check.

You should be able to send me one of the following responses:

The order, Invoice No. _____ was shipped on (date _____).

Certain parts are backordered. You will receive them on (date _____).
The backordered parts are: (Please list).

The order, Invoice No. _____ lacks several items which we expect to receive
soon. The order will be shipped no later than (date _____).
The items we still need are: (Please list).

If you cannot complete the order or assure me of a shipping date, please return
my check for $2070.00 within 10 days.

Very truly yours,

Joseph Rachiele
General Manager

Encl: copy of Jan. 25, 1988 letter placing order

Starting a Record. It is hard to know exactly when a letter conducting a routine transaction will become a letter of record—documentation at the bank or evidence in court. But Joe Rachiele, who orders supplies regularly from companies that may appear and then disappear, knows that he, not RGB Electronics, is responsible for the "paper trail" that will keep his business solvent—as the trail will help to keep his suppliers honest. For this reason, Joe writes very factual letters that describe specific products, actions, or requests. He avoids guessing about actions that may have occurred. He does not express his growing frustration because he wants to keep his message clear and direct. As you prepare your correspondence, remember that you are creating the record of your efforts and interactions. That record should reflect favorably and factually on you and your employer. This objective, reasonable attitude toward the receiver becomes quite important when you must take the final step in creating a paper trail, drafting a letter to file a complaint.

Complaints

You have studied letters of complaint: Joe Rachiele's complaint about undelivered manuals and Angela Knauss's about obtaining an itemized bill. Such letters exist to build a record that will help to solve a problem. Joe's letter, Figure 10.9, works toward a solution by its reasonable attitude, organization, and direction. Joe, a survivor of business competition, knows that angry—not happy—associates are less likely to write checks.

Attitude. Note that Joe's letter assumes a factual, nonthreatening attitude until its last paragraph. At the start, Joe tries to build a personal approach by bringing Mr. Chips into the events with the repeated uses of "You" to start paragraphs and "your firm" to begin lists of events. In other words, the message is, "This is your firm, Mr. Chips, and you should be concerned about what it did to me." Without pandering, Joe attempts to retain some good will in the third paragraph through a frank expression of fear about his $2070.00. Then Joe introduces the idea of collection as a mutual problem that they both can solve, to prevent worse problems.

Organization. Joe uses the first paragraph to outline the scope of the problem and to explain its unpleasant consequences. The facts of the case then appear in the second paragraph. The third and fourth paragraphs explain that Mr. Chips can act—he can write a check—to solve the problem Joe has outlined in paragraph one and described in paragraph two. This three-part organization is elegant, for it allows Joe to present several levels of details as well as summaries, feelings, and a request—all in a one-page letter. Thus, anyone who must help Joe Rachiele to collect (his attorney, for example) can glean the facts of the case in a one-page reading. Finally, note that Joe again uses visual cues, as he enumerates the facts of the case in paragraph two, to create easy reading.

Direction. Direction is an important concept in drafting letters of complaint if you feel these works may become letters of record. Joe Rachiele's letter moves from the facts, to his feelings, to the future in a logical arrangement. One idea builds on another inductively. The facts suggest that Joe's feelings of concern voiced in the third paragraph are reasonable. The facts combined with Joe's statement of his feelings make it seem logical that litigation could result from Mr. Chip's failure to write a check. Joe makes no idle threats, but the outcome is not left to Mr.

**FIGURE
10.9**

A Typical Letter of Complaint for the Record.

RIVERSIDE COMPUTER COMPANY

208 Georgia Avenue
Bowling Green, Ohio 43402
Phone: (419) 354-2219

March 10, 1988

Mr. Larson Chips, Sales
RGB Electronics
1234 Memory Lane
Silicon Valley, CA 96807-1228

Dear Mr. Chips:

You should have received two letters and a telephone call from me during the past six weeks. I write to inform you that I have placed an order in good faith with your company, and I have received neither a confirmation statement, nor a shipment, nor my returned check.

My records show the attempts to contact your firm that are listed here.

1. On January 25, 1988, I placed an order with your company, and I confirmed that order with a bank check of $2,070.00.
2. On February 10, 1988, I sent a request for a confirmation of my January 25 order because I had received none.
3. On March 8, I called your firm to see if I could speak with you about obtaining my check for $2,070.00 which your firm has had for one month. Ms. Aguilla stated that you were gone for the weekend, and that I could not reach you over the weekend because you were in the mountains. When I asked to book a time for a telephone call this week, she stated that your schedule was difficult to follow and that she could not book an appointment.

You probably can understand that I feel concerned about the $2,070.00 I sent you six weeks ago. I have checked with Buckeye Savings and found that the check has been processed.

Your firm has not acknowledged my order, and I have not received any of the parts I ordered with the check. Please issue a check promptly in the amount of $2,070.00, so that both our businesses can avoid the costs of litigation.

Very truly yours,

Joseph Rachiele
General Manager

Encl: copies of Jan. 25 and Feb. 10 letters.

Chip's imagination, either. Note that Joe combines active and passive voice verbs to make his letter sound both action-oriented and thoughtful. Thus, the letter outlines a clear course of action which RGB Electronics will not be able to avoid if Mr. Chips fails to return the money. The letter shows that Joe does not try to act as his own attorney, but rather, that he knows how to build a trail of events. An honest company can trace and honor its commitments or an attorney will use these letters to pursue Joe's claims.

Recommendations

Joe Rachiele writes two types of recommendations, for products and for employees. Since product recommendations are often short reports rather than letters, we concentrate on Joe's recommendation letter for John Thieu, a technician who plans to move to Fort Collins, Colorado.

By now, you should notice that our models share a common outline:

1.0 State the case or problem — define your scope.
   1.1 State your summary.
   1.2 State your approach.
2.0 Write a summary sentence about Event, Fact, or Example One.
   2.1 Describe the Event, Fact, or Example.
   2.2 Tell the readers what you or they should think about it.
3.0 Write a summary sentence about Event, Fact, or Example Two.
   3.1 Describe the Event, Fact, or Example.
   3.2 Tell the readers what they, not you, should think about it.
4.0 Call for action by describing the action needed.

This outline applies to almost any one-page letter we write. If you use it for recommendations, your letter will evaluate the candidate positively, but with detail instead of general, vague terms. Joe Rachiele uses the outline to be specific about his favorite employee.

Introduction. The scope and problem are the recommendation of Mr. Thieu. The summary is, "He is one of the best employees I have ever hired." Joe's approach is his decision to talk about John Thieu in terms of his attitude and his skills.

Fact One. Joe's first sentence summarizes John Thieu's attitude. He then offers four examples of Thieu's positive attitude. His conclusion suggests that Newman should agree with the other employees that Thieu is outstanding.

Fact Two. The first sentence is Rachiele's summary of Thieu's technical skills. Example One contains the description of Thieu's training. Example Two stresses how Thieu uses his skills to solve systems problems. The paragraph's last two sentences tell Newman how to evaluate Thieu.

The *conclusion* calls for action by stating, "I would try to keep him here in Bowling Green." Such a conclusion now convinces the reader, in this case Mr. Newman, because he will have adequate details for Joe Rachiele's glowing recommendation.

FIGURE
10.10

A Letter of Recommendation.

RIVERSIDE COMPUTER COMPANY

208 Georgia Avenue
Bowling Green, Ohio 43402
Phone: (419) 354-2219

November 30, 1988

Mr. Isaac Newman, Manager
Personnel
Pioneer Computer Repair
Fort Collins, CO 80342

Dear Mr. Newman:

I write to recommend Mr. John Thieu to you for employment as a Senior Repair Technician. Mr. Thieu has worked for me for slightly over two years; he is one of the best employees I have ever hired because of his attitude and his skills in electronics.

Mr. Thieu displays an outstanding attitude toward his work. He volunteers consistently for overtime work when repairs stack up here, even though this work has cost him whole weekends away from his studies or his family. He displays dedication to my business and to me personally that I will find hard to replace. Also, Mr. Thieu is quite polite to customers. Sometimes, the customers arrive upset because they need immediate repairs, but I can rely on Mr. Thieu to express interest in their concerns and to calm them down. His polite manner when we do installations has, without doubt, won repeat sales for Riverside Computer. The other employees here think he is easy to work with, too.

In the area of training and curiosity about electronics, Mr. Thieu is among the best technicians I've seen. He obtained his initial training at Control Data Institute; then he gained an associate's degree in electronics at the University of Toledo. This fall, he started back for his B.S. He understands computers both in theory and in use. Mr Thieu can integrate his informal knowledge of programming with his understanding of mainframes and tabletops, so that he can solve problems that involve the system as well as the hardware. In this sense, his skills and his curiosity about the new developments in electronics approximate those of an electrical engineer. Simply, he can repair a computer more quickly than anyone else in my shop can.

You would be fortunate to hire Mr. Thieu since I understand he and his family are relocating in Fort Collins. I know him well, and I understand what an asset he is to a business. If I could keep him here in Bowling Green, I would try to do so. If you need further information about Mr. Thieu, please call or write.

Very truly yours,

Joseph Rachiele
General Manager

A major problem with recommendation letters is that writers often think it is enough only to recommend an employee highly or to state that an employee's performance was marginal. Our outline builds in the facts, examples, and reasons that will make your view seem worthwhile. Again, a hierarchy of information from general to specific, concluding with a recommendation for action, becomes most important when you offer your highest recommendations of candidates, for readers are least likely to believe these recommendations. Only a hierarchy of information can change skepticism to belief.

Bad News

You may wonder how to deliver these messages. In any business, you must deliver bad news: denial of applications for credit, notices of termination, and notices of unwelcome changes in your company's policies. In businesses that deal with complex technology, bad news often involves explanations of the technology and its breakdown. So remember that your receiver often struggles with several levels of difficulty simultaneously. Here are some suggestions that result from research by credit companies and from our own experiences with clients.

Offer a Reason. Explain that an event or a process caused the difficulty the receiver now perceives. Sometimes, such an explanation can alleviate the receiver's feelings that he or she caused the difficulty. However, with technical subjects this is not always possible. For example, when Mrs. Klinmeister calls Joe Rachiele, he will tell her about the damaged "motherboard" that contains her computer's central circuits. How can she understand this figure of speech? Simply, Joe must explain that motherboards wear out the way shoes do, only very suddenly. Also, you can say little about an expired warranty. But Joe has tried to ameliorate the situation with this approach.

Use the "Personal Approach." Note that Joe's first paragraph talks about the receiver of the letter, about the prompt examination of the machine, and about the attempts to reach Ms. Klinmeister quickly. This emphasis on Ms. Klinmeister may enable her to feel that at least Riverside Computer offers reputable, courteous service, something to feel happy about these days. Joe also reassures her by explaining that the machine can be repaired quickly. He also uses visual cues as part of his "personal approach" in paragraph two when he prepares Ms. Klinmeister for bad news by saying, "We regret to tell you," and "misinformed" instead of "wrong" or "uninformed."

Write a Longer Letter. Note in paragraph two that Joe takes time to explain that his workers called the company to check the warranty. He also takes a little longer to develop paragraph one than he might, and really, paragraph three could be rewritten to "Please call us and tell us what to do." But as a smart salesman, Joe Rachiele has learned to take longer telling bad news. A longer, but not unreasonably long, letter makes the receiver of bad news feel better.

Don't Apologize. Note, though, that Joe does not claim responsibility for what happened to Ms. Klinmeister's machine. Apologies can quickly sound insincere, because your reader, although ignorant about technology, may not lack experience. Here again, an explanation, if possible, of the difficulties that caused the bad news, can comfort the receiver more than an apology.

FIGURE
10.11
A Letter Delivering Bad News.

RIVERSIDE COMPUTER COMPANY

208 Georgia Avenue
Bowling Green, Ohio 43402
Phone: (419) 354-2219

January 4, 1989

Ms. Sophie Klinmeister
811 Springdale Boulevard
Dayton, OH 45419

Dear Ms. Klinmeister:

The RGA Electronics Zoom 200+ Pro Pack computer that you brought to our
Bowling Green store can be repaired easily. In fact, we examined the machine two
days ago, and we were unable to reach you by phone, so we decided to write to
give you the details of our estimate. Your machine looks as though it is in good
shape, so we should have it ready for you to pick up in seven days.

We regret to tell you, however, that you may have been misinformed about the
warranty. We called the company in Oklahoma, and their records show that your
warranty on parts and labor expired in November, 1988. We have enclosed an
estimate for parts and labor, because we will bill those costs directly to you, not
the company.

Would you please go over the estimate and call us to tell us whether or not we
should begin to repair the machine? Although our costs compare favorably with
other local estimates you might obtain, we realize the expense of this repair was
not one you expected, so we will hold the machine until you call us.

Very truly yours,

Joseph Rachiele
General Manager

Stress the Positive. Note that Joe turns quickly to the repairs at hand, which can be done, but at a price. He now directs Ms. Klinmeister's attention to the repair work that must occur, warranty or no warranty, if she wants a working computer.

FOUR TECHNIQUES FOR EFFECTIVE LETTERS AND MEMOS

Read Figure 10.12, Barbara Haggler's letter seeking a refund or a repaired necklace. Barbara's letter shows that you should incorporate four tips into your style and message to create successful correspondence. A strong, effective letter or memo exhibits:

☐ A personal orientation.
☐ A concise style.
☐ A correctness of usage and appearance.
☐ An orientation toward action.

How do these four tips appear in Barbara Haggler's letter? How can you use these techniques in your own work?

Personal Orientation

Researchers have discovered how to make your text easy for the reader to understand. First, involve the reader in your message by stating a problem that the two of you share and can treat mutually. For example, Barbara starts by admitting how much she likes The Gallery, as we assume the manager, Alexis Notting, also does. Note that the letter begins with a discussion of The Gallery, not of Ms. Haggler's problem, because The Gallery is much more interesting to Mr. Notting than is Ms. Haggler. Some authors refer to this attitude as a "shared purpose" with the reader or a "goodwill builder." It is important not to exaggerate or pander, but in paragraph one you should define your scope in terms that will interest or motivate the receiver to continue reading. Barbara Haggler attempts to retain Alexis Notting's goodwill with a factual—not accusatory—description of the problem.

What else does she do to involve the reader? Study the visual cuing of this letter. Note that the letter narrows its focus until Mr. Notting's two choices stand alone, visible, and quickly readable. Your expository writing professor might not like this format in a theme, but a series of researchers have shown that Mr. Notting is more likely to read this format that points to his choices rather than embedding them in a paragraph. Practice visual cuing to highlight the important information in your letters. Use indentations, tabs, columns, tables, underlining, boldface type, freestanding grammatically parallel lists, and anything else that is in good taste to keep your reader interested.

You can read more about the personal orientation in the literature of business communications, but initially, remember to create a reader-directed message with a tone that defines the problem in the reader's terms and a format that emphasizes this tone with chunks of readable text.

Concise Style

In correspondence, this means one page. We are really only interested in teaching our students how to write one-page letters, because we know that reader response to letters of

FIGURE 10.12   An Effective, Personal Orientation.

460 S. Summit, No. 31
Bowling Green, OH 43402

February 23, 1988

Alexis Notting
The Gallery
171 Railroad Street
Bowling Green, OH 43402

Dear Mr. Notting:

Because your store offers the most unusual and attractive jewelry in Bowling Green, I shop there regularly. I have referred my friends to The Gallery, too, because you and your employees have always treated me courteously and sold quality merchandise. I believe you are the sort of manager who will understand and want to help with my concern about a purchase that hasn't worked out.

On February 6, 1988, I purchased a necklace and earring set for $73.00. The set, style number NR10663, was made of ivory and brass. Unfortunately, the necklace broke the first time I tried it on at home. I saved all of the beads and returned them to your store with the earrings on February 10. The sales clerk informed me that I would have to speak with you directly to obtain a refund.

Although I left a message with my phone number, I have not yet heard from you. Since it is difficult to contact me by phone, I would appreciate it if you could:

   Have the necklace repaired,

   or

   Issue a store credit for $73.00, and consider the merchandise returned.

I will call you during the week of March 1 to discuss our problem with the necklace. Please take care of this matter, so I can continue to shop at my favorite store.

Very truly yours,

Barbara Haggler

more than one page drops off considerably. Do you think Mr. Notting cares about page two of Ms. Haggler's complaint? Plan your message, using a hierarchy of details, so that it fits on one page. If it doesn't fit, rewrite the letter. To keep the messages to one page, consider breaking up long messages with these strategies:

☐ Turn a long letter into two items—a report with a dynamic letter of transmittal.
☐ Cover a multi-topic letter by drafting several short letters—one for each item that requires action.
☐ Use enclosures and attachments dynamically—have the message describe, but not summarize, them.
☐ Select your message with utter precision—one, and only one, message per letter.
☐ Draft every letter as an inverted pyramid—scope leading to details, leading to a call for one, mandatory action before closing.

Correct Usage and Appearance

A hidden grammar teacher of the most critical type lurks within every letter reader. How often have you read a letter and then said "Look at that. He wants me to write a check for this, but he can't even spell! What kind of outfit does he run?" More recently, we watched during a hiring committee meeting as highly qualified applicants for a professional position were disqualified by committee members who joked about the spelling errors in the applicants' letters. A slightly sadistic committee? Yes, readers respond with acid when you send them defective work, and an incorrect letter is a defective product. Proofread your letter for these items:

Spelling of the receiver's name.
Spelling of the receiver's place of business.
Spelling elsewhere—including your own name and firm.
Correct typing throughout, including:
    Numeric text—figures, money, measurements.
    Dates—especially dates as parts of letters of record.
Correct usage, especially:
    Subject-verb agreement.
    Pronoun reference.
    Noun-pronoun agreement.
Punctuation, in particular:
    Colon—nothing else—after the salutation.
    Comma after the complimentary close.

Finally, check for the five errors our students make most often by answering these questions:

Did you type in a dateline to validate the letter as record?
Check again: did you put a colon after the salutation?
Did you place a comma after the complimentary close?
Did you sign the letter below the complimentary close?
Did you initial or sign the memo to the right of your name?

Orientation Toward Action

Write your letter to ask the receiver to do something. Make your request simple, direct, and as easy as possible for the receiver to do. For example, Barbara Haggler asks Alexis Notting to perform either of two actions that are easy for him: ship the necklace for repair or write a check. She requests a specific action, not a "prompt adjustment" of the problem or "immediate attention to" her concerns. These phrases are trite because they lack detail. Ask the receiver to perform an act rather than to think about your respect. Acts are visible—tangible to busy, preoccupied readers; thoughts can always wait. Your letters and memos have better results if they are requests for action.

Maintain, as well, your own orientation toward action. Do not waste correspondence opportunities by sending routine forms instead of dynamic messages. Each year, our new technical writing students hand in their final reports with letters of transmittal attached in front. Novice students write dull letters that say, "Here is the report you assigned. Gee I hope you like it and give me a good grade." These students work hard, but fail to sell their reports with dynamic cover letters that explain why the report is important to us, why we should look forward to reading it, and what it will ask us to do or teach us.

Don't make that mistake. Whether you write a form letter, a routine letter of transmittal, a letter of record, or a persuasive appeal, use the opportunity to deliver a message and to ask the receiver to act in your behalf. If you adopt this approach, you will write better letters.

SUMMARY

In this chapter, you learned that technical communicators and personnel in technical businesses write letters to make requests, confirm agreements, start and keep records, file complaints, make recommendations, and record bad news that calls for action. You have learned that the format of your correspondence shapes its message and that you can draft messages that obtain results by involving the reader, keeping the message to one page, proofreading for correctness, and using communication routines dynamically. After reading the chapter, you should be able to:

- Format a full block, modified block, or alternate block letter.
- Format a memo manually or with your computer software.
- Format second and subsequent pages, although you will first try to draft one-page messages.
- Format a letter when you have a centered or left-placed letterhead.
- Format a letter on plain bond.
- Draft a letter of request that places an order.
- Write a confirmation letter that initiates a record.
- Draft a complaint that calls for action without making threats.
- Create a recommendation letter that has adequate levels of details.
- Compose a letter to deliver bad news that calls for action.
- Design drafts of letters that incorporate elements of the personal orientation: shared purpose, chunked format, visual cues, persuasive diction.
- Write letters that fit on one page and maintain a hierarchy of details.

- Proofread letters to avoid the most common mistakes of students.
- Use routine channels of correspondence to deliver dynamic messages that call for action.

ASSIGNMENTS

1 Make an informal assessment of the word processing software that is available for student use on your campus. Write a one-page letter to your instructor recommending a software package that you find helpful for correspondence.

2 Find an example of a poorly written piece of correspondence. Such an example might arrive with your daily mail. Save the original letter and rewrite the message, using as many as possible of this chapter's objectives. Staple your revised version to the original letter, and bring your paper to class.

3 In class, exchange letters with a partner. Read your partner's letter to see which objectives of this chapter he or she tried using in the rewrite.

4 Write your partner a one-page memo describing the one action that can most improve the rewrite.

5 Have a conference with your partner and explain all the changes you recommend for the rewritten letter.

6 Examine the memo you receive from your partner and take notes about the other changes he or she recommends.

7 Rewrite your letter using the written and spoken advice you have received. Staple your latest version to the top of the other two drafts and submit your letter for a grade.

8 Draft a letter of complaint about a professional, business, or recreational situation you seek to correct. If you have nothing to complain about, perhaps you can help someone in your family or in the office. Draft a one-page letter that outlines the problems and calls for action without making threats. Bring the letter to your writing class.

9 Find a new partner and exchange letters. Prepare a reply to the complaint you have in hand. Remember, you might not be able to take the action the writer requests; if so, you must write a bad news letter that explains why.

10 Do you ever wonder what others say about you in their letters of recommendation? Now you can find out. Select another, new partner in your writing class. You and your partner should interview each other and take notes about each others' professional capabilities. Write a letter of recommendation that will help your partner to obtain his or her next job. You may assume any role: supervisor, teacher, friend of the family, or some other position that allows you to make such recommendations.

CHAPTER 11

Resumes, Letters of Application, and Interviews

This chapter introduces you to the preparation of resumes, letters of application, and interviews for employment. After reading this chapter, you will be able to:

- Decide whether to write a chronological or a functional resume.
- Compose a resume that sells your talents effectively.
- Draft a letter of application that persuades personnel representatives to call you for an interview.
- Design your letter of application so that you can modify and rewrite it quickly.
- Research and prepare for your interviews.
- Participate in your interview with poise and confidence.

CASE STUDY   **Siblings Prepare for a Competitive Job Market**

Randy Goetz, a new graduate of Bowling Green State University, must find a job. A journalism major, he has a minor in biology, a subject he enjoys. In fact, Randy's interest in technical and scientific writing, particularly in the reporting of science, stems from his background in biology and journalism. Too late to change his major, Randy learned he could qualify as a technical writer. Instead, he has tried to put together the journalism and science credentials necessary for him to enter technical writing with a journalism degree.

Randy's sister, Linda Harris, is a job seeker of a different sort. At 32, she offers more than 10 years of work experience. She has worked her way through college to obtain a liberal arts degree in psychology. By the time she had earned the degree, her employer, the telephone company, paid her more than the available, entry-level psychology jobs would. She has risen to the top of the salary scale as a test desk operator. As the single parent of two children, she felt she should stay with the telephone company. But she does not like her work. Furthermore, the increasing automation of her technical area may soon lead to her termination.

Why the Current Market Presents a Challenge to Applicants

Randy and Linda feel overwhelmed by the competition in the current job market. Randy, who worked for his journalism degree, is angry that liberal arts majors seem to work harder for jobs. He is frustrated by the $9,000 to $11,000 salaries offered to fellow journalism graduates lucky enough to obtain entry-level newspaper jobs. He realizes he must put aside his love of biology to move into technical writing, a more lucrative field. But how can he convince employers that he, a liberal arts major, can learn the specific skills and concepts of technology and science? How can he compete with the technical communication majors already on the job market? How can he stand out in this year's crop of graduates as the one worth hiring?

Linda faces the typical problems of the adult, experienced job applicant. She worked for ten years to get her degree and other training. She has proved she can master both liberal arts and technology. Only her leadership in the employees' union has relieved constant boredom on the job and problems with certain supervisors. She now commutes three hours daily because the company has phased out two offices closer to her home. She wants very much to move into either human services, her area of training, or a technical position with a challenge and some variety. She is discouraged by starting wages in human services or technology.

Linda, too, is angry. With all of her education, her technical experience, and her excellent employee record, why, at 32, must she start the job search again? Nothing in her experience has prepared her for the depression, anxiety, and lack of confidence she feels.

Randy Goetz and Linda Harris are two cases this chapter examines and develops. Randy is entering the working world for the first time. To do so, he must make others notice that he offers talents they need. He must convince an employer that he is worth the time and effort needed to train him. Linda, on the other hand, begins the second stage of her career with the job change she anticipates. Like many adults, she must complete one phase of her work experience, prove that she created multiple successes for her previous employer, and convince a prospective employer it is not too late to train an experienced adult as a new worker.

Randy and Linda remind us that job insecurity characterizes today's work world. If a new graduate, your first job search is actually step one of a career, a lifetime of job searching. If you are a middle-aged worker or an adult returned to school, then you have learned that education and training do not end with graduation and that continued education and training increase your likelihood of continued employment.

FIGURE 11.1   A Resource List for Starting the Job Search.

Have you checked all the available resources listed here? Most are free and available to all job hunters.

Your College or University Placement Service May Offer
Career counseling
Self-analysis programs, surveys, and brochures
Seminars about resume preparation
Interview schedules of corporate recruiters on campus
List of companies who hire graduates of your college
Library materials so you can research the companies you select

Your State Employment Agency May Help with
Lists of companies that have stated they are hiring
Leads about specific jobs
Information about companies' attitudes toward minorities
Small group instruction in resume preparation and the completion of job applications

Your College, University, or Local Public Library Can Provide
Directories of companies
Reference materials with detailed financial information about companies
Newspapers in which you can read about the companies' recent advances, concerns, or setbacks
Telephone directories of major cities in the United States
General works that discuss how to improve your job search, resume-writing, and interview techniques.

Your Network of Work Associates, Friends, and Club Members Sometimes Assist with
Specific leads from within the firms where they work
Specific knowledge of the type of individuals their firms will hire
Telephone calls which act as introductions and references to obtain the interview for you
Coaching in the finer aspects of managing the job interview

In this chapter, we examine materials typical of both groups, the new and the experienced workers, because you will need to prepare your job application materials as your career matures and as you move from one category to the other.

RESUMES

Since everyone seems to be a resume expert these days, we sometimes read contradictory advice. For example, some experts claim that no resume should be longer than one page; others, that you should have one page per 10 years of work experience; and still others claim that quality, not length, is what you should develop. All this advice means simply that your resume must conform to the standard formats for presenting experience, and, at the same time, it must be individualistic enough to detail your talents convincingly. Well-prepared resumes usually have most of the sections described here.

FIGURE 11.2	**Should You Prepare a Chronological Resume?**

Should you use a chronological format for your resume? If you answer yes to these questions, consider the chronological pattern.

Do you intend to stay in your previous field?

Does your employment background show growth, progress, or wage increases?

Do you want to emphasize your last employer's name in the field; for example, have you completed an internship with a major firm?

Are you training for a "traditional" occupation such as teaching, health care, or government service?

Are you a "typical" graduate: early twenties, newly trained, obtaining first degree after high school?

When should you avoid the chronological resume? Be cautious of the following employment problems:

☐ Redirection of career objectives, for example, from sales to personnel.
☐ Frequent change of employers.
☐ Extended or repeated absences from the workforce.
☐ Too long in the same job—no promotions.

Name and Address

Usually horizontally centered as a block of text, this portion should offer enough detail for an employer to call you without consulting your file or directory assistance. Minimize the use of abbreviations. Include full area codes for telephone numbers and ZIP codes for mailing addresses. If you are an on-campus student, supply both your campus and home addresses and phone numbers.

Professional Objectives

In one or two sentences, state your entry-level, short-term goal, and then expand it into a feasible long-term management goal for which a company could train you. New graduates beware: recruiters complain currently that everyone wants to enter the workforce as a manager and that no one wants to work. In your short-term statement, use verbs such as "work," "improve," "develop," or other verbs that suggest growth and training on the job. (Please see our list of Action Verbs.)

FIGURE 11.3

Should You Prepare a Functional Resume?

Perhaps the functional resume suits your needs. If you answer yes to these questions, then consider the functional pattern.

Are you changing careers, maybe leaving teaching for computer science?

Are you returning to the workplace after a lengthy absence, for example, after starting a family?

Have you moved about often, resulting in different or apparently unrelated work experiences?

Have you owned or operated your own or your family's business for any length of time or did you grow up working in a family business?

Do you need to demonstrate talents or achievements that your current employment does not require?

Should you avoid the functional resume? Be watchful of the indicators listed here:

☐ A work history with steady growth, promotions, or raises.
☐ A work history lacking diversity; jobs with limited responsibilities or functions.
☐ A pattern of working for the same company.
☐ A history of work in a "traditional" occupation such as teaching, health care, or government service.

FIGURE 11.4

Action Verbs for Resumes and Letters of Application.

ACTION VERBS FOR RESUMES AND LETTERS OF APPLICATION

accompany	educate	judge	promote
accelerate	effect	justify	prompt
achieve	eliminate	launch	propose
acquire	enact	lead	prove
act	encourage	license	provide
administer	engineer	locate	recommend
advise	enhance	maintain	reconcile
analyze	establish	manage	reduce
arrange	evaluate	manufacture	regulate
assemble	exceed	market	reinforce
assist	execute	master	relate
budget	exhibit	mediate	reorganize
build	expand	merit	report
clarify	facilitate	monitor	research
command	form	motivate	revamp
complete	finalize	negotiate	review
compose	finance	nominate	revise
conceive	formulate	normalize	satisfy
conduct	found	obtain	schedule
construct	generate	officiate	secure
control	govern	operate	serve
convert	graduate	order	service
cooperate	handle	organize	simplify
coordinate	hire	originate	solve
correlate	identify	participate	spark
create	implement	perceive	stimulate
decide	improve	perfect	streamline
delegate	increase	perform	structure
demonstrate	induce	pilot	succeed
design	initiate	pinpoint	supervise
determine	inspire	pioneer	support
develop diversity	install	place	teach
devise	instruct	plan	train
discover	insure	prepare	transfer
display	integrate	preside	transform
direct	intensify	procure	use
double	interpret	produce	verify
earn	invent	progress	write

Courtesy of Bowling Green State University Placement Office.

**FIGURE
11.5**

Resume of an Entry-Level Technical Communicator.

LOUIS BABCOCK
1200 S. Carpenter Rd.
Cleveland, Ohio 44211
Phone: (216) 225-3211

Objective
: To work as a technical writer or editor and ultimately manage a technical writing staff.

Experience
: Allen-Bradley Systems Division
Highland Heights, Ohio

Time: 1/14/88 to 6/15/88
Supervisor: Bill Jacobs

Wrote manuscripts dealing with hardware for industrial automation. Have written six manuscripts, including two manuals.

Falcon's Nest, Bowling Green State University
Bowling Green, Ohio

Time: 5/1/87 to 1/10/88
Supervisor: Rita Fowls

Student Manager duties included cashiering, ordering, and training. Also developed a manual on scheduling procedures.

Education
: Candidate for Bachelor of Arts in Technical Communication,
Bowling Green State University,
August, 1988.
GPA—3.74.

Writing classes include technical writing, professional editing, advanced writing, journalism, and advanced technical writing. Technical classes include twelve credit hours of computer science, six credit hours of electronics, and a course each in training, management, and accounting.

Awards
: Outstanding Student Paper, Technical Communication, 1987, BGSU.
McMillan Scholarship for Minority Students, College of Arts and Sciences, 1987.
Dean's List, Junior and Senior years.

FIGURE 11.6   Randy Goetz's Resume for an Entry-Level Position as a Science Writer or Editor.

<u>RANDY GOETZ</u>

Present: 480 Lehman #208
Bowling Green, OH 43402
(419) 354-1808

Permanent: 4839 Anderson
Akron, OH 41075
(216) 687-2312

Career Objective: To work as a science writer or editor and ultimately manage a scientific or technical writing staff.

<u>Education</u>	Bowling Green State University (BGSU), Bowling Green, Ohio. Bachelor of Arts degree in journalism expected in May, 1988. GPA of 3.38 with two terms on Dean's List.
<u>Technical Writing Experience</u> 1985 (summer)	Gulf Coast Research Laboratory (GCRL), Ocean Springs, Mississippi. <u>Organized, wrote,</u> and <u>edited</u> scripts for two slide shows on invertebrates to be shown at the Marine Education Center in Biloxi, Miss. First slide show designed for the general public and second designed as a teaching aid. <u>Researched</u> and <u>wrote</u> a pamphlet on fish kills in Mississippi Sound to be printed at GCRL.
1983	Technical Writing 488 (<u>abstracts, proposal letters,</u> and <u>feasibility studies</u>) at BGSU.
<u>Extended Education</u> 1982 (summer)	<u>Project Ocean Search: Atlantic/The Cousteau Society.</u> Program directed by Jean-Michel Cousteau. Introduction to the marine sciences and study of an estuary.

Education

You have great discretion in drafting this section of your resume. For example, if you have an extensive work history, as Linda Harris does, you may wish to place your experience before your education in the resume. On the other hand, new graduates usually place their education right below their professional objective because their new degrees are their strongest assets. You may also determine the length of this section.

Some applicants simply state their school's name, dates of graduation, and degrees obtained. Others write a few lines describing their special courses, internships, or other educational credits to strengthen their credentials. For example, Randy Goetz mentions his outstanding grade point average. Nicholas Marone and Li Ming list their strongest courses. Lucille Winchester, a nontraditional student, explains the curriculum that her proprietary school experience has covered. Here are some items this section might include: degrees, academic institutions and majors, grade point average, attractive courses, academic honors,

FIGURE 11.6	Continued.

1984 (summer)	Six-week field study of the physiology, behavior, and taxonomy of marine invertebrates at GCRL.
Work Experience 1979–1985	Carrier Recruiter and Open Route Supervisor for the <u>Sun Messenger</u>. Inventoried boxes, tested parts and painted offices at Anchor Industries. Busboy at the Inner Circle.
Offices Activities 1982–1985	<u>Publicity Officer</u> for student chapter Society for Technical Communication: Wrote summer letter to incoming technical writing students to recruit new members. Worked with media for publicity.
	<u>Publicity Officer</u> for literary/arts journal <u>Prairie Margins</u>: Wrote letter to university faculty to solicit support for the journal. Worked with media for publicity.
	<u>BFA Representative</u> on Creative Writing Committee: Provided undergraduate input in decisions made by Creative Writing Department.
	<u>Undergraduate Representative</u> on the Undergraduate Curriculum Committee: Provided undergraduate input in curriculum planning for the English Department.
	<u>Sigma Phi Epsilon Fraternity—Corresponding Secretary</u>: Member of Executive Board. Wrote letters and cards to parents, university officials, and other organizations. Responsible for publicizing fraternity functions.

cooperative education experience and internships, service activities, and the percentage of college expenses you earned.

Experience

Arrange this section in reverse chronological order, starting with your most recent experience. The chronology is important because employers expect to see continuity.

If you cannot present continuous employment, consider using the functional resume. This resume enables you to arrange your work experience by categories such as salaried work, volunteer work, political work, or special training. The functional pattern presents only the experience relevant to the job you seek.

As in the education section, general guidelines can help you customize your resume's employment section. Remember to include:

☐ Your most recent experience, with some details emphasizing your responsibilities and accomplishments.
☐ Description of employment responsibilities that readers may not understand. For example, Linda Harris translated her technical experience into terms the lay audience can understand in her liberal arts resume.
☐ Promotions either from within a firm or when changing jobs.
☐ Accomplishments that show your talents on each job.
☐ Specific training that your company provided to promote you or upgrade your skills.

Service Activities

This section is optional. You should list here the campus or civic activities that enhance your competence. For example, Randy Goetz lists such activities under the heading, "Offices/Activities," because as a new job applicant he wants to show that he understands the importance of service to one's community and working cooperatively with groups of people. This record makes him seem dependable. If you have experience with budgets or planning because of your service commitments, list these accomplishments. They also help recruiters understand that you are responsible.

Interests

In this optional section, list hobbies or personal pastimes that may interest an employer. For example, if you have your pilot's license, this accomplishment may interest an employer who owns a small plane used for business travel.

Organizations

List here any professional or civic memberships which enhance the picture of your stability and your ability to serve the public. For example, if you belong to Toastmasters International, you might list this membership because it indicates that you care about effective public speaking.

Publications

Professionals such as professors, medical doctors, lawyers, and engineers should pay close attention to this section. This section of a resume is a complete list of published books, chapters, and articles. It should also list public presentations, preferably in a separate section which appears immediately after the list of publications.

References

You may either list your references—the names, addresses, and phone numbers of people who have agreed to recommend you—or type "Available upon request." Before you send your resume to a possible employer, ascertain whether to include a list of references. If the answer is yes, list them in the resume to keep your application simple.

Chronological Organization

This simple pattern employs a reverse-order chronology. Its purpose is to show continuity in your education, employment, and other activities. This method of organization lends itself well to:

☐ New college graduates, 21 or 22 years old, with a short work history.
☐ Professionals with advanced degrees in education, law, medicine, or engineering.
☐ Highly specialized personnel in business careers who have taken a direct, uninterrupted path upward through regular promotions.
☐ Persons who have worked entirely for Fortune 100 firms whose names offer maximum prestige value.

If you use a chronological pattern, account for every year of your career from high school graduation to the present. For this reason, young people find this pattern useful. Many people over 30 combine a chronological format for the most recent 10 years with capsule descriptions of all previous years.

Functional Organization

This format requires that you organize your work experience by job type (for example, by management duties, technical jobs, or personnel functions). This format minimizes chronology, and it emphasizes the broad view of an individual's talents. Use it to show a diversity of talents or accomplishment. The functional format works well for:

☐ Persons over 30 who want to confine their resumes to two pages.
☐ Applicants trained in two or more useful careers.
☐ Parents returning to the workforce after an extended absence to care for their children.
☐ Former military service personnel who should translate their military training and experience into civilian terms.
☐ Rehabilitated criminals, substance abusers, or others with gaps in their work history.
☐ Dependent spouses who have moved frequently.
☐ Owners of independent businesses or farms who also work in industry or business.

If you choose a functional format, use logical, general categories for grouping your experience as Linda Harris does. Place your experience first and your education second. In the cover letter, explain your experience and create some continuity in your work history. Use the cover letter to prove that you are reliable because the gaps in your work history result from normal or socially acceptable absences such as those necessitated by family life.

Tips for Successful Resumes

Appearance. Your appearance should look neat and attractive. If possible, have it typeset and printed on fine quality stock, approximately 20% rag, 20-lb bond. Use white, pale grey, or off-white paper. Avoid colored papers and unconventional typefaces. Each resume should look like an original. If you have typeset resumes, you can create this effect. You can also create this effect with a daisy-wheel, letter-quality, or laser printer. Avoid dot-matrix printers and near-letter-quality printers.

Usage. Your resume must be letter-perfect, with no spelling or grammar errors. We have seen applicants disqualified immediately because of such errors.

**FIGURE
11.7**

Resume for Entry-Level Job in Computer Programming.

NICHOLAS MARONE
1600 University St.
Toledo, OH 43478
(419) 536-8247

OBJECTIVE

To obtain a computer programmer a position with a progressive, growing firm. Have a solid background programming in four computer languages. Am dependable, hardworking, willing to travel and relocate.

EDUCATION
6/87

University of Toledo, Community and Technical College, Toledo, Ohio, Associate of Arts Degree, computer programming.

Instruction involved writing, documenting and debugging programs in ASCII, COBOL, Assembler 360/370, FORTRAN IV, RPG II. Gained hands-on experience with batch processing, DOS and MSDOS. Participated in testing the IBM laboratory as part of a federal grant.

9/85–12/86

Lakeland Community College, Mentor, Ohio.

Earned 48 credits toward an Office Management major. Completed courses in accounting, finance, technical communication, and word processing.

EXPERIENCE
2/86–present

Blue Cross, Blue Shield, Northern Ohio, Toledo, Ohio.

Started as receptionist, answering telephones, taking messages, doing some word processing. Currently processing third-party payments for union contractors.

5/86–9/86

Cleveland Museum of Art, Cleveland, Ohio.

Temporary office worker. Substituted for vacationing employees. Worked in every department of the museum on word processing, filing, accounting, and public relations.

9/85–5/86

Pizza Hut, Incorporated, Lakeland, Ohio.

Night manager of carry-out restaurant. Hired, trained and supervised store personnel. Doubled the store's evening sales within a ten-month period.

COMMENT

I have worked forty hours per week throughout my education, and I have earned a 3.74 GPA while doing so. I manage time effectively and complete my assigned tasks efficiently.

REFERENCES   Available upon request.

**FIGURE
11.8**   Resume of a Nontraditional Student in Electronics.

LUCILLE WINCHESTER
1000 E. Manhattan
Toledo, OH 43611
Home Phone: (419) 286-6509
Business Phone: (419) 289-9099

OBJECTIVE   To join a growing company as a computer/electronic technician.
Have demonstrated ability to test, maintain, and repair central
processors and peripheral equipment. Am hardworking, enjoy
working with people, and am willing to travel.

EDUCATION
1/86–6/86   Control Data Institute, Toledo, OH
Earned certificate in Computer Technology. Successfully completed
an 8½ month course in 5 months, with above average grades,
covering basic electronics, central processing units (hardware
and software), peripherals (display terminals, line printers,
card readers, tape and disk drives), and microprocessors.
Gained hands-on experience operating, maintaining, and repairing computer systems.

2/86–6/86   Barbizon School, Toledo, OH
Successfully completed a 20 session course in modeling/personal
development program. Gained knowledge in dealing with people, social savvy, appearance, and personal development.

12/81   Davis Business College, Toledo, OH
Associate of Science, Computer Science. Subjects covered data processing, typing, keypunching, records management, flowcharting, Cobol and BASIC programming.

EXPERIENCE
3/79–5/85   Toledo Lucas County Public Library, Toledo, OH
Gained 6 years of experience in the area of customer relations
from working with the public. Had hands-on experience in
editing and entering data into the computer. Revised and
purged filing system that resulted in efficient utilization of
space and reduction of time in locating information. Responsible for typing, collecting agency cases, and first and second
notices. Other responsibilities included calculating payroll, inventory, ordering supplies, scheduling, balancing monthly
statistics, and cash control.

10/74–2/75   Home Modernizers, Toledo, OH
Responsible for training and supervising other solicitors. Produced
top sales in telephone soliciting.

| **FIGURE 11.9** | Resume for an Entry-Level Job in Medical Records Management. (This resume also shows how to accentuate positively your ethnic American status.) |

<u>LI MING</u>
1670 Dixie Highway
Dayton, OH 45449
(513) 426-4580

OBJECTIVE — To work in a major hospital on the management information systems aspects of data storage and dissemination. To qualify for promotion to the management system of a major research hospital.

EDUCATION — Bowling Green State University, B.S. Medical Record Administration, May 1987. GPA: 3.9.

Special Courses: Advanced Personnel Administration, Health Care Law, Legal Writing, Advanced Technical Writing, Trends in Medical Record Administration.

EXPERIENCE

1/87–5/87 — Internship, Medical College of Ohio, Toledo, OH.
Medical Records Department. Updated existing records, answered physicians' requests for information, specialized in information for oncology patients.

8/86–12/86 — Directed Practicum, Flower Hospital, Toledo, OH.
Assisted the Director of Medical Records with accounting pro-

Style. Your resume should look conventional. At first glance, a reader should be able to define it as chronological, functional, or a combination of both styles. It should be concise and easy to read. Use the list of action verbs included in this chapter to write brief but descriptive summaries of your employment experiences. Avoid the passive voice.

Ethics. It is acceptable to translate your technical experience into terms a lay audience can understand, but never lie about your experience. Some investigators suggest that 60% of the resumes currently on the market may contain fabrications about applicants' experiences. Therefore, personnel managers now closely check resumes.

Multiple Resumes. If you have a letter-quality or laser printer, consider storing the sections of your resume in separate files on disks. You can then merge the files to create a custom resume for each application. For example, Randy Goetz would like a job as a scientific writer, but he might also qualify for a position as an entry-level documentation specialist in a com-

FIGURE 11.9	Continued.	
		cedures, collection notices, and third-party payments. Revised Medical Records Handbook (Accounting Section, 50 pages).
	5/15–8/15/86	Summer Employment, Dr. Elsa Marwick, Sylvania, OH. Worked full-time, assisting with patient contact, records management, accounting, and insurance.
	5/15–8/15/86	Summer Employment, The Monroe Clinic, Toledo, OH Worked full-time for three specialists in geriatric medicine. Worked primarily on third-party payments, state and federal contracts.
	1983–1987	Student Employment, Bowling Green State University's Health Center. Worked throughout college in all areas of medical records-keeping and administration: answered phones, scheduled appointments and minor surgeries, contacted insurance companies, maintained and created records. Earned 75% of my college expenses.
	HONORS	1986, Toledo Chinese-American Foundation Scholarship for outstanding grades—$1,000 1986, Inducted into Medical Records Honor Society's Bowling Green Chapter Dean's List throughout college
	REFERENCES	Available by request

puter company. Randy can use his computer to generate two distinct resumes, one slanted toward scientific writing and indexing and the other toward technical documentation.

Resumes and Nontraditional Students

If you are over 21 and attending school, you are a nontraditional student, that is, one with special needs and concerns. For nontraditional students, particularly persons in and out of college as they earn their expenses, the sample resumes shown in this chapter offer some effective strategies. For purposes of job strategies, we like to consider the needs of women and minorities in particular. We also have specific advice for applicants with no continuous employment record.

Women. Various reports show that women usually receive substantially lower wages than men. Sometimes women may start at a lower position and lower wage because of an inadequate level of detail on their resumes. These suggestions may help as you prepare resumes.

**FIGURE
11.10**

Typical Resume of a Working Person—Designed to Focus on the Worker's
Technical Competence.

LINDA S. HARRIS

400 S. Main St.
Defiance, Ohio 46910
(419) 452-2216

EDUCATION

State University, Muncie, IN. Bachelor of Arts in Psychology, 1979. GPA: 3.25.
Minor in sociology. Completed 41 hours toward Bachelor of Science in Electronic Technology.

United Telephone Company, Courses in Labor Relations, AC DC Theory, Defensive Driving, Pole Climbing, 82A and 84A Carrier, Throw Crew, Telephone
Installation and Maintenance.

Illinois State University, Labor Law, Grievance Analysis and Arbitration,
Steward's Training.

AFL-CIO Summer School, Labor Law, Women in the Labor Movement.

American Arbitration Association, Current Trends in Arbitration.

EXPERIENCE

1970 to Present General Telephone Co. of Ohio

Nov. 1981 to Present   Position: Test Desk Operator
Duties: Operated Automatic Electric (AE) Model One and Badger Centra-Line
612A Remote test systems with AE LEICH, NORTH, and STEP switching
centers. Operated Service Version Release (SVR) 1124 System using a Televideo
950 (hold circuit) and 212 type coded data set on a 2-wire dial-up facility for
the AE GTD-5 computerized switching center. Tests were done in cooperation
with field repairpersons to locate and clear trouble. Conducted tests in connection with movement of service, regrades, conversions, area transfers, and toll
and trunk circuits. Operated computerized Trouble Administration System
(TAS), dispatched installation and repair forces and maintained appropriate
records.

May 1980 to Nov. 1981   Position: Installer-Repairer
Duties: Climbed poles in all kinds of weather to install, maintain and remove
single and multi-line rotary dial and touch-call instruments at business and
residential locations. Bonded and grounded telephone lines in accordance with
Public Utilities Commission and company standards. Engineered efficient

**FIGURE
11.10**

Continued.

system cutover for local bank. This included a survey to determine amount and routing of distribution cable; selection of placement sites for multi-line instruments and the preparation of a narrative and diagrams of the survey. Have experience in use of Simpson 260 multimeter, GTE Lenkurt 8410A Station Carrier Test Set, Comdev ANI-3B module, Wilcom 136 BGM Circuit Test Set, Black and Decker power drill, and the GTE Lenkurt 82A Station Carrier Systems.

Mar. 1979 to May 1980   Position: Lineworker
<u>Duties:</u> Performed aerial and ground work in connection with line and cable transfers. Worked extensively with prints and engineering.

Aug. 1977 to Mar. 1979   Position: Frameworker
<u>Duties:</u> Placed, removed and rearranged jumper wires. Tested on distributing frames. Cleaned, inspected, and oiled apparatus, handled routine maintenance and testing of frame equipment. Performed Automatic Number Identification (ANI) strapping and testing.

Jan. 1970 to Aug. 1977   Position: Switchboard Operator
<u>Duties:</u> Operated switchboard and direct-dial equipment. Assisted public with local and long-distance calls from business, residential, and pay station telephones. Worked directory assistance as well as rate and route.

<u>HONORS</u>

National Honor Society and Psi Chi National Honor Society in Psychology.

<u>MEMBERSHIPS</u>

Local 986, International Brotherhood of Electrical Workers. Seven years as Bowling Green Chief Steward with four stewards reporting to me. As Chief Steward, I am responsible for processing and settling grievances through District level for 115 people. Seven years as Treasurer for Local 986, representing 900 people. As Treasurer, I am responsible for issuing checks, maintaining accurate check records, auditing all financial records, and participating in policy decisions as a member of the Local Union Executive Board. Appointed four times to Union Bargaining Committee to negotiate contracts between Local 986 and United Telephone Company.

<u>PERSONAL BACKGROUND</u>

Born in Sumter, South Carolina. Lived in Davenport, Iowa; Elmendorf Air Force Base (Anchorage), Alaska; Seward Air Force Base (Smyrna), Tennessee; Terre Haute, Bloomington, and Muncie, Indiana. Interests include music,

FIGURE 11.10

Continued.

literature, needlework, and working with people. Worked 40 to 48 hours per week with a full class schedule to earn my degree in Psychology. Willing to relocate.

REFERENCES

Available upon request.

FIGURE 11.11

Typical Resume of a Working Person—Designed to Focus on the Worker's Management Capabilities.

LINDA S. HARRIS

400 S. Main St.
Defiance, Ohio 46910
(419) 452-2216

EDUCATION

State University, Muncie, IN. Bachelor of Arts in Psychology, 1979. GPA: 3.25. Minor in sociology. Completed 41 hours toward Bachelor of Science in Electronic Technology. GPA: 3.31.

J.I. Shank and Associates, Davenport, Iowa. Employee Involvement Facilitator/Coordinator Workshop.

United Telephone Company, Courses in Labor Relations, AC DC Theory, Defensive Driving, Pole Climbing, 82A and 84A Carrier, Throw Crew, Telephone Installation and Maintenance.

Illinois State University, Labor Law, Grievance Analysis and Arbitration, Steward's Training.

AFL-CIO Summer School, Labor Law, Women in the Labor Movement.

American Arbitration Association, Current Trends in Arbitration.

EXPERIENCE (Uncompensated)

1976 to Present—Local Union 986, International Brotherhood of Electrical Workers, Norwalk, Indiana

| **FIGURE 11.11** | Continued. |

Positions: Steward, Chief Steward, Treasurer, Executive Board Member.
Duties: Appointed Chief Steward in 1978. Assist union members with problems unrelated to contract disputes. Problems include: alcohol and drug abuse, personality adjustment difficulties, stress problems, and career counseling. Charged with enforcement and administration of labor contract for 115 employees of United Telephone Company. Responsible for investigating possible contract violations. Investigations include interviews with management and craft employees to ascertain relevant facts, analysis and determination of whether a violation does, in fact, exist, and, if appropriate, filing of proper documentation to present the violation to management for redress. Elected to office of Treasurer and Local Union Executive Board Member. As Treasurer I am responsible for the issuance of checks, maintenance of check records, and auditing of all financial records. As Executive Board Member I represent 900 Local Union members through participation in budget development, expenditure review and approval, and policy decisions that directly affect the course of the Union.

EXPERIENCE (Compensated)

Jan. 1970 to Present   United Telephone Company

Dec. 1984 to Present   Member, United Telephone Company State Steering Committee, Employee Involvement Program.
Duties: Appointed as member of committee to develop and implement an Employee Involvement Program for the Ohio operating company.

Feb. 1985 to Present   Facilitator, United Telephone Company, Employee Involvement Program.
Duties: Training of craft and management employees in techniques of effective problem solving, monitoring of progress of employee involvement program on a local level, and serving as a resource person to help resolve problems in local programs.

Nov. 1981 to Present   Position: Test Desk Operator
Duties: Conducted tests in cooperation with field repair personnel to locate and clear telephone trouble. Also performed tests in connection with service, equipment and circuit changes. Maintained appropriate records.

May 1980 to Nov. 1981   Position: Installer-Repairer
Duties: Climbed poles to install, maintain and remove telephones. Bonded and grounded telephone lines in accordance with Public Utilities standards. Have experience in the use of test equipment and hand tools.

**FIGURE
11.11**

Continued.

Mar. 1979 to May 1980   Position: Lineworker
<u>Duties</u>: Worked extensively with prints and engineering to perform aerial and ground work in connection with line and cable transfers.

Aug. 1977 to Mar. 1979   Position: Frameworker
<u>Duties</u>: Placed, removed and rearranged jumper wires. Tested on distributing frames. Cleaned, inspected, and oiled apparatus and handled routine maintenance and testing of frame room equipment.

Jan. 1970 to Aug. 1977   Position: Switchboard Operator
<u>Duties</u>: Operated switchboard and direct dial equipment. Worked directory assistance and aided public with local and long distance calls.

<u>HONORS</u>

National Honor Society and Psi Chi National Honor Society in Psychology.

<u>MEMBERSHIPS</u>

Local 986, International Brotherhood of Electrical Workers. Chosen by membership as delegate to the Union's International Convention. Appointed three times to Union Bargaining Committee to negotiate contracts between Local 986 and United Telephone Company.

<u>PERSONAL BACKGROUND</u>

Born in Sumter, South Carolina. Lived in Davenport, Iowa; Elmendorf Air Force Base (Anchorage), Alaska; Seward Air Force Base (Smyrna), Tennessee; Terre Haute, Bloomington, and Muncie, Indiana. Interests include music, literature, needlework, and working with people. Worked 40 to 48 hours per week with a full class schedule to earn my degree in Psychology. Willing to relocate.

<u>REFERENCES</u>

Available upon request.

☐ Study the resumes of women in comparable jobs and how they describe themselves.
☐ Show your resume to other women who work full-time. Listen to their suggestions for improvement.
☐ Join the groups in your field who support professional women. For example, technical communicators can join Women in Communication, Inc., an advocacy group for women

throughout the communications industry. Such groups offer special job leads, information about companies that hire women, and special seminars to assist women with their job search strategies.

☐ If you are a woman returning to the workplace after an absence for child care, highlight your motherhood and your community or church service as the responsible positions they truly are. You may have managed budgets, solicited funds, or written grants for these organizations. If you have not joined community service organizations, doing so will allow you to build the record that can lead to employment.

☐ Perhaps you have earned supplemental income for your family as you have raised children. For example, if you have operated a child care facility, place an accurate dollar value on your work and include this information in your resume.

☐ Overcome any feelings of false modesty about your accomplishments. The resume exists to sell your skills; list your talents and accomplishments in full detail. Do not avoid doing so by saying, "Oh, anybody could have done that," about your domestic accomplishments.

Ethnic Americans. The Civil Rights Amendment prohibits discrimination on the basis of race or ethnic background. Nonetheless, members of visibly ethnic groups are still under-represented throughout virtually all employment sectors. As a member of such a group, actively research companies for an environment in which you can achieve success.

☐ Join the groups that support native and ethnic Americans in your professional field. Use your memberships to obtain leads and advice about your job search.

☐ Research the companies that interest you. Use your campus placement service or cooperative education office to find out about the companies' track records with recruitments and promotions of ethnic Americans. Some companies have programs to encourage their managers to hire specific groups of Americans.

☐ Although it is illegal to discriminate against applicants on the basis of race or ethnicity, some companies do. Avoid these companies and concentrate on the firms that hire ethnic Americans.

☐ Once you have identified the companies with an interest in ethnic applicants, highlight your special status on your resume. Because it is illegal to ask applicants if they belong to an ethnic or racial group, employers must rely on you to disclose that information. Do not state your race or ethnic background directly, but if you have won an ethnic scholarship, for example, describe that scholarship as, "Awarded to the outstanding Black senior in engineering annually." Develop your ethnic status as an employment asset. Notice that Li Ming's resume makes an asset of her Chinese heritage.

Interrupted Employment. Very few persons have records of continuous employment. Interruptions such as for retraining, sudden family relocation, child care, or military service are common. Sometimes a record might contain less savory interruptions such as time spent on rehabilitation after substance abuse or mental illness. These situations are of course difficult to address in a resume, but some guidelines have emerged as people have become more mobile.

☐ Present your experience in a positive light. For example, if as a dependent spouse you relocated regularly, state that you managed all the business affairs of the moves including finances, logistics, and the rest. Present the image of your personal stability and effective management.

☐ Translate your military technical experience into civilian technical terms. Do not use the military names for the systems on which you worked when the generic or civilian equivalents will be more descriptive. State how often you relocated, and emphasize your personal flexibility. Emphasize as well any special training you may have received or any higher education you gained while in the service. List any security clearances you have obtained.

☐ If your absence from the workforce resulted from a social or legal problem, avoid a direct discussion of the situation by using a functional format. If the absence is long enough to require an explanation, do so in your cover letter, again emphasizing your accomplishments and successes as you rehabilitated yourself. For example, if you are a rehabilitated alcoholic, tell the reader how often you speak in public as a result of your leadership in Alanon or a similar organization.

LETTERS OF APPLICATION

At some point in the hiring process, you will need to send a prospective employer your resume and a formal letter of application. Use these documents as persuasive tools to convince a personnel manager to hire you. However, remember that personnel managers have stated that the letter and resume method of applying for jobs is not the most effective job search strategy unless you use this approach very specifically. For example, a "confetti" job search campaign, one in which you send hundreds of resumes and letters across the nation, will offer rather limited success. Use the letter of application and resume method only in the following cases:

☐ To respond to the positions advertised in your campus placement office.
☐ To answer advertisements for jobs in your local newspaper.
☐ To follow up on leads provided by your friends or co-workers.
☐ To answer national ads that call for special expertise or licenses, such as an ad for an airplane mechanic.
☐ To follow up on any contacts made through the more successful job search strategies suggested by your placement experts.

The letter accompanying your resume should contain three types of information: a statement of the facts—how you heard about the job and what materials you are submitting with the letter, a description of your interest in the firm, and a persuasive conclusion that urges the receiver to contact you at a particular time for more information. In short, this resembles the typical sales letter because, indeed, you should use the letter of application to "sell" your skills. As we discuss each of these three elements, read through the first and revised drafts of letters of application included in this chapter. Note that throughout the revised letters we seek to establish the personal orientation that we defined in our correspondence chapter.

**FIGURE
11.12**

Randy Goetz's First Draft of a Job Application Letter.

480 Lehman #208
Bowling Green, Ohio 43402

May 15, 1988

Ms. Elaine Hoff
Manager, Personnel
Arlington Tools
Bryan, OH 46218

Dear Ms. Hoff:

I am a very energetic and determined individual who enjoys a variety of tasks. I
am very interested in your organization and feel that my personal characteristics
may be a match with the qualities you look for when hiring a new employee. I
am extremely interested in a position in your communications group. I spoke to
you at a career day seminar at Bowling Green State University in 1987.

I believe that my past work history indicates that I am capable of learning about
technical material and writing about it. I understand the importance of paying
close attention to technical details, so that manuals will be accurate. Also, my
professional courses such as Management and Production Techniques have
increased my awareness of administrative and communications skills which are
also essential in this field.

I am very interested in Arlington Tools and would like to arrange a convenient
time to meet. I can be reached at the address above. I look forward to hearing
from you.

Very truly yours,

Randy Goetz

Introduction—What You Heard About the Job

In the first sentence, explain how you heard about the job. If possible, allude to a contact with
the personnel manager, a recruiter, or a firm member who told you about the position. It is important to use this information to create a picture of continuous contact with the company.
Note that our three rewritten drafts all state clearly how the applicants heard about the
positions.

**FIGURE
11.13**

Randy Goetz's Rewritten Draft of a Job Application Letter Showing a
Personal Orientation.

480 Lehman
Apartment 208
Bowling Green, OH 43402

May 15, 1988

Ms. Elaine Hoff
Manager, Personnel
Arlington Tools
Bryan, OH 46218

Dear Ms. Hoff:

Your firm came to my attention when we spoke at Careers and the Future, the
career seminar held at Bowling Green State University, December 11, 1987. You
mentioned that Arlington Tools recruits as technical communicators writers who
are willing to learn how to use power tools. Since I use power tools regularly, I
may be of use to your communication group, and I am enclosing my resume to
start my application to work with your firm.

Arlington Tools is a well-known company in the Goetz home, because my dad and
I both use your equipment for our hobby of repairing and restoring old furniture. We have found that your tools allow us the maximum of control when we
remove old finishes; this type of experience during the last six years has allowed
me to learn about your product from the user's point of view. I have encouraged
other restorers to purchase Arlington products, and I know I can explain your
tools from a technician's point of view. Along with my technical hobby, I have
obtained at BGSU a journalism degree with special emphasis on technical communication. I have a solid background in technical writing, editing, and print
media production and management. My internship at the Gulf Coast Research
Laboratory taught me the importance of accuracy in the publication of technical
specifications, and my minor in biology, another of my hobbies, shows my
appetite for learning and writing about technical material.

May I call to make an appointment to show you my portfolio? Such a discussion
can focus on how my training can prove beneficial to Arlington Tools as you
develop new tools that woodworkers will purchase. I shall call the week of June 1
to make an appointment to review your needs in technical communication.

Very truly yours,

Randy Goetz

**FIGURE
11.14**   Lucille Winchester's First Draft of a Job Application L̶

Lucille Winchester
1000 E. Manhattan
Toledo, OH 43611

May 15, 1988

Mr. Henry Slavin
Aztec Computer Repair
200 Federal Highway
Columbus, OH 49917

Dear Mr. Slavin:

I graduated recently from Control Data Institute, and I am seeking a career as a
computer or electronics technician. In addition to my recent technical training,
my work background includes six years of customer service experience in the
areas of complaints, second billings, and orders. I also have gained experience
with payrolls, inventories, and electronic filing systems.

I sincerely appreciate your time and consideration concerning my desire for
employment with your company in which I would incorporate all of the above
qualifications as well as sincerity and enthusiasm in building a career in the
computer industry.

I am available to come in for an interview at your earliest convenience. You may
contact me at the address on the letterhead.

Very truly yours,

Lucille Winchester

In the first paragraph, develop a mutual purpose with your audience. For example, Randy
Goetz mentions not only the career day at his college, but also his extensive experience with
the Arlington Tools products. Similarly, Lucille Winchester mentions her telephone call, so
the receiver will remember her as an individual. She then uses this opportunity to suggest
that she understands and can help correct the firm's shortage of mainframe computer technicians. Since she has no previous contact, Linda Harris's very short introduction enables her
to discuss immediately the employer's — not just the applicant's — concerns and interests.
Finally, note that all three writers use the word "you" and avoid the word "I" as much as possible. They use active voice verbs to create a dynamic and businesslike tone.

FIGURE
11.15

Lucille Winchester's Rewritten Draft of a Job Application Letter Showing a Personal Orientation.

Lucille Winchester
100 E. Manhattan
Toledo, OH 43611

May 15, 1988

Mr. Henry Slavin
Aztec Computer Repair
200 Federal Highway
Columbus, OH 49917

Dear Mr. Slavin:

Our telephone conversation on Monday, May 12 helped me to understand your personnel needs. Since you already have good technicians to work on desktop computers, and you need a person with training on mainframes, I would like to apply for the job.

The increased volume of calls to service mainframes, which you described on the phone, requires a technician who can work specifically on those machines and who can work at the job site with a minimum of supervision. My resume shows you that I can assist you with the technical area and that I can work independently to produce the results you want. For example, I have just completed a special training course in mainframes at Control Data Institute in Toledo, so I can offer you the most current skills in computer repair which I obtained through self-paced instruction and weekly testing. Please note, too, that in my recent work experience at the Toledo Library, I worked entirely alone and received continuous promotions because of my responsible management of my duties and my skill in working with the public.

Your firm requires mature, stable technicians who can work on-site without disturbing your customers' routines. I offer that experience; I work well with others, and I can perform the specialized technical work that you require. May I call you the week of May 22 so we can review my application for your advertised job? You will find my technical competence and my stability can increase profits for Aztec Computer Repair in time saved at the site and in satisfied customers.

Very truly yours,

Lucille Winchester

Body—Why the Company Appeals to You and What You Can Do for Them

The body of your letter should answer the receiver's unwritten question, "How will I benefit from hiring you?" In short, write the body of the letter from the employer's—not your—perspective.

For example, Randy Goetz answers with an overview of his family's experience with Arlington Tools. He uses the rest of the paragraph to not simply state his experience but to explain how his experience can benefit the company. Thus, each fact becomes a piece of persuasive evidence explaining that Arlington Tools will gain an asset when they hire him. Lucille Winchester uses the same approach. She, too, tells the prospective employer how to read her resume. After defining Mr. Slavin's needs, she selects from her resume those experiences that demonstrate her technical competence and trustworthiness. In doing so, she turns a liability, age or "overqualification," into an asset, a responsible independence.

Linda Harris, also an experienced worker whose credentials might price her out of the competition, uses a slightly different approach. In paragraph two, she translates her previous experience so that Ms. Singleton can see its possible application to her own firm. Instead of referring to her positions with terms specific to labor unions, Linda Harris uses generic terms that apply both to labor and management. By doing so, she selects the experiences which make the strongest case for her employment, and, again, she tells the receiver how to read her resume and evaluate her experience. (Any working adult or former member of the military should practice Linda Harris's method.)

Conclusion—Why You Should Call the Receiver

Since the applicant in each case states that he or she will call for an appointment, you may think that the three rewritten letters contain very aggressive comments. Assuming that you have contacted the prospective employer previously or that you offer substantial qualifications for the position, you should take the initiative in following up with a telephone call. Remember, the purpose of your letter is to create a mutual purpose with the prospective employer. An interview is a natural outcome of that shared purpose.

Consider the alternative—no phone call. How long will you wait for the personnel manager to shift through the stacks of resumes and "blind" applications that arrive daily? We have observed many cases in which a manager ignores files of mailed resumes and letters because a telephone call forces him or her to pay attention to a particular application. Therefore, follow up with a telephone call within two weeks of your packet's expected arrival at the manager's office. You risk only rejection, and you know that rejection can result if you fail to call proper attention to your materials in the first place.

To review, consider some "dos" and "don'ts" of applications.

Do

1. Contact employers before mailing resumes and letters of application.
2. Tailor your resume to the particular job you apply for—even though you qualify for several types of jobs.

FIGURE
11.16

Linda Harris's First Draft of a Job Application Letter.

400 S. Main St.
Defiance, OH 46910
(419) 452-2216

May 15, 1987

Ms. Donna Singleton
Manager, Personnel
Electronics Unlimited
1700 Ajo Way
Phoenix, AZ 86219

Dear Ms. Singleton:

I write to apply for the position of employee representative that you advertised
recently in the Frontier News. I offer all of the qualifications for which you
advertised, and I have excellent recommendations which attest to my maturity
and technical ability.

I am a graduate of the State University in Muncie, Indiana where I obtained a
B.A. degree as I worked my way through school. I majored in psychology with an
emphasis on industrial psychology and training, so that I could apply my learning to a business setting. At the same time, I obtained extensive training through
United Telephone, my present employer. I have been sent to school to study law,
arbitration, labor relations, and the various technical skills necessary for my job
at the telephone company.

My resume shows that I have earned steady promotions with United Telephone.
Starting as a switchboard operator, I also volunteered for "nontraditional" jobs
such as frameworker and lineworker. Because I have worked to obtain four promotions from the telephone company, I understand how large companies operate
and how they manage their employees. Therefore, I can speak effectively for your
employees.

After you review my resume, you may have questions about my career, or you
may need further information about my skills as an employee representative.
Please feel free to call me at the number listed on my resume to obtain any further information or the names and numbers of my references.

Very truly yours,

Linda S. Harris

FIGURE 11.17

Linda Harris's Rewritten Draft of a Job Application Letter.

400 S. Main St.
Defiance, OH 46910
(419) 452-2216

May 15, 1987

Ms. Donna Singleton
Manager, Personnel
Electronics Unlimited
1700 Ajo Way
Phoenix, AZ 86219

Dear Ms. Singleton:

I am applying for the job of employee representative that you advertised in the May 11 Frontier News. I enclose a resume; my transcripts and letters of reference are available upon request.

Your position caught my interest immediately because for the last 10 years, I have been involved in representing the employees of United Telephone Company, and my experience can help you to cut through the red tape of federal requirements and routine labor and management negotiation procedures. For example, through my positions, first as a facilitator, then as a supervisor, and finally as the manager of some sixty persons, I have learned how to address labor/management conflicts, file grievances, prevent nuisance complaints, and motivate employees. As a labor manager, I have reduced grievances in my shop by 30% as I have increased our productivity by 12% despite the continued layoffs and relocations of our employees. Your firm, which is in a right-to-work state, has more flexibility in labor negotiations than United Telephone does, so an experienced labor representative can help you to reduce overtime, product rework, and the unfounded complaints that can eat up your managers' time with costly negotiations.

You should note that I offer more than a simple "hands-on" talent for resolving labor and management issues. My resume shows that I have special training in psychology with an industrial emphasis, labor and management negotiation, and communication. Your firm will also make good use of my coursework in electronics, and the extensive training in electronics systems that I obtained at United Telephone.

I shall call you the week of May 22 so that we can arrange to discuss your specific needs. Since travel and relocation present no problems, I can schedule a visit to your Phoenix facility on short notice.

Very truly yours,

Linda S. Harris

3. Show in your letter of application that you know something about the company.
4. Write a letter of application that tells the receiver how to interpret your experience.
5. Write a letter that persuades the receiver you can do something for his or her company.

Don't

1. Mail out hundreds of "blind" resumes and letters.
2. Send the same resume for each application.
3. Use "To Whom It May Concern," "Dear Sir," or "Dear Madam" as the salutation of your letter.
4. Emphasize what you hope to learn on the job.
5. Dwell on what your personal experiences have meant to you.

FIGURE 11.18	Fifty Questions Asked by Employers During Interviews with College Students.

1. What are your long-range and short-range goals and objectives? When and why did you establish these goals? How are you preparing yourself to achieve them?
2. What specific goals, other than those related to your occupation, have you established for yourself for the next 10 years?
3. What do you see yourself doing five years from now?
4. What do you really want to do in life?
5. What are your long-range career objectives?
6. How do you plan to achieve your career goals?
7. What are the most important rewards you expect in your business career?
8. What do you expect to be earning in five years?
9. Why did you choose the career for which you are preparing?
10. Which is more important to you, the money or the type of job?
11. What do you consider are your greatest strengths and weaknesses?
12. How would you describe yourself?
13. How do you think a friend or professor who knows you well would describe you?
14. What motivates you to put forth your greatest effort?
15. How has your college experience prepared you for a business career?
16. Why should I hire you?
17. What are your qualifications that make you think you will be successful in business?
18. How do you determine or evaluate success?
19. What do you think it takes to be successful in a company like ours?
20. In what ways do you think you can make a contribution to our company?
21. What qualities should a successful manager possess?
22. Describe the relationship that should exist between a supervisor and those reporting to him/her.
23. What two or three accomplishments have given you the most satisfaction? Why?
24. Describe your most rewarding college experience.
25. If you were hiring a graduate for this position, what qualities would you look for?

INTERVIEWS

Prepare for two types of interviews, screening and intensive. Screening, or preliminary interviews, usually last about 30 minutes, and typically occur at your college placement office. During this discussion, the recruiter and you determine your suitability for employment with a particular firm. Intensive interviews occur later, usually at the employer's place of business. At this stage, you are usually one of fewer than six finalists, and the employers and you should seek to know each other better.

In either case, the interview process consists of three stages—preparation, interview, and follow-up—so we discuss these points in general terms. To prepare specifically for either type of interview, use our suggestions here to work intensively at your placement office or with your career counselor to prepare for your specific requirements.

FIGURE 11.18	Continued.

26. Why did you select your college or university?
27. What led you to choose your field of major study?
28. What college subjects did you like best? Why?
29. What college subjects did you like least? Why?
30. If you could do so, how would you plan your academic study differently? Why?
31. What changes would you make in your college or university?
32. Do you have plans for continued study? An advanced degree?
33. Do you think your grades are a good indication of your academic achievement?
34. What have you learned from participation in extracurricular activities?
35. In what kind of work environment are you most comfortable?
36. How do you work under pressure?
37. In what part-time or summer jobs have you been most interested? Why?
38. How would you describe the ideal job for you following graduation?
39. Why did you decide to seek a position with this company?
40. What do you know about our company?
41. What two or three things are most important to you in your job?
42. Are you seeking employment in a company of a certain size? Why?
43. What criteria are you using to evaluate the company for which you hope to work?
44. Do you have a geographical preference?
45. Will you relocate? Does relocation bother you?
46. Are you willing to travel?
47. Are you willing to spend at least six months as a trainee?
48. Why do you think you might like to live in the community in which our company is located?
49. What major problem have you encountered and how did you deal with it?
50. What have you learned from your mistakes?

From the *Northwestern Lindquist-Endicott Report*, by Victor R. Lindquist. Northwestern University, Evanston, IL ©. Used by permission of Northwestern University Press.

Preparation

Most applicants make their first major error in their failure to prepare for the interview. Plan to spend a minimum of three hours preparing for each interview; use the steps we list here.

1. **Research.** Use the materials in your college placement office to read about the company. Many of these same materials are available in your public or college library. Use news and magazine indexes to look up the recent company news in papers such as the *Wall Street Journal* and magazines such as *Business Week*. Call the company to obtain brochures and literature about the firm. Many companies publish magazines and brochures primarily for job applicants' research and understanding. As you read about the company, write questions you would like to ask the interviewer.

2. **Plan.** Prepare a statement that tells the recruiter about your experiences, education, and activities. As with the job application letter, focus your presentation on how your experience can benefit the company. At the same time, prepare samples of your work for the recruiter. If you are a technical communicator, advertising major, or artist, prepare a portfolio. In any other field, prepare attractively mounted, carefully collected writing samples you can discuss with the recruiter. Not only is there an increased emphasis on written competence in today's job market, but the samples also show that you can complete work satisfactorily.

3. **Practice.** Use the questions in this chapter to practice your interviewing techniques. Have friends test you with these questions until you feel quite comfortable with your answers. Practice as well the difficult questions you will need to ask in the interview such as the questions about wages and benefits. Practice asking the recruiter if he or she intends to offer you a job. Practice asking when you can expect to hear about the job offer that you request.

4. **Review.** Review the relaxation techniques we outlined in the oral technical communication chapter. For the interview, place in an attractive leather or vinyl folder or briefcase your interview materials, your notes, the questions you will ask, and blank paper for note-taking.

Interview

As with public speaking, you simply cannot practice interviewing too much. Therefore, we refer you to your placement office for extensive instruction in this area. We can, however, offer guidelines for effective interviewing.

1. **Keep Track of the Time.** A preliminary interview usually lasts 30 minutes. Wear a watch you can easily see so that you can accomplish everything you need to. Use the first five or ten minutes to present your accomplishments, couched in terms of their interest and benefits to the company. The interviewer may then spend about ten minutes asking questions or explaining various company policies. During the final ten minutes, you should have a dialogue with the interviewer that allows you to ask if you will be hired, how you will be hired, and how the company treats its employees.

    For an intensive interview, request an itinerary. If time does not permit a written itinerary, obtain one by telephone, so you can plan to follow or alter slightly the preliminary schedule for each group you meet.

2. **Take the Initiative.** Recruiters report regularly that new graduates fail to take control of the interview at the start. Typically, they sit passively awaiting the recruiter's questions. Remember, the interviewer will respect the applicant who takes the initiative. Do so by greeting the interviewer by name, shaking hands, and beginning the dialogue with some polite conversation about the interviewer's visit to your campus or region of the country. Then indicate that you are prepared to describe your qualifications for the specific company and that you would like to explain your capabilities. Answer all questions, but do not simply sit there and await them. After finishing your five- or ten-minute presentation, ask the recruiter if he or she would like to ask questions or discuss your experiences in detail.

3. **Avoid Negatives.** Do not discuss your weaknesses, gaps in your education, or any other "negatives" the recruiter may bring up. Many recruiters are taught to ask, "What do you feel is your greatest weakness?" or "What professional problems are you working on overcoming presently?" Do not lie, but use such questions to stress your involvement with the work ethic. For example, you might state that you have worked all of your life to overcome bad habits. You are currently working to increase your concentration on the particular job at hand. Note that the entire answer is cast in a positive light. These are "trick" questions, and you should manage them effectively.

4. **Ask for an Offer.** In a sense, we describe another test the interviewer may give you. The recruiter may talk about the company in detail, but he or she may not mention whether or not the company intends to offer you employment. Ask, "Do you intend to hire someone today?" Or you might state, "I understand this is a preliminary interview. Please outline the steps you will follow to hire one of today's applicants. Could you also tell me your schedule, so I can understand whether or not you can offer me employment." You should ask these questions toward the end of your interview, when you and the recruiter discuss your mutual interest in the company. Ask as well about the pay range and benefits. Avoid stating the pay range you expect. If you cannot avoid answering this question, then remember that companies plan on offering you 20% less than the salary you mention.

Remember to exit from the interview as assertively as you entered. Shake hands with every company representative present and thank them for the interview. State that they will hear from you and tell them that you will expect to hear from them.

Follow-Up

Anytime you interview, send a thank-you note to demonstrate that you can follow through on scheduled projects, and, more importantly, to remind the personnel manager that you still retain an active interest in the company. Examine Randy Goetz's thank-you note to Arlington Tools as a model. Use the note to confirm the event of the interview, and to include any new information, such as a change of address or a time when you cannot be reached.

Finally, after sending the thank-you note, file your materials for this particular job application for future reference. For example, Linda Harris, an electronics specialist also searching for a job in human services, might at any time apply seriously for 25 different positions. Without a system for her correspondence and telephone calls related to those applications, she would be quite confused. Purchase a supply of letter-sized file folders and label each with the company name to which you apply. Store all your dated correspondence with that company there. Keep notes about your interviews and names of contacts in the file, too. This method will prevent the embarrassing mistake of sending a thank-you note to the same per-

**FIGURE
11.19**   Randy Goetz's Thank You Note After a Preliminary Interview.

480 Lehman
Apartment 208
Bowling Green, OH 43402

June 2, 1987

Ms. Elaine Hoff
Manager, Personnel
Arlington Tools
Bryan, OH 46218

Dear Ms. Hoff:

Thank you for giving me the opportunity to talk with you this morning.

Your company manufactures the finest tools available in the medium price range, and
you certainly have convinced me that you will maintain your leadership position. I
would welcome the opportunity to be involved with your continued growth and success.

Thank you again for your time and consideration.

Very truly yours,

Randy Goetz

sonnel manager twice or applying for the same position twice. In short, a system which tracks
your job search will save you hours of work.

SUMMARY

In this chapter you learned to compose various types of resumes and letters of application; to
prepare for screening and intensive interviews; to have your resumes, letters, and interviews
reflect a strategy toward the job search that results from your research in your placement office, library, and among your friends and co-workers. After reading this chapter, you should be
able to:

■ Study and understand the current, unstable job market.
■ Decide whether your experience should appear in a chronological, functional, or combination resume.

■ Compose several types of resumes that define your qualifications for various jobs.
■ Select the typesetting, paper, and correct design for your resumes.
■ Show that you know when to use a letter of application.
■ Compose several types of personally oriented letters of application that explain your qualifications as benefits to be acquired.
■ Prepare for screening and intensive interviews with research, planning, practice, and reviewing of successful interview techniques.
■ Conduct any kind of interview by controlling the conversation.
■ Ask the kinds of interview questions for which you need answers.
■ Ask a recruiter or a manager if he or she intends to hire you.
■ Draft a thank-you note that shows you are punctual.
■ Keep records of your job search, so you avoid mistakes.

ASSIGNMENTS

1 Informally assess your college's placement office. Find out what seminars and research resources it offers to students and alumni. Prepare a five-minute report for the class.

2 Take reference notes while at the placement office. Prepare a one-page handout that the other students in the class can use to find resources for their job search.

3 Work with a partner for this assignment. One should survey the public library, and the other the college library. Look for resources students can use to obtain information about companies and professions. Prepare a handout that synthesizes your information and that will accompany the five-minute oral report you will give to the class.

4 Using the nearest metropolitan newspaper, find a job advertisement that interests you and prepare an application for the job. Prepare a letter and resume and bring each to class. Bring the advertisement to class with you, too.

5 Find a new collaborator and give your materials to your partner to rewrite. First, your partner should interview you to determine your strengths; then the partner should work on improving the persuasive aspects of both the resume and the letter.

6 Rewrite your letter and resume using your partner's most useful suggestions. Submit all three versions, your original, your partner's rewrite, and your revision, to your instructor for review and discussion.

7 Collaborate on this assignment. Compose five questions you need to ask in an interview in order to develop a true sense of a company's purpose and your ability to fit into that organization. Write down the questions and prepare to share them with your classmates. By the way, how would you answer these questions if you were the recruiter?

8 Work alone on this one or work with a trusted family member or friend. Make a list of your weaknesses or other unfavorable traits you would prefer to avoid discussing in an interview. For example, you may have a problem with tardiness. Develop a strategy for each weakness should it come up in an interview. Remember, you cannot lie, but you can emphasize the positive aspects of the situation.

CHAPTER
12

Proposals

This chapter will introduce you to the preparation and writing of formal and informal proposals for business and institutions. After reading this chapter, you will be able to:

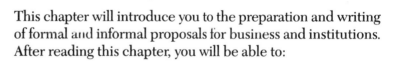

- Identify situations in which proposals will be useful.
- Decide whether to write a formal or informal proposal.
- Obtain guidelines for proposals with dictated formats.
- Choose the appropriate formats for proposals that call for individual approaches to style and layout.
- Develop the best persuasive and descriptive appeals for your proposal.
- Draft a convincing proposal with a clear introduction, body, and conclusion.

Remember Steven and Elaine Miska of Chapter 4? As owners of a small insulation business, they have to research and develop new business opportunities. They wonder if they should expand into the sound insulation business. Since their county has adopted a new ordinance to control noise pollution, the Miskas want to know if their business can easily use its existing procedures for insulation and construction to move into the sound-proofing business.

Bill Dorsey, a junior technical writer with a small communications firm, views the Miskas as a source of future business for his company. If he can provide the study the Miskas need at a reasonable fee, he can use that same information to generate similar contracts in other cities that enact noise pollution ordinances.

This opportunity offers Bill a potentially rewarding challenge, because he can perform most of his research for the first study, the Miska feasibility report. With such data, he can then generate "boilerplate," or reusable portions of the report, to combine with local needs assessments in other markets. First, he must convince the Miskas that his firm, Research Associates, Inc., can provide some of the information they need. To win this new account, Bill must start with a proposal.

WHY WRITE PROPOSALS?

Bill Dorsey's case is not unusual. A typical technical writer can expect to draft proposals related to almost any field of commerce or technology. In fact, proposals are written for any of three reasons—reasons which you should understand so that you can draft a proposal with its true purpose in mind.

Often, proposals are used to obtain new business or to renew and upgrade regular business contacts. A firm or an institution publicizes its call for bids on a product or service in a request for proposals, often referred to as an "RFP." The RFP gives the specifications of the product or service the purchaser requires. It outlines, as well, all the terms and conditions a supplier must meet in order to do business with that firm or institution. An RFP can be as informal as a purchasing agent's telephone calls to a firm's regular bidders on its contracts, or as formal as a multivolume set of Defense Department specifications. RFPs always call for written bids in the form of proposals.

You might also be called upon to write a proposal to obtain funding from a public agency, a philanthropic foundation, or a corporation for a nonprofit activity. For example, community agencies, school systems, universities and colleges, businesses, and churches all compete regularly to obtain financial support for their ongoing nonprofit activities. Your local fire department may submit a proposal to the state for new equipment, or your library may compete for federal funds to enhance its collection. You may be involved in community proposal-writing projects in the areas of public support for the arts, litter control, sewage management, equipment for the local schools, conservation, and church-related charities. At some point in your career, you will write a narrative to obtain funds for such an activity. Proposal writers are in great demand in government and in the arts.

Proposals also exist as the records of the transactions between suppliers and businesses. Sometimes, all the parties to a contract have agreed to work together before a proposal is even written. Still, the proposal must be written and submitted for review to create a record of the agreements between the clients and vendors. For example, a large printing company may hire the same contract writers to help with deadlines on a regular basis. Nonetheless, a proposal is needed for each hiring so that the company and the contractor can have records of the work promised, the negotiations that occur, and a basis for determing quality assurance. Thus, the proposal, once accepted and signed by a firm's purchasing agent, becomes a part of their contract, a legally binding document to which both the company and the supplier must adhere or be liable for litigation.

In short, proposals, whether written to obtain new business, to expand existing budgets, or to confirm informal negotiations, are the backbone of business. The clarity, organization,

---

FIGURE 12.1   Definitions for Proposal Writers.

THE VOCABULARY OF PROPOSAL WRITERS

Benefit—This is a favorable result, improvement, or gain that a customer can expect as a result of your services. For example, a benefit might be that the customer will acquire a complete training evaluation system from your firm.

Boilerplates—These are standard passages of text your firm or institution always includes in proposals, as might be, for example, their equal employment opportunity statement. These are written, stored, and reused with each proposal, and not drafted each time.

Goals—Goals are broad, guiding principles (and they are generally *worked toward*, not usually *met*). Example of a goal: To help improve morale among student commuters.

Objectives—Objectives are definite and measurable (and you will be evaluated by them). Example of an objective: Seventy percent of the participants in the training will pass the exam to gain licenses.

Outcome

Terminal—Prescribed in the statement of objectives, very specifically to be met at specific points in the schedule.

Results

Enabling—Required to achieve terminal results. These are milestones allowing the customer to supervise you, such as, for example, the design of a questionnaire. **State these clearly.**

Serendipitous—''Freebies'' or extras included that will result in improvement, such as, for example, a study of employees' benefits that might also improve their productivity.

and appearance of your proposal can increase your firm's earnings, just as poorly executed proposals can cost an employer clients and profits.

SOLICITED AND UNSOLICITED PROPOSALS

Solicited proposals are bids to which suppliers have been invited to respond. For example, the printing company that you represent may have "preferred bidder status" with a state agency, meaning that your company qualifies to do business with the state government. Your company, then, is invited—solicited—to submit a cost estimate whenever the state needs documents printed. Although your company may compete against six other firms for the job, you don't have to compete against every printer in the state. Obviously, solicited proposals have a greater rate of acceptance than nonsolicited ones. The authors of solicited proposals usually have their firm's history with the client to draw upon as they compose plans. Since the estimators and writers usually know exactly which companies they are competing against, they can estimate with greater accuracy. The purchaser has a record of success with the vendor upon which the proposal writer can draw for persuasive examples.

On the other hand, unsolicited proposals must be extremely competitive in order to obtain attention and success. Some not-for-profit funding agencies estimate that only 8 percent of their funded proposals are unsolicited. You may be asked, for example, to write an unsolicited proposal after your supervisor receives a direct mail announcement of funds available for particular training or research activities. If your firm has no previous experience with the targeted funding agency, your proposal must be perfect. It must be visually pleasing, written exactly to the specifications of the RFP, and most importantly, contain a perfectly outlined plan of action for the concept for which you seek funds.

FORMATS FOR PROPOSALS

Most funding agencies issue format specifications for proposals with their RFPs. You may have to complete a formal application on that company's or agency's form and attach a written statement of your firm's plans and policies. When a business or agency issues specific guidelines for the formats of proposals, you must follow these guidelines exactly, or your bid may be disqualified from the competition. Obtain all the forms needed to complete the proposal, and before you start the bid, design a checklist that you may use to make certain you have supplied every bit of information the agency has requested. Such a form will help you to define the task's scope as you plan a proposal. As you identify the areas in which you feel less than adequately informed, you can obtain the assistance and information you need from your associates. For example, you may not know of your company's policy on equal employment opportunity, but if you bid on a Federal contract you must supply a statement of compliance. If you review the proposal forms and compose the checklist early, you can inform your legal department of your needs, and they can prepare a statement of equal employment opportunity for you.

More commonly, you will work for a firm that designs its own business proposals. Of course, if you are self-employed, you will have total control of the proposals you write, but in

FIGURE 12.2	The Proposal-Writing Process.

PLANNING YOUR PROPOSAL

Since your proposal is both persuasive and descriptive, you must convey a clear picture of your plan to the customer—without explaining how to do the job. So that you can write clearly, try the following composition process. After all, planning and rechecking details are the most time-consuming aspects of proposal writing.

Make a Jot List—one idea on one piece of paper, so you can shuffle the papers later to create sequence.

Sketch a Plan—create a flowchart of your work plan much as Bill Dorsey did.

Create a Time Frame—use a timeline to map out exactly how long each phase of your plan will take. Mark as well the phases which will overlap.

Analyze and Debug—look at your rough sketches and ideas, and ask if your plan is realistic. Can you deliver it all?

Draft a Rough Budget—remember to include all your expenses including paper, printer ribbons, etc. Add 10% to cover your errors and unforeseen price increases.

Make a Checklist—of every item you will need to include in this particular proposal to comply with your client's format.

Select Your Persuasive Appeal—from our list for the introduction, or perhaps you have an appeal selected.

Select Your Descriptive Details—you will develop these in the middle of the proposal. Remember, tell them that you have a plan for doing the job, but don't tell them how to do it themselves.

Write the First Draft—and seek reviews from co-workers.

Rewrite and Refine Your Draft—and use the checklist so you are sure to include everything.

Proofread Your Final Copy—three times.

Inventory Your Final Copy—use the checklist to make certain you have included every sheet.

both cases, you must be aware of the conventions of appearance and content that govern proposal writing so that you can produce truly competitive bids.

Appearance

Because your proposal introduces the potential buyer to your company, appearance is critically important. Experienced proposal writers refer to the ten musts of a winning appearance:

1. Use a handsome cover of heavier stock than the rest of the proposal. Have a graphic design that conveys your company's identify or message. Your company's logo should appear prominently. Use a spot color if possible to involve readers with the message.

2. Use good paper for the proposal itself: a 20 lb. white paper (no colors) with at least a 25% rag content. Such paper survives abuse better than cheap stock, and it is your responsibility that your proposal stays attractive through the many readings you hope it receives.
3. Use wide margins—1½ inches on the left, right, top and bottom margins—so the proposal is readable after it is bound.
4. Type or print on only one side of the page even if your copy machine has duplex capabilities.
5. Double space throughout to create readable copy.
6. Start each section with a fresh sheet of paper; again, this maintains readability, as does double spacing.
7. Proofread carefully so that your proposal is clean, articulate, and literate. You can be sure that the competition's proposals will be perfect.
8. Use a checklist to check your proposal for completeness. Does your company usually attach an abstract at the front of the proposal? Have you included one? A checksheet can help you detect such errors.
9. Look at copies of other proposals submitted to the firm or agency you approach. See what others did with their formats.
10. Make the budget look attractive. This is the page the decisionmakers will read first. You must do everything possible to make it attractive, correct in appearance, self-explanatory, and readable. This section alone makes or breaks the proposal.

Content

Your proposal must have a clearly identifiable introduction, middle, and conclusion. This content convention is so important that we devote this chapter's entire next section to it. If you fail to observe the content conventions—for example, if you put your timeline in the wrong section of the proposal—the client may identify you as a novice and decide on that basis not to hire you. Even in proposals phrased as memos and letters, the three sections should be readily identifiable.

Before we discuss the sections of the proposal, a word about redundancy may be useful. Well-conceived proposals tend to repeat the same ideas several times. In this sense, they are quite different from the essays your instructors may have described as well-written. Economy of words is not characteristic of all successful proposals. The very structure of the proposal requires you to present your concept in the introduction, describe it in detail in the middle, and rephrase it in the conclusion. You might feel a bit awkward with this procedure, especially if you are writing a proposal to the format of a funding agency. Remember, different members of your audience may read only the introduction, the conclusion, or even the budget. Your concept in each section of the proposal must be as clear as possible. Although we caution against redundancy as a rhetorical device, you may find that effective proposals—that is, those that are funded—repeat themselves to some degree throughout. In short, the winning proposal is not always a masterpiece of concise writing or elegant style.

INTRODUCTION

Your introduction must persuade the prospective client to hire you. This usually translates into several related objectives. First, state your understanding of what caused the client to

| FIGURE 12.3 | Organization of the Proposal. |

ORGANIZING YOUR PROPOSAL

Well-written proposals of any length contain three sections. You can modify the amount of detail in any section to suit your proposal's length and purpose.

Front Section—leads to main section
1. Establishes your understanding of the client's needs
2. States objectives and purposes governing your work and the accomplishments to result from it
3. Arouses urge to act

Main Section
1. Flowchart or pictorial description of the work
2. Timeline tied to flowchart
3. Narrative detailing results, accomplishments, and benefits

Conclusion
1. Experience and capability—yours and your staff
2. Procedures for utilizing your services
3. Organized conclusion leading to your retention

seek your services. Second, offer an overview of the goals and objectives of the services you propose to perform for the client. Finally, instill in the client the "urge to act," that is, the desire to purchase your services. For example, you might emphasize the client's need to comply with existing laws or to increase efficiency or profits. A prospective client may want to avoid negative publicity or to seek consumer recognition in a particular area. All of these ideas may suggest appeals you can use to encourage the client to hire you. Figure 12.9 lists other approaches.

Depending upon your proposal's length and style requirements, the front or introduction should contain the eight items we describe here in capsule form.

Letter of Transmittal

In a formal proposal, this document should be on your firm's letterhead, and it should state correctly the proposal's name, the persons' names to whom it is submitted, and any other facts that will assure its reaching the appropriate audience. This information is most important in proposals submitted by sealed bids or by mail. In an informal or letter-format proposal, this information should appear in the first paragraph.

The letter of transmittal should also briefly discuss the experiences that led to the submission of the proposal. Did you speak with the purchasing agent on the phone? Were meetings held at which you were told of the need for your services? Did you learn of the company's needs through recent publicity? Mention any of these facts in your attempt to establish your record of successful transactions with the client.

Cover

Of heavier stock than the inside of the proposal, the cover introduces your firm's image to the client. Its appearance conveys your attitude about the client and your style or approach to the services you offer. If possible, use color conservatively to enhance the cover's appearance. The cover should clearly display the title of the service or project you propose, your firm's name, and its logo. Remember, the cover suggests what will follow.

Title Page

The first page after the cover, the title page, lists all of the pertinent facts. Accuracy is important here because your client will start a record of the proposal by creating a file from the title page. This page should contain:

- ☐ The name or title of the proposed service.
- ☐ The date on which you submit it.
- ☐ The name of the individual to whom you submit the proposal.
- ☐ The name of your firm or your own name if you work as an independent proposal writer.
- ☐ Any restrictions on the proposal's distribution such as your copyright or a distribution list.

Abstract

The abstract should reflect your proposal's length, but if possible, on no more than one page. This short, single-spaced summary should describe the results, outcomes, or advantages the client will gain from your services. Although the abstract can mention your methods, it should touch on them lightly, emphasizing instead the gains the client can expect.

Table of Contents

The table of contents should interest and involve the reader in your proposal. Instead of a cumbersome list, write titles that will interest the client. For example, instead of using "Results" for that project section, write "Magnetics, Inc. Will Improve Profitability," or similar phrases describing your promised result. This list should be inviting and user-friendly; avoid jargon and words that may need defining.

Assurances

Some writers call this section the "boilerplate," and others call it the policy statement. Either way, this section explains your working conditions and policies. For example, your company may use this section to outline its cost-containment procedures or to explain that it is not involved currently in any controversial litigation. In this section, your civic organization might include its standard statement of nondiscrimination in hiring. Other proposals your firm has submitted are valuable sources for the types of assurances or procedures to include in this section.

Statement of Need

This statement should convince the client to hire you or your firm. Advertising specialists refer to creating a need, that is, convincing the public that a certain product is indispensable. Your proposal does the same through your clear presentation of your assessment of the client's needs. When possible, use numbers to convince the client that your assessment is accurate. Then, be sure to define a need consistent with the plan of action you offer. This section must answer the question, "Why does this client need your services so badly that he or she should be willing to purchase them?" That answer rests not with your description of competence, but instead with your accurate presentation of the client's unique problem that only you or your firm can solve.

Goals and Objectives

In this section, state the goals, that is, the guiding principles of your effort, and the objectives, the measurable work, you will do to achieve the goals. (See Figure 12.1 for detailed definitions.) For example, Bill Dorsey talks about the goal of improving the quality of life through soundproofing. This is a true goal because it is attainable, and probably immeasurable. Specifically, Bill Dorsey backs up his goals statement by attaching six objectives — six actions he would perform, such as analysis, research, and legal investigations — to his statement. All his objectives are results that the purchaser can own, within the eventual report that Bill hopes to sell with his proposal.

More to the point, this section creates the transition between the end of your introduction and the start of the middle of the proposal. At the beginning of the middle section, your goals and objectives will appear on a time line and plan of work. Introduce the reader to these ideas in the introduction so you can show a plan of organization next, and outline a plan of accomplishment throughout. Note again the proposal's built-in redundancy.

MIDDLE

Because the introduction of the proposal concentrates on persuading the reader to hire you, the middle should persuade by describing the plan and process that will make your services indispensable. This section should contain two illustrations: the flowchart that shows your plan of work and the timeline that shows when the phases of your project will begin, end, and overlap. The middle contains the narrative that outlines your goals and objectives, and reiterates benefits of hiring you.

The Flowchart

The sample student-designed flowcharts, Figures 12.4 and 12.6, serve two purposes: to illustrate the process as a collection of defined steps beginning and ending at specified times, and to define for a client the exact completion dates of the job's major components.

The students in our class were given a scenario from which to develop the flow diagram and timeline:

> You, as an independent contractor/technical communicator, have been
> asked to submit a proposal to conduct a training seminar which will help

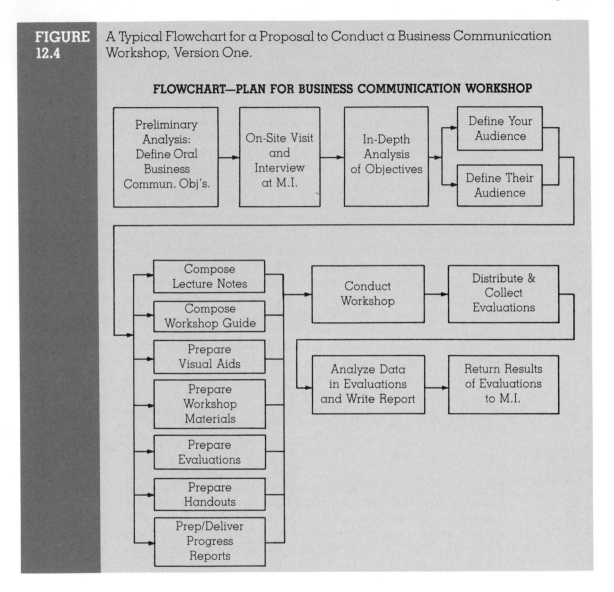

**FIGURE
12.4** A Typical Flowchart for a Proposal to Conduct a Business Communication Workshop, Version One.

FLOWCHART—PLAN FOR BUSINESS COMMUNICATION WORKSHOP

engineers improve their oral technical communication. You will have about five or six weeks to prepare for the workshop upon acceptance of your proposal.

Note what students did. First they used large rectangles to suggest the project's major components: preliminary analysis, on-site visit, in-depth analysis, and audience definition. They then used smaller rectangles to suggest ongoing activities that would be part of the au-

FIGURE 12.5	The Timeline for the Proposal to Conduct a Business Communication Workshop, Version One.

A SAMPLE TIMELINE
Tentative Schedule of Contract Project

Task	9/22–31	10/1–10	10/13–17	10/17–27	10/27–31	11/3–14
Initial Contact w/Co. (Proposal)	▓					
Needs Assessment (Or Get Interview)		▓				
Status Report of Co. (Specific Objectives)		▓	▓	▓	▓	
Final Preparation (Collect Materials)			▓	▓	▓	
One-Day Workshop					▓	
Report Project Competition (Evaluation)						▓

dience analysis: compose lecture notes, compose workshop guide, prepare visual aids, prepare in-session workshop materials, prepare evaluations, prepare handouts, prepare interim report. Then large boxes are used again for conducting the workshop, distributing evaluations, analyzing data, and returning results.

These students chose to include a detailed plan for audience analysis because so many activities of the project itself are included there. But they also could have omitted this section and shown only the major activities with large rectangles. In Figure 12.7, you can see another

FIGURE 12.6 A Flowchart for a Proposal to Conduct a Business Communication Workshop, Version Two.

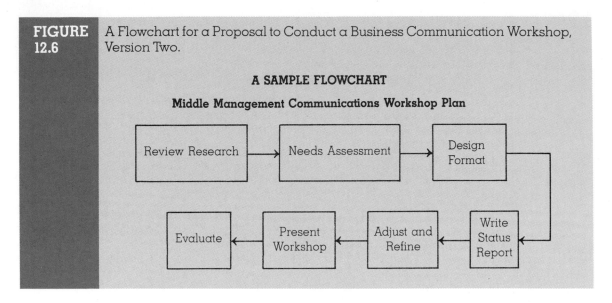

student-designed process chart for the same assignment. Note that they showed only the major activities of the project and a slightly different terminology. In fact, you have a great deal of choice in the type of flow diagram you design because you, the proposal writer, define your project's central activities. However, the flowchart must appeal to the reviewer's logical sense and conform to the rules of flowcharting. (See our graphics chapter.) In short, the flowchart's size and complexity should reflect the size and complexity of your plan of work.

Timeline

This illustration should display each of the flowchart's major activities, each activity's start and finish date, and the overlapping activities in your timeframe. Note that in the class assignment, the two groups of students designed quite different timelines, and both are correct. Figure 12.5, which accompanies the more detailed flowchart, uses blocks of time and shading to communicate each phase's start and finish dates. Simple horizontal lines in Figure 12.7 show the project's phases. Note also that these four illustrations differ markedly from the flowchart and timeline in Bill Dorsey's proposal to the Miskas. All three sets of illustrations are correct. These illustrations are more emphatic when color and shading are used to suggest similar activities or to highlight timeframes. The degree of detail used will vary according to what is appropriate for your audience.

Some conventions about the flowchart and timeline are worth reviewing.

1. These drawings are referred to generically as flow or process charts, process plans, timeframes, and timelines. Don't use any of these terms in the titles of your drawings. Instead of "Process Chart," write "Workshop Planning Phases" or something descriptive of the particular project at hand. Call your timeline a "Schedule for Workshop Preparation," for example.

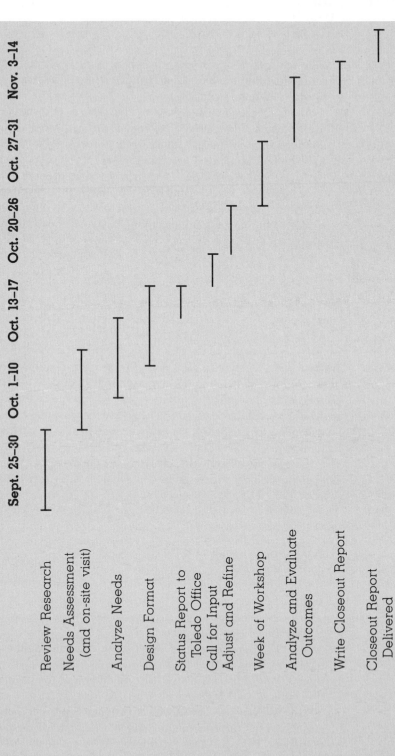

FIGURE 12.7 A Timeline for a Proposal to Conduct a Business Communication Workshop, Version Two.

A SAMPLE TIMELINE

MIDDLE MANAGEMENT COMMUNICATIONS WORKSHOP SCHEDULE

Sept. 25–30   Oct. 1–10   Oct. 13–17   Oct. 20–26   Oct. 27–31   Nov. 3–14

Review Research

Needs Assessment
(and on-site visit)

Analyze Needs

Design Format

Status Report to
Toledo Office

Call for Input

Adjust and Refine

Week of Workshop

Analyze and Evaluate
Outcomes

Write Closeout Report

Closeout Report
Delivered

2. Be sure your timeline is realistic. Nothing hurts your credibility more than a project behind schedule. Build extra time into your timeline by allowing some of the major phases to overlap where they can.

3. If others, such as your supervisor or co-workers, are involved with the project, show them the flowchart and timeline for review and discussion before drafting the proposal. You cannot meet your deadlines without your co-workers' cooperation.

4. Keep both illustrations simple. For example, if you propose a project with 20 phases, don't try to fit them in a one-page flowchart. Instead, prepare a major phase chart of the four most important activities and detailed charts of the less important phases which can be shown in appendixes. Keep your timeline to one page, too. You want to impress your readers with your organizational skills, but you do not want to intimidate with the project's size.

Narrative

This section includes the results, accomplishments, or benefits that will occur from adopting your proposal. This section should describe your methods, but certainly not how to do the job.

For students, the narrative is the proposal's most difficult section since they always wonder how to describe the task at hand without revealing how to do the job. The "Breakdown of Tasks" section of Bill Dorsey's proposal lists quite a few actions—a market survey, library research, legal research, and a package of designs—to be studied. But note that Bill reveals neither his definitive library sources nor his interviewee for legal information, and so on. Those are Bill's trade secrets, so he gives only a paragraph about each activity and the path he will follow to accomplish the task. Obviously, only he and his firm can accomplish the job he outlined.

Here are the objectives worked into the narrative for the oral technical communication skills seminar about which the students in our class wrote their proposals.

Plan of Work

A one-day training seminar for ten persons will consist of a lecture session, a demonstration, and the participants' involvement in four modules, with a lunch break. A summary outline shows that the workshop leads from the discussion phase, which will introduce the concepts to be stressed, to the actual participants' presentations to be analyzed and improved.

Session 1. The needs of middle managers, basics of oral technical communications, basics of persuasion and demonstrations.

Session 2. The requirements of visual supports including the appropriate uses of low- and high-technology aids.

Session 3. Workshop session for the preparation of the participants' three- to five-minute discussions.

Session 4. Workshop session in which participants present their discussions, receive support and advice on their technique, and self-evaluate. If the conditions permit it, the participants will be videotaped to improve their oral technical communication skills.

**FIGURE
12.8**   Motivational Appeals from the Advertising Industry.

MOTIVATING AND PERSUADING READERS

Here is a list of some human motivations, or needs. Although people make decisions for various reasons, the appeals listed here are the tried and true ones of the advertising industry. Can you think of ways to use them in your proposal?

Making Money
a. Improve job skills, increase pay.
b. Learn new job skills, obtain better paying jobs.
c. Earn outside, spare-time income.

Saving Money
a. Sales, closeouts, discount operations, factory-direct, introductory offers, pre-publication offers, charter member offers.
b. Longer wearing, less replacement cost.
c. Elimination of expense (do-it-yourself, reduce overhead, cut consumption).

Winning Praise
a. Improve homemaking, gardening, etc., skills
b. Receive awards, citations, certificates for generous or meritorious actions.
c. Get better grades in high school or college.

Helping Children and/or Family
a. Health appeals (food, vitamins, medicines, exercise).
b. Education.
c. Recreation.

Saving Time and Effort
a. Avoid physical drudgery.
b. Instant results, overnight benefits.
c. Carefree maintenance.

Impressing Others
a. Specific status symbols (possessions).
b. Indirect status symbols (memberships, certificates, awards, speeches, leadership, etc.).
c. Education and knowledge (traveling, speech, courses, degrees, books, art, music).

Having Fun
a. Traveling.
b. Family outings
c. Nightclubs, race tracks, movies, bowling, dancing, etc.

Self-Improvement
a. Physical (weight, muscles, hair, clothes, makeup, sex appeal).
b. Mental (education, philosophy, psychology, controlling others, influencing others).
c. Spiritual (religious, helping others, self-control).

**FIGURE
12.9**

Persuasive Appeals from Advertising.

ADVERTISING APPEALS PROPOSAL WRITERS CAN USE

You can learn a great deal about persuasive writing from the formulas used to write copy in the advertising agencies. Here are some old standards. Can you work them into your introduction?

A-I-D-A

This is the oldest of the formulas. AIDA simply means:
Attract attention,
arouse interest,
stimulate desire,
ask for action.

D-D-P-C

Your copy starts by being DRAMATIC, continues by being DESCRIPTIVE (of your product or service), continues further to be PERSUASIVE (about the benefits the reader receives if he or she accepts your offer), and ends by CLINCHING the sale.

PICTURE-PROMISE-PROVE-PUSH

This is one of the simplest formulas ever invented. It is good because you can picture it mentally.

You start by painting a word PICTURE of what the product or service will do for the reader. Then PROMISE that the picture will come true if the product is purchased. Offer PROOF of what the product has done for others. Finally, end with a PUSH asking for immediate action.

PICTURE—Headline or first sentence to evoke desire as well as get attention.
PROMISE—Definition of product or service. Success story.
PROVE—Testimonials and endorsements. Special feature. Statement of value.
PUSH—Action closer. Maybe a restatement rephrasing the headline.

Note that any manager reading this narrative has a very good idea of the nature of the service purchased with the seminar, but we have shown none of our materials, training films, or any part of the self-evaluation process that we teach. Since those items are proprietary, we do not distribute them with the proposal. In a short proposal, write one or two paragraphs about each activity that will lead to your objectives. In a lengthy proposal, write a page or more for each activity.

The narrative should also reassure the customer that he or she will be able to supervise you or the firm you represent. We speak here of the feedback mechanisms that keep the

FIGURE
12.10

How to Promise Your Proposal Will Bring Benefits.

PROPOSING AND DESCRIBING BENEFITS

A benefit is a favorable result, improvement, or gain that the customer can expect as a result of your services. The customer expects a benefit because you were hired. Here's how you can define and explain the benefits your firm offers.

1. Promise a benefit in your lead or first paragraph—**your most important benefit.**
2. Immediately **enlarge** upon this benefit.
3. Tell the reader **specifically** what he or she will receive.
4. Back your statements with **proofs** and endorsements.
5. Tell the reader what he or she **might lose** if he or she **fails to act.**
6. Request action **now.**

FIGURE
12.11

Six Questions That Test Your Persuasive Appeal.

TESTING YOUR PROPOSAL FOR LEVEL OF DETAIL

Do your proposal's persuasive and descriptive passages answer these six questions used by sales personnel? If not, your proposal might require more details.

1. What will you do for me if I listen to your story?
2. How are you going to do this?
3. Who is responsible for the promises you make?
4. For whom have you done this?
5. What will it cost me?
6. How do I order from you?

customer satisfied with your overall performance. You also can profit from this requirement by soliciting feedback so the project presents the customer with no unpleasant surprises. Here are some examples of mechanisms you can use to involve the customer in the project without inviting excessive customer interference.

Example 1.

For the oral technical communication seminar, we worked in the on-site visit to assess the corporation's requirements. At this meeting we asked the managers to tell us what they thought the engineers needed to learn. Two weeks before the seminar, we also presented a two-page report so the employer could review and comment on our final plans before the actual seminar.

Example 2.

> A client recently won a three-month contract to perform a program evaluation for a major military defense contractor. He was concerned that the plant manager would find his costs excessive, because no study results would be submitted for three months. He wrote into his proposal a clause that called for a two-page status report during the third week of each month. This report is more work for the evaluator, but with it he can solicit his manager's comments and prepare the manager for the investigation's results. In other words, his manager has a mechanism for supervising the expensive consultant he has hired.

Example 3.

> Bill Dorsey's firm provides a consultant weekly for an informal progress report on the noise pollution research project. These efforts, although time-consuming, create two-way communication that prevents the complaints of an inadequately informed customer.

At the end of your narrative, reiterate the primary benefit to your employer for hiring you. For example, Bill Dorsey points out that his firm offers an extra eight-hour special consulting session that no other bidder offers. The benefits vary with the nature of your practice or firm. Carefully analyze what only you and no other competitor can offer. Here are some examples of benefits.

Expertise. Two Ph.D.s who consult regularly refer to their extensive credentials in the theoretical aspects of their profession. Businesses love to hire Ph.D.s for this reason.

Experience. You can have a great deal of this even if you are young. Our young students often do internships with companies that purchase the latest robotics equipment. These students have that experience. Perhaps you have published an article or news story while attending school.

Fiscal Knowledge. If you have worked with budgets, mention it. Again, you may have acquired this background through your philanthropic or charitable activities. Nonetheless, you have the knowledge, and you should advertise it.

Reputation. Perhaps you work or have worked recently for a firm with a major reputation in your field of competition. Mention it.

Special Training. You may have attended a special seminar to obtain the latest instruction in your field—for example, a special training workshop in the development of industrial videos. If you have any such training, cite it.

These are just a few of the benefits you can offer an employer. You should be able to list many others if you interview co-workers and analyze your competitors. In the redundant method of the proposal, use this section to sell your competence once again to your prospective employer before directing his or her attention to your proposition's evaluation and costs.

CONCLUSION

Your conclusion should contain the following six sections: the evaluation plan and procedures, reporting and distribution plan, management plan, budget, personnel qualifications statement, and preliminary agreement to hire.

Evaluation Plan

A requirement in most proposals submitted to the Federal government, this plan should explain who will examine your work and how the work will be examined to assure accuracy and quality. For example, you may elect to bring in a team of external evaluators known for their expertise in your area. You may administer a test or survey instrument which measures the participants' responses to your objectives, as we planned for the oral technical communication seminar. The purpose of this activity is to allow the client to examine the quality of your services. This procedure shows that you are willing to be accountable for your efforts.

Reporting and Distribution Plan

Again, this activity suggests that you will be accountable for your promises. Offer the client or customer some sort of report associated with the completion of the project. This report should reiterate all the goals, objectives, results, and benefits associated with your services, and it should have a substantial appearance. The same guidelines suggested for the proposal format apply: good paper, heavy cover stock, typesetting if possible, and a binding. This report will be your customer's final impression, and when properly written and produced, can help convince him or her that your services were well worth their costs. Also offer to give a closeout group presentation for your supervisor and his or her associates using the best visual aids you can obtain. This proposal section should detail all these promises. Remember, the report is a product the customer actually receives as a result of your efforts, so it should be substantial in content and appearance.

Management Plan

This section demonstrates your ability to work closely with the client without disrupting his or her office system or staff. Your plan should include a description of the structure of your project, showing key positions, their definitions, their responsibilities and their position in a management structure. An organizational chart may strengthen your narrative. This section should also describe how you and your staff will work with the customer: What mechanism will be used for contact? Will you have weekly consultations, three on-site visits, and so on? This section should end with short statements affirming the qualifications of all the key personnel assigned to the project. These statements should emphasize the experiences that make these individuals desirable. Similar statements should be provided for all firms or contractors you agree to hire for assistance.

Budget

There are many ways to present budgets. Yours should conform to the specifications of your desired funding agency, if such specifications are available. Obtain help from the budget experts in your corporate or institutional office, so that the budget will reflect the true costs your firm will incur completing your project.

If bidding on a project as an independent contractor, determine your true costs, the profit you need to stay in business, and the costs you feel your customer is willing to incur. We know of writers, for example, who cover all of their operational costs in one hourly fee. We know of others who prepare itemized budgets that cover everything from halftones to airplane tickets, to printer ribbons. Here is a budget a consultant recently submitted with a winning proposal to write a user manual for a small business.

Budget

If you accept my proposal by February 1, I will deliver you the Service Manual for the Automatic Garage Door Opener Device, Model S-6000, as well as the copies that I will use to reproduce the manual, computer diskettes containing text and graphics, and any pictures or drawings that I have used by June 15. I have figured my professional fee for the project as follows.

Professional Services	$10,800.00
(360 hours × $30.00 per hour)	
Personal Expenses	$ 300.00
(office supplies, telecommunication, postage, etc.)	
Travel	$ 200.00
Production Materials	$ 295.00
(printing, paper, diskettes, ribbons, halftones)	
TOTAL	$11,595.00

Note the unintimidating quality of this simple budget that has only four line items. If the writer had developed this proposal for a state agency, for example, she would have needed many more line items and a series of descriptions for each line item. But this budget is appropriate for the proposal because its recipient, an owner-operator of a small business, has many jobs to do and a great deal of paperwork to complete. He wants to know what the manual will cost and whether he should proceed with the project. This budget answers his questions as it saves him time. Remember one important fact: the budget is the first thing the decisionmaker for your project will read. The person with the authority to accept or reject your proposal often reads the budget first to determine the project's affordability and then the narrative to determine the project's interest. Your budget must appeal to that individual's fiscal sense.

Personnel Qualifications Statement

This section should contain brief, standardized descriptions of all the personnel who will work directly with the client. A standard resume format, such as Figure 12.12, creates the impression of a competent team. Figure 12.12's format is fairly conventional. Perhaps your company or institution has a standard, required one-page format.

If you compose an informal proposal, a resume may be too lengthy. In an informal, memo-format proposal, we described an associate this way.

FIGURE 12.12

Here Is a Typical Format Companies Use When Their Proposal Requires the Resumes of Several Employees.

A TYPICAL PROPOSAL RESUME FORMAT

Name_____

Title_____   Dept._____

Years of Professional Experience (count college and graduate):_____

Prior Employment Record:

Employer	Position	Responsibility

Academic Achievements:

Institution	Degree	Dates

Publications:

Appeared In	Date(s)	Pages

Current Responsibilities:

Project Experience:

Honors and Awards:

Edward J. Gurnick, Training Support Materials, Dana Corporation. Specialist in high-technology visual supports, author of video scripts for training support and demonstration materials. Expert cost estimator of video scripting, planning, shooting, editing, and production.

With the resumes, you may choose to include any appropriate materials that advertise or promote your firm or services. For example, you can include a company brochure or an annual report. If an independent contractor, you may want to include a short (less than five pages) writing sample, or a list of satisfied clients who are willing to discuss your services. For this section, consider including any materials—provided they are short samples—that present your company's successes, its philosophy, and its approach to work. Long samples should always be in an appendix.

Preliminary Agreement to Hire

Appearing after the resumes and supporting materials, this statement, in the old sales jargon, "asks for the order"; that is, the preliminary agreement enables the reader to hire you upon finishing your proposal. Again, if you write a proposal with a dictated format, the application packet may include all the terms of your retention and the forms you must complete to be hired. But in business, the consultant often determines the terms and method of retention, and you don't want to appear unaware or afraid to negotiate these matters. It is a good idea to attach a statement the reader can use to hire you on the spot. This statement should be the proposal's last page and, ideally, perforated or at least marked so the client can complete the form to retain your services.

For example, one successfully funded proposal concluded with the following form.

EDITING AGREEMENT

I have read this proposal and would like to retain you to complete the editing described according to the terms outlined in Section Two of this proposal.

Please check one of the following:

_____ I have enclosed a $210.00 nonrefundable check as your retainer.

_____ A nonrefundable retainer of $210.00 will be mailed to you within the next ten days.

_____
signature of company's representative

_____
company

_____
date

This form should have a stamped self-addressed envelope attached so the reader can easily complete and mail you the form. Remember, your proposal's whole structure leads to your retention as an employee or as a provider of a service the customer needs. The form you design should be attractive, easy to use, and unambiguous. Experienced sales associates will tell you that new sales representatives tend to make the same mistake: they describe the product or service adequately; they arouse the customer's interest; they stimulate the customer's desire to purchase the product or service; and then they fail to ask directly and clearly for the sale itself. Avoid this mistake by ending your proposal with an orderly form that saves the client time in hiring you.

WHY WRITE SHORT PROPOSALS?

A long proposal such as Bill Dorsey's is an art form in itself, because each section has unique and mandatory content and style requirements. A long proposal has, typically, one objective—obtaining a contract for a major project involving months of work and several types of personnel.

You may also need to write short proposals for various reasons. Short proposals are used to make sales, to create records of decisions made informally in meetings, to confirm financial agreements, and for many other reasons. Because the short proposal, a document of about five pages, is so ubiquitous, it is hard to define. We suggest you follow the general guidelines stated here, but because brevity is the key strength of such a document, tailor the short proposal to its audience's requirements.

GUIDELINES FOR SHORT PROPOSALS

Typically, short proposals take the form of letters or memoranda.

Use a letter for a short proposal that you send to another company or institution. Write a letter if the proposal is for a sale. Bonnie Knapp's proposal shows how to use this form. On the other hand, you may write a memo, if a sale has occurred during discussions, and you write to firm up the details. For example, Leonard Horse hired Stan Kuzak to do his documentation during Stan's routine sales call, so Stan followed up with the memo-style proposal shown in Figure 12.15 (see page 295) to confirm the details of their agreement. Write a memo if the proposal will be distributed internally to another department of your company. For an example, see Figure 12.16 (see page 296), Matthew Gatliff's equipment request to the purchasing agent.

Organization

The short proposal uses a more subtle, less formal plan of organization than does the long proposal. The text often has a three-part plan of organization: 1) thesis, 2) rationale, 3) details. For example, Matthew Gatliff's confirmation or justification proposal follows this pattern. A sales proposal such as Bonnie Knapp's may also follow the thesis—rationale—details pattern. Her thesis is to persuade Mr. Kim to hire her; her rationale is that she will eliminate many time and cost complications for him. Her details are shown in the list of tasks on page two of the proposal. Stan Kuzak's confirming memo uses a two-part plan of organization. First, Stan's thesis confirms the agreements they have reached. Second, he lists the choices Leonard Horse will have. The simple approach is effective because the sale has already occurred; Stan is merely proposing the methods of producing the manual.

(Text continues on page 296.)

**FIGURE
12.13**

Bill Dorsey's Proposal.

A TYPICAL FULL PROPOSAL

PROPOSAL FOR A SOUNDPROOFING RESEARCH STUDY

Prepared for
Steven and Elaine Miska, Managers
Miska Insulation Company

by
Bill Dorsey
Research Associates

January 15, 1988

**FIGURE
12.13**

Continued.

Research Associates, Inc.

480 Pheasant Way • Bowling Green, Ohio 43402

January 15, 1988

Steven and Elaine Miska
Managers
Miska Insulation Company
856 Holland Sylvania Rd.
Holland, OH 43528

Dear Steve and Elaine:

I am sure that you have been following the progress of proposed county ordinance PL101.7. The ramifications of its passage seem critical to the contracting business. We, at Research Associates, have also been concerned with this new ordinance. Ours is the information business. Past experiences have shown that our information-gathering processes can add to our clients' understanding of new technologies. As these processes are combined with your production methods, you will be in a better position to capture an entirely new market—the home soundproofing market.

Please review the proposal we have prepared to meet your unique needs as owner/operators of a small business. I shall call you within the next two weeks, so we can discuss this offer in greater detail. I look forward to your response to this proposal.

Very truly yours,

William "Bill" Dorsey
Research Associates

FIGURE 12.13   Continued.

CONTENTS

**FIGURE
12.13**   Continued.

WILL PL101.7 CHANGE THE BUSINESS CLIMATE?

Proposed county ordinance PL101.7 presents a challenging opportunity to the Miska Insulation Company. Should this ordinance be passed, it appears that new markets for noise control structures will be created. In addition to the obvious challenge to contractors, an obligation also exists. Government will have initiated their legal obligation. Contractors will need to follow suit.

What You Must Know About PL101.7

The following questions seem relevant:

Will the ordinance be passed?
What are the potential markets?
Is the outlook favorable for the residential market?
Will the ordinance apply to only new structures or will existing
    buildings need to be modified as well?
Can Miska Insulation's existing facilities accommodate the new markets?
Will the costs for this accommodation be unreasonable?
How can we determine the best design and materials?

Public sources indicate that the answer to the first question is a definite "yes." We wish to demonstrate that a comprehensive study and report by Research Associates will place you in a position to answer the remaining questions and prepare you to be in the forefront of the subsequent markets.

What Research Associates Can Do About PL101.7

As my colleagues and I sat down to discuss our goals in this matter, we realized how closely they must resemble your own goals. The application of noise control devices will improve the quality of life for our entire community. As a more pleasant environment is created the public will acknowledge the part we have played. Our ultimate goal is to provide our community with safeguards against the plague of noise pollution.

RESEARCH ASSOCIATES CAN OFFER YOU SIX TYPES OF SERVICES

Some specific objectives can lead us to this quiet future. You may consider these objectives our responsibility to you upon the award of your contract.

1. We will give a history of noise pollution and a background of noise control.
2. We will provide documentation that establishes the need for noise control devices in our community.

1

FIGURE 12.13

Continued.

3. We will determine the scope of all markets.
4. We will supply market research to determine the fiscal outlook (i.e., how much will each market be willing to invest?).
5. We will outline present legal considerations and investigate possible future changes.
6. We will examine how your present facilities can be adapted to the new markets.

Obviously, the task before us is immense. Be assured that Research Associates is properly prepared to help you wage the battle against noise pollution. We are a small firm and can provide personalized service. More importantly, my partners and I represent a wide range of skills and experiences. Mr. Art Ellison will be your personal contact with our firm. A background in civil engineering makes him most qualified to meet your needs. More information about Art and our communications with you will follow.

Our Action Plan in a Picture

Allow Research Associates to share in the goal of noise control with you by awarding us a contract for a feasibility study, and only approximately five weeks will be needed to prepare the information that you will use. In addition, the ordinance is expected to become effective at about the time our work concludes. Therefore, there will be approximately one month from the time that you receive our report and the time that the market need will become evident. With our help, you can gain an advantage over the competition and join us on the leading edge of the new construction technology.

I turn your attention to the flow diagram included here. This chart plots the major phases that will control our contract. We believe that this strategy will be comprehensive yet cost-effective.

How We Do the Job for You

Research Associates studies five different areas of interest in the analysis of your special concerns.

Community Need. Assessing the need for noise control devices may appear futile. After all, if the ordinance passes, the devices will be required whether they are needed or not. We need to consider the following concerns, however. First Miska Insulation may be eligible for government loans to cover start-up costs for the new markets, and the government will require some type of needs assessment for the loan application. Second, the assessment will place us in a better position to determine the scope of all the markets.

**FIGURE
12.13**   Continued.

PLAN OF WORK FOR MISKA INSULATION

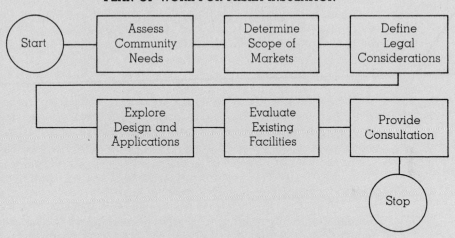

Scope of Markets.   We plan to unite library research with our understanding of the area to assess need and to determine the scope of the various markets. It appears that possible markets exist on the residential front, the industrial front, and the public transportation front. Only a market analysis can determine the depth of need in each of these areas. During this phase, we will discover how much each market will invest in the final product. For example, we may determine that the typical Toledo homeowner will pay no less than X dollars for a soundproofing insulation installation.

Legal Considerations.   The library research is useful in determining the legal considerations also. Some government and state regulations already exist with respect to insulation and sound barriers. Our study will review these requirements and tell you how to keep abreast of changes. As Miska Insulation conforms to these regulations, public trust will build. This will lead to more business, at the same time quality is assured.

Design and Applications.   Our next major step will be the exploration of design and product applications. Mr. Ellison will be invaluable here. He has already done preliminary research on sound-buffing designs. He can rely on his experience and draw from his connections in the engineering field. We are prepared to offer you exclusive rights to these designs should your contract be awarded to Research Associates. Mr. Ellison will be able to adapt these designs to both sound barriers and insulation. A summary of designs and applications can be requested with the final report for this project.

3

FIGURE 12.13

Continued.

Existing Facilities.   Finally we will evaluate your present equipment and training to prepare an analysis that will show how these facilities can be adapted to the future market. Mr. Ellison and I will conduct several on-site assessments of your facilities and processes to determine any necessary changes.

How We Use Your Time

Some specific activities and proposed deadlines are given in the timeline below. Each week from March 5 to April 5, Mr. Ellison will be available for a weekly progress conference upon your request. Or, you may prefer a weekly, written progress report. As you can see, we plan to submit the final report to you during the week of April 20-26.

Why We Offer You Personalized Consultations

We have made a tradition at Research Associates of giving our client an additional bonus. Note that we have made ourselves available during the week of

TIMELINE FOR COMPLETING MISKA PROJECT

	Time Period	Library Research	Market Survey	Background Analysis	Design Research	Engineering Consultation	Data Interpretation	Needs Analysis	Market Analysis	Legal Analysis	Design Analysis	Facilities Examination	Facilities Analysis	Report Submission	Consultation
March	5-8	x												x	
	9-15	x							x					x	
	16-22		x	x	x									x	
	23-29		x	x	x	x		x		x	x	x		x	
April	30-5		x	x	x		x	x	x	x				x	
	6-12						x	x							
	13-19														
	20-26												x		
May	27-3													x	

4

**FIGURE
12.13**

Continued.

April 27–May 3. During this week, Mr. Ellison will be available to you for an eight-hour consulting session. We have found two benefits in doing this. First, we are able to evaluate our work. Thus, each new client earns dividends from previous clients. Second, we are able to clear up any remaining questions you may have. In your case, the session may provide an additional asset. Mr. Ellison could use this time to discuss possible sales and marketing strategies with you.

RESEARCH ASSOCIATES' PERSONNEL OFFER UNIQUE SPECIALTIES

Many of the above steps could be completed by other communications firms. But, could they be completed as effectively? At Research Associates we believe the answer is "No." We keep our firm small but maintain a highly talented staff. In this strategy, we are cost-effective and efficient. We utilize interns to cover some of our library research, so Mr. Ellison and I are free to deal individually with each client. These benefits add up to a final report that prepares you for a challenging market.

A Glance at Our Staff

Please review the short background sheets attached. They offer you an overview of Mr. Ellison's and my experience. These informal resumes show that Art brings a solid engineering, theoretical background to our efforts. I have extensive training and authorship experience in technical and scientific communication. Although we may have the assistance of our interns with the library research, Art and I will be responsible for the entire project including the drafting of your report.

A Sample Cost Estimate

We have broken our budget projections for the final work into the following categories and amounts.

Item		Amount
Labor		$1665.00
Interns	$365.00	
Writer	$500.00	
Engineer	$800.00	
Travel		$ 150.00
Surveying Fee		$ 350.00
Engineering Consulting Fee		$ 300.00
Design Research Materials		$ 150.00
	TOTAL	$2615.00

5

FIGURE 12.13

Continued.

Terms: Contract amounts are payable by check in three monthly installments of $871.66 to Research Associates, 480 Pheasant Way, Bowling Green, OH 43402. We will submit cost estimates monthly by the fifteenth of each month, and checks must be delivered by the last day of each month.

These figures have enabled us to maintain our position as low bidders with our last five clients. We can assure that our figures are accurate, and we can offer, as well, our list of satisfied clients who are willing to discuss our fees and their evaluations of our services.

I shall contact you during the next two weeks, so that we can discuss our proposal. Such a discussion can lead to productive modifications of the plan of work we propose for Miska Insulation Company. Since time is of the essence in your obtaining a competitive lead, the contract will commence within two weeks of your letter informing us of your acceptance.

6

FIGURE 12.13   Continued.

APPENDIX A

Name: Arthur D. Ellison

Title: Engineer
               Research Associates, Inc.

Education: B.S., Civil Engineering, The University of Toledo, 1978

Experience: Art started with Research Associates three years ago. His previous
               experience with the Ohio Department of Transportation, his aca-
               demic standing at Toledo and his specialized coursework brought
               him to our attention.

Project   Four years of experience with ODOT and eighteen months with
Experience: Portland Cement have given Art a solid background in civil con-
               struction and noise pollution. He has worked on five noise reduc-
               tion projects resulting in extensive technical reports for corporate
               and small business clients. He is a chapter officer of the Society of
               Civil Engineers.

APPENDIX B

Name: William R. Dorsey

Title: Technical Writer
               Research Associates, Inc.

Education: B.S. Technical Communication, Bowling Green State University,
               1985.

Experience: After an internship at Westinghouse Corporation, Bill joined
               Research Associates in 1986 to develop noise containment projects
               for municipalities. He has written almost exclusively in this area
               for corporate and small business clients since joining Research
               Associates, authoring some eleven confidential reports on the
               subject.

Project   While attached to Research Associates, Bill has authored reports
Experience: for the Budd Company, Marathon Oil, Cains Snack Company, and
               the Jeep Division of Toledo regarding noise abatement ordinances.
               His article on the topic, "Winning Noise Abatement Proposals,"
               appeared in the February 1988 Technical Communication.

7

FIGURE
12.14 A Concise, Letter-Style Proposal That Confirms an Agreement Reached in
Discussions.

A SHORT LETTER-STYLE PROPOSAL

A PROPOSAL FOR THE TS-425 SERVICE MANUAL

prepared for
Younghee Kim, Vice President
Northwest Security Equipment, Inc.
P.O. Box 844
Arlington, Ohio 43517

by
Bonnie J. Knapp
Technical Communicator
266 Chestnut Street
Bowling Green, Ohio 43402

January 20, 1988

**FIGURE
12.14**

Continued.

266 Chestnut Street
Bowling Green, Ohio 43402

January 20, 1987

Younghee Kim, Vice President
Northwest Security Equipment, Inc.
P.O. Box 844
Arlington, Ohio 43517

Dear Mr. Kim:

After reviewing the materials you gave me, I believe I can provide you with a
manual that will save your company time and money. At the same time, you will
be able to provide your customers with a concise, usable guide for the installation, operation, maintenance, and troubleshooting of the Automatic Security
System, Model TS-425. Ideally, those twenty minute phone calls from frustrated
customers who are trying to communicate a part name and number will be
eliminated as a result of the drawings and the parts index.

The manual, which will set the standard for the others that you need, will combine text and graphics with thorough reference guides so that anyone using it,
whether it is your technician, the system's electrician, or an outside contractor,
will be able to find the problem quickly and easily and decide what to do about
it. This will eliminate unnecessary communication between your clients and your
company, thus freeing you to do other things. It will also save you money by
eliminating unnecessary visits to the institutions that purchase the system.

You will find my expertise as a technical writer and editor, as well as the professional resources I have available to me here in Bowling Green, invaluable in coordinating the project. I am accustomed to working quickly, but efficiently, under
the pressure of deadlines, and my organizational skills will provide the basis for
your project. My experience with the Macintosh computer, the LaserWriter, and
desktop publishing software will save you the expense of having the manual
typeset.

The work you have already done in producing your current manual is notable
and will simplify my job, which will include the following:

☐ Researching your current documents
☐ Extracting useful information and drawings
☐ Writing a concise manual that meets the literary standards necessary for
   clear communication
☐ Organizing the information in an easily used format targeting installation,
   operation, servicing, and troubleshooting

**FIGURE
12.14**

Continued.

Younghee Kim
January 20, 1987
Page 2

☐ Creating a usable parts index
☐ Developing an easy method for updating the manual
☐ Consulting your engineers for their technical expertise
☐ Establishing regular communication for your feedback

Not only will your customers find satisfaction in its ease of use, but also you will benefit in savings of time, frustration, and cost. That, of course, should help to make your job easier. Furthermore, the availability of such a manual will help promote the credibility and good reputation of Northwest Security Equipment, Inc. as it continues to expand.

I look forward to working with you on this project should you accept my proposal.

Sincerely,

Bonnie J. Knapp

Attachments: Flowchart
              Timeline
              Plan of Work
              Budget

**FIGURE
12.14**   Continued.

<u>Proposed Plan</u>

My main objective then is to provide a concise, organized, well-indexed service manual for the TS-425. I will achieve this objective by doing the following tasks:

☐ Assimilate and refine current information.
☐ Write new information according to the literary standards necessary for clear communication.
☐ Index information in an orderly, useful manner.
☐ Communicate regularly with you and your technicians.
☐ Format the information in a manner that appeals to your users.

I will produce the manual on a Macintosh Plus using the Apple LaserWriter to print text. The photographs and drawings that you provide will be incorporated in the manual as I find it necessary.

The manual will be no more than 150 pages long and will be bound in a three-ring binder for easy updating. I propose to use an updating format similar to the one used by the military. You will send full pages to your clients, dated, with a symbol signifying the changes you have made for that page. When you must add a new page or a drawing, you will be able to adjust the numbering system without redoing the entire manual. Consequently, you will have a boilerplate to which you can make changes efficiently unless you introduce a new product that is totally different.

The following Table of Contents reflects my perception of the manual's structure. I included only some of the subheadings to clarify certain areas. The parts index will be located under "Ordering Parts" and will index parts by name, number, and illustration.

1

FIGURE 12.14   Continued.

Table of Contents

Budget

If you accept my proposal by February 1, I will deliver the Service Manual for the Automatic Security System, Model TS-425, as well as the copies that I will use to reproduce the manual, computer diskettes containing text and graphics, and any pictures or drawings that I have used by June 15. I have estimated my fee for the project in the following way:

Professional Services 360 hours × $20.00 per hour	$7200.00
Personal Expenses (office equipment, phone calls, mailing costs, etc.)	$ 300.00
Travel	$ 200.00
Materials (printing, paper, diskettes, ribbons, halftones, etc.)	$ 250.00
TOTAL	$7950.00

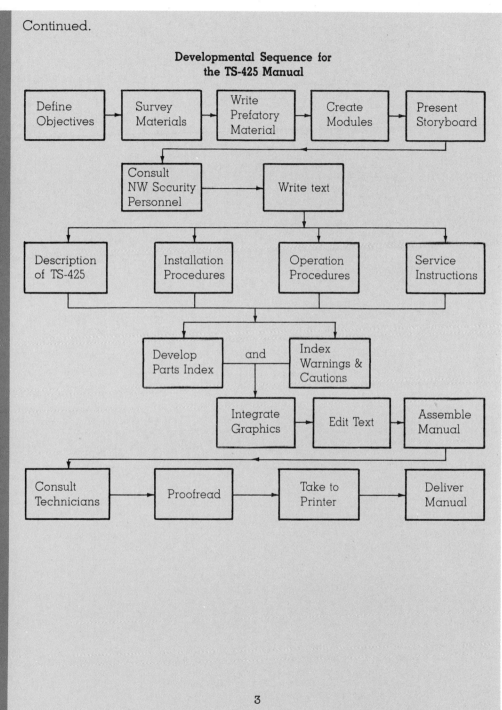

FIGURE 12.14

Continued.

Developmental Sequence for the TS-425 Manual

**FIGURE
12.14**

Continued.

Schedule for TS-425 Manual

	February		March		April		May		June	
Week of	1	15	1	15	1	15	1	15	1	15
Define objectives	▓									
Write prefatory material		▓								
Create module specs			▓							
Present storyboard				▓						
Write description of TS-425				▓						
Create pre-installation checklist					▓					
Write installation procedures					▓					
Write operation procedures						▓				
Write service and troubleshooting procedures						▓				
Develop parts indexes						▓				
Index warnings and cautions							▓			
Formulate evaluation section							▓			
Develop update procedure							▓			
Integrate artwork and graphics						▓	▓			
Edit text								▓		
Assemble manual								▓		
Edit and proofread									▓	
Take to printer									▓	
Deliver manual										▓

| FIGURE 12.15 | A Concise, Memo-Style Proposal That Confirms the Choices Discussed at a Business Meeting. |

A SHORT MEMO-STYLE PROPOSAL

TO: Leonard Horse
        Cornerstone Computing, Inc.
        Toledo, OH 43614

FROM: Stan Kuzak
        Technical Communications Specialist
        7512 Lock St.
        Maumee, OH 43537

DATE: October 17, 1988

SUBJECT: Three options for editing your manuals.

Here are the three options we discussed for the production of the reference manual for your new software. Please select one. I will call you next week to learn of your choice and begin writing the manual.

Option 1—Certain standards have evolved within the software application industry. Option 1 assumes your compliance with these standards so the end product would meet the highest editorial and production expectations. Please refer to the samples enclosed for examples. This option includes high-quality paper, offset printing, color illustration, professional binding, and the other trademarks of excellence, such as a professionally designed cover. Your key savings under this option would be my overseeing the production process at no extra cost.

Option 2—This option excludes professional printing and production where possible. We would use my computer, software, and printing capabilities to prepare the document. I would prepare camera-ready layouts. We would send the document to the printer for spot color, offset type for the cover, and binding. The advantage here would be cutting the cost of some of the production "frills" as we produce a professional-quality document.

Option 3—This option assumes your manual will remain in its present form—computer produced, letter quality, loose-leaf copy. I would perform what is called a "Level Two edit." Essentially, I'll check for errors of grammar and syntax, the matching of illustrations with text, and the matching of table numbers and chapter numbers.

FIGURE 12.16 A Memo-Style Proposal (Internal Distribution) to Justify or Confirm an Order for Equipment.

A CONFIRMATION OR JUSTIFICATION PROPOSAL

TO: Megan McCurdy

FROM: Matthew Gatliff

DATE: October 12, 1988

SUBJECT: Balance in Project Account for Communications Equipment

Thanks for your call informing us that the balance of our equipment money would be released on November 1. Although we placed our annual order before your July 1 deadline, we still need supplemental equipment. Each item we request is explained below followed by a summary budget.

Software: Our original request included both upgrades on our existing software and new desktop publishing software, which we have not completed purchasing.

Tables: To construct the workstations properly we need a sufficient number of tables of the right height and dimensions. We also need tables to create spaces for layouts and paste-ups.

Mice: Much of the new software we use is menu driven in ways that make keyboarding without a mouse cumbersome. We need to equip our writers and graphics personnel with mice.

Hard Disk Drives: Our complex software uses large amounts of RAM and disk space; we find it difficult to operate effectively without hard disk drives in the computers. We can improve our productivity by equipping selected workstations with hard drives.

Thus, in the short proposal, a simple plan of organization is best, because you write the short proposal to save a busy reader time in analyzing and solving a problem. Keep the message simple, and its details to a minimum. For example, Matthew Gatliff develops his proposal with only two types of details: definitions (his rationale) and an extended definition (his budget). He selects his material with great economy. After all, what does the purchasing agent care more about than budgets? Thus, Matthew Gatliff has selected the best persuasive strategy as well.

Tone

Speak directly to the reader, using the personal orientation defined in Chapter 10. In a sales appeal, use the respondent's name where possible. In the first paragraph, refer to your

<table>
<tr><td>**FIGURE**
12.16</td><td>Continued.</td></tr>
</table>

Laser Printer: As external printing costs continue to escalate, we must consider the benefits of a laser printer. We estimate that 75% of our publications can be produced in-house, thus shortening the payback period estimate of your March 4 guidelines.

Please note that our proposed budget allows $2730.00 to remain as an emergency fund. I will call you next week to see if you need further information before you place our order.

BUDGET

Quan.	Item	Unit cost	Cost
Software			
1	WordStar 2000+ update (release 2.0 to 3.0)	$ 240.00	$ 240.00
1	Word Perfect update (release 4.2 to 5.0)	100.00	100.00
1	Ventura Personal Publisher 1.1	895.00	895.00
Office Furniture			
12	Tables 36" x 72"	85.15	1021.80
Hardware			
14	Bus Mouse (serial port)	82.50	1155.00
5	Seagate HDD internal 30 M	320.00	1600.00
1	Hewlett-Packard Laser Jet II	988.00	988.00
	w/2 M memory board	746.25	746.25
	and cables	40.00	40.00
	TOTAL		$6734.25

previous contact with the recipient if space allows. For example, Bonnie Knapp refers to her previous positive conversation with Younghee Kim as does Stan Kuzak with Leonard Horse. This technique reminds the reader that he or she, not you, the author, is the decisionmaker.

Design

Place the narrative or text at the front in a letter or memo. Place the graphics and "extras" afterward either at the end of the communication, as Matthew Gatliff does, or in attachments, as Bonnie Knapp does. This arrangement reinforces the rhetorical strength of the short proposal — its brief, efficiently delivered message — so remember to separate the message and the text. Finally, plan your illustrations so that they complement the text. Note that uncomplicated, one-page or half-page illustrations appear in each short proposal shown here. Avoid multiple-page or difficult-to-read illustrations. Concentrate on the flowchart, timeline, list of tasks, and, above all, the budget.

Because proposals are persuasive, make your short proposal attractive. If you compose a short sales proposal, place an attractive, heavyweight cover on it as Bonnie Knapp did. Also, use the technique of "chunking" throughout your text to break the communication into short, easily read segments. Use visual cues such as boldface type, italics, and hanging indentations to emphasize the proposal's strategic sections. The visual cues are at least as important as the content. After all, the visual cues help to lead your reader through the short proposal to the important decision or authorization you hope to obtain.

SUMMARY

In this chapter, you learned how to design formal and informal proposals for business and not-for-profit institutions. You have learned that certain proposals—for example, those of the Federal government—have a prescribed format, and that others—for example, certain business proposals—have highly individualized formats. You have learned that proposals are solicited and unsolicited. After studying this material you should be able to:

- Read an RFP so that you can tell whether a proposal is to have a prescribed or individualized format.
- Show an understanding of your proposal's origin—solicited or unsolicited.
- Sketch a flow diagram of your proposal's plan of work.
- Sketch a timeline tied to your flow diagram.
- Select your persuasive and descriptive appeals.
- Draft a set of goals and objectives.
- Draft a needs assessment for a prospective client.
- Write an appeal that makes a prospective client want to hire you.
- Draft a narrative explaining your plan of work and how it will benefit your client.
- Compose an appropriate budget for your proposal.
- End the proposal with a form that results in your retention.
- Draft and revise short, letter-style proposals.

ASSIGNMENTS

1. Investigate the operation of your college's grants or proposals office. Collect samples of the forms and RFPs they have. Write a report to your instructor about your investigation.
2. After visiting your school's proposal office, select and redesign the form you understood least to improve its readability and ease of completion.
3. Find an individual in your community who has written proposals funded by the state or Federal government or by a foundation or philanthropic institution. Ask that writer what made his or her proposals successful. Write a report to your classmates about what you learned regarding proposals from this person.

FIGURE 12.17   A Short Proposal in Need of Revision.

PROPOSAL NEEDING REVISION

A Proposal: Cooperative Education for the Community

The Cooperation for Employment Program (CEP) has been designed to provide the school systems in Davenport with a progressive system of instruction affording its citizens "the opportunity to improve their own quality of life and contribute to the overall economic growth of the metropolitan area."

The Retail Education Program can be conceptualized within the framework of the Universal Cooperation Model (Lincoln, Nebraska Project, 1984). All systems (educational, social, and technological), are comprised of inputs, processes, and outputs. These components can be identified in the Retail Education Program: the input has been provided by the concentrated effort of the Retail Education Program Curriculum Committee to establish the program's goals; these goals have been processed into a scope and sequence. Subsequently, the necessary program modules were developed to insure successful implementation of the course, Retail in the Shopping Mall. While it is premature to fully realize the output, the thoroughness of the plan supports positive results.

Any system, regardless of its environment, should contain a fourth component, feedback, in order for the system to continually adjust to changes within that environment. Evaluation of the system through the feedback component provides an opportunity for modification and redesign insuring constant alignment with community needs.

For this reason, we propose an evaluation of the Retail in the Shopping Mall course. The evaluation process will provide data to insure consistency with the purpose of the coursework—providing a viable foundation and tracking for specific retailing activities at the high school level.

The introduction of courses in Cooperative Education continues to be a radical and innovative approach in the evolution of Business and Marketing curriculums. Consistency between the philosophy and content has eluded many programs with the noted exception of the New Jersey Plan, the California Model, and a derivative, the Cooperation for Employment Program. Consequently, in conjunction with the analysis of Retail in the Shopping Mall, we propose to investigate the question, "Is there a measurable gain in the sales ability of students in the Cooperative Education course, Retail in the Shopping Mall, as compared to students in traditional sales and marketing courses?"

4 Find a topic related to your community, about which you think someone should develop a proposal for funding. For example, one student of ours has written a litter control proposal that was funded. Compose a flow diagram, timeline, and checklist such as ours for your project.

5 Investigate funding agencies for your proposal. Write a memo-report to your instructor explaining how and where you will try to obtain funds for your project.

6 Write the first draft of your proposal for submission to your classmates for review and suggestions.

7 Assume your firm agreed to hire Bill Dorsey during a recent meeting. After studying Figure 12.13, his full proposal, draft a short, confirming version of it for your instructor to review.

8 Figure 12.17 is a proposal needing rewriting and attention to its style. How would you revise it? Revise or rewrite the proposal. Submit your version to a classmate for discussion and review.

CHAPTER
13

Short Reports

This chapter introduces you to the organization and composition of short reports, specifically progress and overview or short summary reports. After reading this chapter, you will be able to:

■ Use a generic outline for a short report to draft a progress report, a short summary report, or a recommendation report.
■ Compose a short report in either of two formats.
■ Use a checklist to review and improve your report.
■ Discuss the uses, strengths, and limitations of short reports.

CASE STUDY   **A Consultant Writes a Progress Report**

Carlos Andrade, a statistician with a firm of training specialists, Consultec, Inc., must supervise and complete the external evaluation of the training program that Evco Electronics has conducted, using a grant from the federal government.

As a gifted, well-educated program evaluator, Carlos Andrade is confident with his assignment—despite the hard work involved in transferring the hard copies of the corporation's training records to usable collections of data for a computer's statistical analysis. But he has never become comfortable with the technical reports he must compose to continue and complete his work according to his firm's contract. For example, for this external evaluation he must submit a progress report for each two-week period, a book-length closeout report, and a summary of the closeout report for distribution among Evco Electronics' managers.

Because of the densely technical nature of the subject and the mathematical method of analysis, Carlos Andrade has never worked well with a hired writer. In the end, he usually composes and edits his own reports because very few editors know the particular types of abbreviations a statistical analysis requires.

In this chapter, you will study the first progress report Carlos Andrade writes to fulfill his firm's contract with Evco Electronics. In the next chapter, you will examine the longer report, Andrade's summary of his book-length closeout report. As you examine these and other documents, we will explain the requirements and challenges of writing technical reports in a business environment.

THE USES OF SHORT REPORTS

Short reports—that is, documents of five pages or less—facilitate communication in a corporation, because their length encourages flexibility. Typically, short reports create and maintain a record of progress, confirm oral or informal transactions, or announce the immediate details of a policy change.

A Record of Progress

Carlos Andrade's progress report is a typical example of an effectively used short report. He composes a two-page "self-report" of his training evaluation activities for each two-week period. The report enables him to inform the Evco Electronics vice-president about the actual work the evaluation process may involve. He can use the reports to educate the client about the evaluation process the company has purchased. He can reassure the client about the normal activities of the evaluation, for example in this case, the project's schedule. He can also prepare the company for any possible changes or problems. His progress report makes a request for the data processors he needs to maintain the project's schedule.

A Confirmation of an Informal Transaction

Short reports in a memorandum format create records of informal transactions. For example, a manager may write a short report of a conference with an employee after a routine performance evaluation. Depending upon the company's policies, the manager may include that report in the employee's personnel file.

Similarly, an employee may draft a short report to remind a supervisor of a decision he or she has discussed but not confirmed. The author of the report on decision support systems shown in this chapter shares very specific assumptions with the reader, an instructor and consultant in decision support systems. Thus, the report commits to writing previously discussed concerns about Applia Corporation's decisionmaking process. The supervisor and employee have a record of their mutual concerns and goals for their project.

An Announcement of a Change in Policy

Companies routinely use short reports to announce policy changes to thousands of employees. For example, a company may announce a change in its employees' health insurance benefits in a memo report that first states the news and then lists the new policy's terms. The report enables the company to communicate the new policy's major points in an easily readable, cost-effective manner.

THE ADVANTAGES OF SHORT REPORTS

Short reports offer four distinct advantages.

Quick Reading

Because the short report is no more than five pages, it is quickly read. Therefore, the short report is the natural means of communication for decisionmakers, who haven't much time to study problems. Short, easily read reports also provide natural channels of communication for mass distribution, as in the case of announcing a change in a company's benefits policy. When the audience analysis reveals that the audience does not want to read a lengthy document, distribute the information in a short report.

Quick Record

The short report enables you to create a record quickly by using a fairly informal channel of communication. Thus, if members of a communications group decide to add new information to a manual under development, a memo-style short report can record the decision quickly. The group, then, is not delayed by the need for many levels of approvals for the change. The record allows new employees or temporary personnel assigned to the project to read about and incorporate the change as they study the project's files.

Obtain Action

As Carlos Andrade's progress report shows, the short report creates an opportunity to obtain action or request a decision that enables you to maintain a production schedule. Use short reports that synthesize and define a problem to outline a proposed solution and to request a specific decision or an authorization for special activities. A short report may explain why extra personnel must be hired or describe why an external vendor must complete a particular printing job.

Contain Expenses

Even short reports require time for their planning and composition. In this sense, reports are expensive because they are labor intensive. Thus, the shorter the report, the more you can cut expenses. For this reason, consider using standard formats and forms. The standardized form used, for example, in all travel reports helps you save time on the design and planning of the finished document. Finally, there is the issue of cost-effective distribution. Short reports might be distributed on an electronic mail system. Or, if hard copies must be prepared, the short report saves money on paper, copying, and mailing.

WHAT ELEMENTS BELONG IN A SHORT REPORT?

This question is difficult to answer because a short report should address its audience's informational needs. Because you have limited space, you must be precise. Here are some conventional report elements that you should try to include either directly as individual sections or indirectly in your overall plan of organization.

Purpose. Why are you writing? State your thesis and plan of organization in the first sentence of the first paragraph. Don't keep the readers guessing. Both Carlos Andrade's and Ruth Silk's first sentences are thesis statements.

Call for Action. Also called a request for a decision, your report should, typically, ask the reader to make a decision. Examples of decisions are in all three reports: Carlos Andrade wants more temporary employees, Darlene Deitrick asks for a review and comments, and Ruth Silk indirectly asks the reader to practice soil conservation. What do you want your reader to do? You must ask, and furthermore, why are you writing a report if you don't need something? For strategy recommendations about the call for action, study the next section of this chapter.

Summary. Sometimes archaic or difficult to work in, the summary is still a report requirement. In a short report, your summary should be no longer than one or two paragraphs, giving an overview of the entire report.

History. Usually this section describes the events or conditions that prompted a report. How was soil erosion treated in the past? What were last year's budget concerns? Why did a decision support system become necessary? This section should give the problem a background and a context.

Body. This section is the actual report in the present tense or the continuous past tense. This section lends itself well to all the methods of rhetorical development, but particularly well to cause and effect, comparison and contrast, and spatial organization.

Recommendation. Again, placement of this section depends on the method of organization. Although your recommendation can be less specific than your call for action, readers respond best to a well-detailed statement. Don't recommend, for example, cost-effectiveness. Instead, recommend that the reader reduce travel expenditures, entertainment expenses, or something he or she can visualize.

HOW SHOULD YOU ORGANIZE A SHORT REPORT?

We deliberately phrased these section titles as questions. In the technical communication profession there is little consensus on how to design a short report. Before considering any specific guidelines, understand two important views of this matter.

View One—The Traditional Method of Organizing a Report

Classroom and business customs have called for a short report with a chronological, spatial, or cause-and-effect pattern of organization. After explaining the problem or the progress, the report then calls for a decision or an action. For example, Carlos Andrade's short report uses a spatial pattern that leads to a request for temporary help. Thus, Andrade's sentence outline might look this way:

> Memo Topic: We are making good progress.
> 1. A summary statement suggests we will end on schedule.
> 2. An interviews section states we have interviewed Dayley, Dimplick, Martinez, Hoby, Solomon, Kane, and Werk.
> 3. A data processing statement says we obtained help from Mason and Appleby.
> 4. A conclusion calls for temporary help to maintain the schedule.

Since Andrade's method of organization is perfectly acceptable for a very traditional audience, you could safely imitate his approach. However, his plan of organization has one drawback. You would have to read Andrade's entire report before understanding that he needs your action—the authorization of temporary help to stay on schedule. What if you fail to read the entire report? Because readers usually scan or read selected sections, Andrade risks losing his client's cooperation, even though his report is quite correct.

The Newer Method of Organizing a Report

Now study Figure 13.5, Andrade's reorganized report. Note that he has placed the request for temporary help right below his statement of purpose. This strategy assumes that Franklin Berry will see the request sooner and will grant the request because he can learn at the start of the report what Carlos Andrade expects him to do. Figure 13.5 would draw upon this outline.

> Memo Topic: Your evaluation is progressing well, but you must authorize temporary help so we can finish on schedule.
>   1. A summary statement suggests we will end on schedule.
>   2. An interviews section states we have interviewed Dayley, Dimplick, Martinez, Hoby, Solomon, Kane, and Werk.
>   3. A data processing statement says we obtained help from Mason and Appleby.

Advocates of this more reader-centered plan of organization suggest that the strategy involves the reader at the start, and thus encourages busy readers to cooperate. If writing for an action-oriented audience, consider this pattern of organization. This approach immediately involves the reader in the action and your report's results.

HOW TO WRITE EXCELLENT SHORT REPORTS

This chapter describes short reports' contents and organizational features. However, the short report's brevity requires the writer to make bold decisions about each document's scope and approach. These guidelines, then, can be useful.

Content

In a conservative context, use the traditional plan or organization. For a livelier document, try the newer method. Your report should contain as many of the elements we have named as your content can allow. The unique nature of your topic may require you to omit some of the elements we have defined, and if so, provide an adequate level of detail within the sections you retain.

Headings

Use headings freely in either the traditional or the newer form. Remember, do not use generic names such as "body" or "call for action." Use headlines to suggest how the reader should react to the text. For example, instead of "summary," use, "How the Data Have Been Stored."

Appearance

Any time you are uncertain about the appearance requirement, prepare a title page on heavy paper. Arrange the sheet as you would for a long report.

Design

Prepare progress and business reports for internal distribution as letters or memoranda. As you use the correspondence form, remember to include a complimentary close, signature, and list of attachments. Prepare summary reports and review studies such as Ruth Silk's for a larger audience with various reading levels. Use third-person narrative for descriptions, and move to second person, imperative mood if you must include instructions or procedures.

FIGURE 13.1   A Checksheet for the Writers of Short Reports.

CHECKSHEET FOR A SHORT REPORT

Use this checklist to complete your short reports. Modify it to suit your requirements. (For example, most students forget to initial their memoranda until they have used the checklist a few times.)

Appearance—did you

Include all the titles in your heading or in your cover memo: Memorandum, To, From, Subject, Date?
Initial the memo correctly, to the right of your name?
Use simple headings to divide the text and create white space?
Proofread for spelling and punctuation?
Proofread the text for factual accuracy?
Edit the report for the inclusion of all attachments?

Introduction—did you

Write an introductory sentence that gives an overview?
Use the second sentence to define the scope of the report?
Include a sentence that names the details the report discusses?

Summary—did you

Refer to the time the report covers?
Write a sentence that reviews your accomplishments for this period?
Write a sentence that informs the reader of the problems you have defined?
Alert the reader to any actions you will ask him or her to perform?

Body—did you

Refer to the continuing problem or investigation that required your report?
Enumerate your accomplishments for this period?
Define any problems you have encountered?
Outline solutions for any problems you have encountered?
State as briefly as possible the reasons for your conclusions?

Conclusion—did you

Ask for action as simply and clearly as possible?
Explain why any action you request is absolutely necessary?
Include all the paperwork that will make it easy for the reader to respond to your request or to authorize your expenses?

**FIGURE
13.2**

A "Generic" Outline for Short Reports.

A SAMPLE OUTLINE FOR A SHORT REPORT

All short reports share certain common elements of organization. For example, all short reports contain an internal summary very near the start of the report. This outline stresses the elements that progress, summary, and other short reports have in common, so you may modify the outline to meet the requirements of your reports.

Introduction (one paragraph—preferably a short one)
    Lead sentence states topic of report and type of report
    Other sentences state
        Time covered by the report
        Scope of this discussion
        Plan of organization of this report
Summary (one paragraph—preferably a short one)
    Lead sentence states accomplishments of "positive" information
    Other sentences state
        General problems anticipated
        Crucial sections of the report
        Action or decisions which may be required
Body (several paragraphs—you have flexibility here)
    Lead sentence mentions current problem you work on but have not yet solved.
    Subsequent sentences or paragraphs list all the work you have completed.
    Separate paragraph describes any unique problems you have discovered.
    Include here the solutions you propose for the problems you have investigated.
Conclusion
    Call for action or authorization if needed
    Explanation of action needed
    Attachment of all forms required for action

Illustrations

Attach line drawings, photographs, graphs, and tables at the end of the report. Do not waste valuable narrative space by placing them within the body of the text. Carefully select the visuals so you retain the prominence of the text and the call for action.

Type

A review copy of a report for comments and revisions should be prepared as a manuscript— that is, double spaced or on a computer at three lines to the inch. If preparing a final copy, single space the paragraphs in either full block or modified block style as your instructor or employer requires. If the report is single spaced, do not crowd the page with type. Use "chunking" techniques to create blocks of readable text and allow enough white space to frame the text.

DARLENE DEITRICK'S PROGRESS REPORT

Darlene Deitrick, the author of this report, wrote it as the leader of a group of business students assigned to develop a decision support system for a corporation. This report describes the group's initial efforts with the project, because they had to obtain their professor's approvals for each phase of the effort.

Format

Darlene Deitrick divides her report with very explicit headings: purpose, tools and techniques, findings, and recommendations. These headings allow a supervisor to read the most pertinent sections first, or to avoid sections that do not apply to his or her area. The use of obvious labels divides the text into readable "chunks," and the plan of organization acts as a time-saving mechanism.

The author has chosen to compose the overall report through the process of attachment. That is, she writes a memo to introduce the report. The report, a self-contained document, is attached to the memo, and the questionnaires appear as an appendix. Because each section is free-standing, Darlene Deitrick enables readers to flip to the sections of interest, instead of making them wade through a complicated narrative for the important information. This format is inviting to the reader.

Content

Time is this report's basic theme: how much time the author's team needs to analyze their problem, and how much time the managers need to analyze their sales data. Darlene Deitrick divides the discussion into two parts. The cover memo, part one, outlines the time needed to complete the project. The timeline is illustrated on a Gantt chart.

The report uses a pattern of increasing complication to explain its topic. In the first two sections, Darlene Deitrick outlines the group's problem and their method of analysis. She then uses the "Findings" and "Recommendations" sections to create a greater level of detail for the most important issue — a solution for the time managers spend analyzing overly complicated data. Thus, Darlene Deitrick's emphasis is correct because more than half of the report discusses the problem, while the earlier, shorter section creates its broad framework.

Darlene Deitrick wisely enumerates her group's recommendations because researchers have demonstrated that readers learn more quickly (and retain better) when information is divided and enumerated, rather than presented in the traditional prose narrative form. Since the recommendations are the most important section of the document, Darlene Deitrick has emphasized the right material.

Improvements

Darlene Deitrick might enhance the "Findings" section with more obvious enumerative techniques. For example, she could list and number the problems the group encountered. She could also number and list in a column the results of the questionnaire that are now shown in a traditional narrative. The use of extra white space and the listing would improve the report's readability.

**FIGURE
13.3**

Progress Report.

TO: Alice Philbin

FROM: Darlene Deitrick

DATE: February 25, 1988

TOPIC: Outline for independent study project.

The preliminary analysis report and feasibility study report are complete (pending changes that may be made due to system changes).

I am including with this report a Gantt Chart that shows the schedule for our system development project.

I expect to have the summary of the subsystem function and narrative of the functioning program completed by April 16, because the coding and testing will be complete by April 2 and these two documents are a result of that phase of the development project.

The Operator's Manual, User's Guide, Implementation, Conversion, and User Training Plans will be completed during the weeks of April 16–April 30.

I will submit all materials to you on May 1. I plan to submit all materials to you in the same form as we submit our project to Dr. Yoo, so that you will be able to see how the documentation works within the overall system.

GANTT CHART

Feb. 3   10   17   24   Mar. 3   10   17   31   Apr. 2–30

```
PPPPPPPPP
        DDDDDDDDDD
                 CCCCCCCCCCCCCCCCCCCCCC
                                      TTTTTT
                                            WWWWWWWWW
```

KEY: P —Planning and Analyzing—February 3–10.
          D —Designing (System outline, subroutines, spec)—February 10–24.
          C —Coding—February 24–March 17.
          T —Testing—March 17–31.
          W—Writing (Implementation and user training procedures, operation manual, user's guide)—March 31–April 30.

FIGURE 13.3   Continued.

<u>Preliminary Analysis Report</u>

<u>PURPOSE OF THIS REPORT</u>

The purpose of this report is to state the facts obtained in studying the old decision making system at Applia Corporation. The study was done in response to Mr. Washington's and Mr. Lincoln's request for a DSS. This report also presents alternatives to the present decision making system used at Applia Corporation.

Mr. Washington, Vice-President of the Home Appliance Division of Applia Corporation, and Mr. Lincoln, Director of Marketing Operations, have requested that the feasibility of creating and implementing an on-line, interactive decision support system be studied. Managers at Applia Corporation are finding it increasingly difficult to make decisions involving product mix, introduction of new products, investments in high-tech equipment, demand forecasting, and cost analysis. This report is the result of the initial investigation of the present system and brainstorming for alternatives to the old decision making system.

In this report, the reader will find discussions of the tools and techniques used to complete the initial study, the findings about the old decision making system, and a list of alternatives to the old system.

<u>TOOLS AND TECHNIQUES</u>

We sent a questionnaire to the managers in the Home Appliance Division. The returned questionnaires are included in this report as the Appendix. Basically, the questionnaire's purpose was to discover the present system for decision making and discover how better to satisfy the needs of the managers who will be using the proposed new system in their decision making. In addition to distributing and considering the responses to the questionnaire, this system study group has researched the company and its competitors through various books, magazines, and corporate literature in order to get a better idea of the features that should be included in a DSS for a company like Applia Corp. This system study group has also examined the reports that the managers of the various departments have said they use to make their decisions. We have also observed the managers as they use the data and reports to make their decisions and plans.

<u>FINDINGS</u>

In the current system of decision making, the managers at Applia Corp. obtain data from individual departments (Manufacturing, Marketing, Research and Development, Accounting, and Personnel). The managers then make their decisions based on that data.

One problem with the current system is that it is difficult and time-consuming for the managers to interpret the raw data that they obtain from the various departments. The managers have to use modeling techniques on the data without the aid of a computer, or they must get their subordinates to model the

**FIGURE
13.3**   Continued.

data for them and create reports from the modeling process. In addition, managers often obtain irrelevant data, and sorting the relevant data from the irrelevant data is also time-consuming.

Managers are constrained by the time they must spend with the data from the individual departments. They cannot make as many decisions per unit of time as they could if the information they need were in an appropriate form, and they cannot make decisions as quickly as they need to in the increasingly competitive home appliance environment.

In reviewing the questionnaires, we found that the areas managers would most like to have decision support in are: product mix, introduction of new products, purchase of high-tech equipment, demand forecasting, and cost analysis.

It is the opinion of this analysis group that centralizing the data the managers are now getting and putting that data into one database is most desirable. Once the data are brought together, the data can be processed and the types of support and reports that the managers have asked for (see above paragraph) can be made available to the managers.

RECOMMENDATIONS
After analyzing the current system, we suggest the following alternatives to the present decision making system at Applia Corp.

1. Offer the managers a database which they cannot change, but which is updated monthly by the DP department, and an interactive system for decision making that includes menus and prompts to ask the manager the type of decision he wishes to make.
2. Offer the managers a database which they can change and an interactive decision making system that includes menus and prompts to ask the manager the type of decision he wishes to make.
3. Purchase database, spreadsheet, and statistical analysis packages and modify those packages to suit the needs of Applia Corporation managers.
4. Use the Delphi Method or Nominal Group Method with a consultant on these methods present.
5. Leave the decision making system as it is.

FIGURE 13.3   Continued.

APPENDIX

QUESTIONNAIRE

Name   G. Washington

Department   Vice-President of Home Appliance Division

1) Where do you obtain your data and information on which to base your management decisions? Please be specific.

   I obtain my data from the individual department managers at Applia Corporation's Home Appliance Division. I obtain information on the rest of the corporation from corporate reports and meetings with my superiors. In addition, I obtain information about the environment external to the company from trade journals, newspapers, and magazines.

2) Do you find that these sources are sufficient to satisfy your needs for information on which to base decisions? If not, why are these sources insufficient?

   I do not find that these sources are sufficient to satisfy my information needs. I obtain far too much irrelevant data, and I have to spend much of my time sorting through the data to find information that pertains to the decisions I need to make and that I need to approve. The loss in fourth quarter profits due to the poor decision making with regard to the marketing strategy that we implemented in 1987 is of great concern to me.

3) In what areas are you most interested in having decision making support? For example, product mix and demand forecasting.

   In view of the problems that we had with our marketing strategy, I believe that summarized reports and decision support in the areas of product mix and introduction of new products are called for.

Comments

   Is it possible to have a system developed that would allow me to see reports that would help me make decisions in the above areas? This is what I would like to see in the Decision Support System.

**FIGURE
13.3**

Continued.

<u>QUESTIONNAIRE</u>

Name   A. Lincoln

Department   Marketing

1) Where do you obtain your data and information on which to base your management decisions? Please be specific.

   I obtain my data from individual department reports such as R & D reports and Manufacturing reports. I also obtain information by talking to other department managers.

2) Do you find that these sources are sufficient to satisfy your needs for information on which to base decisions? If not, why are these sources insufficient?

   No, I do not find that these sources are sufficient to satisfy my information needs. I obtain far too much irrelevant data, and I have to spend much of my time sorting through the data to find information that applies to the decisions I need to make. Applia Corporation lost profits in the fourth quarter of 1987 due to the poor marketing strategy based on decisions made in my department. The Home Appliance Division at Applia Corporation lost 15% in earnings due to the marketing strategy we implemented in 1987. When the Home Appliance Division contributes 38% of Applia Corporation's overall earnings, a loss of 15% due to poor strategy and decision making does not make our division, and specifically my department, look good.

3) In what areas are you most interested in having decision making support? For example, product mix and demand forecasting.

   Considering the problems that we had with our marketing strategy, I believe that summarized reports and decision support in the areas of product mix and introduction of new products are of vital importance.

Comments

FIGURE 13.3   Continued.

QUESTIONNAIRE

Name   H. Taft

Department   R & D

1) Where do you obtain your data and information on which to base your management decisions? Please be specific.

I obtain my data through reports from individuals in my department and individual department reports such as Marketing and Manufacturing reports. I also obtain information by talking to other department managers. In addition, I obtain information from trade journals.

2) Do you find that these sources are sufficient to satisfy your needs for information on which to base decisions? If not, why are these sources insufficient?

I find that these sources are sufficient to satisfy my basic information needs, but I find myself wishing that I had ways of modeling situations other than manual modeling methods that consume a lot of my time. I can think of many times when I would have suggested developing new, more technologically advanced products if it wouldn't be for the fact that I just don't have the time to do cost-analysis modeling in order to justify the costs for the benefits I think Applia Corp. would reap by adding more advanced features to some of its products.

3) In what areas are you most interested in having decision making support? For example, product mix and demand forecasting.

Modeling and reports in the areas of high-technology equipment and cost analysis would certainly help my decision making.

Comments

**FIGURE
13.3**   Continued.

QUESTIONNAIRE

Name   J. Kennedy

Department   Manufacturing

1) Where do you obtain your data and information on which to base your
management decisions? Please be specific.

I obtain my data from reports that come from my department and individual department reports such as R & D reports, Accounting reports,
Marketing reports, and Personnel reports. I also obtain information by
talking to other department managers.

2) Do you find that these sources are sufficient to satisfy your needs for information on which to base decisions? If not, why are these sources insufficient?

No, I do not find that these sources are sufficient to satisfy my information
needs. I obtain far too much irrelevant data, and I have to spend much of
my time sorting through the data to find information that applies to the
decisions I need to make.

3) In what areas are you most interested in having decision making support?
For example, product mix and demand forecasting.

I'd like to have support for making decisions on investments in high-tech
equipment. It is always difficult and time-consuming to do a cost/benefit
analysis on those issues. As you have suggested, support in demand
forecasting would be extremely helpful—that would cut my modeling time
down. A cost-analysis model or reporting system would also help me in my
decision making.

Comments

**FIGURE
13.3**   Continued.

QUESTIONNAIRE

Name   T. Roosevelt

Department   Personnel

1) Where do you obtain your data and information on which to base your management decisions? Please be specific.

   I obtain my data from reports that come from my department and individual department reports such as Accounting reports, Marketing reports, and Manufacturing reports. I also obtain information by talking to other department managers. I get a lot of information on the environment external to Applia Corporation through labor statistics and reading the newspaper.

2) Do you find that these sources are sufficient to satisfy your needs for information on which to base decisions? If not, why are these sources insufficient?

   I find that these sources are sufficient to satisfy my basic information needs, but I obtain far too much irrelevant data. I have to spend much of my time sorting through the data to find information that applies to the decisions I need to make.

3) In what areas are you most interested in having decision making support? For example, product mix and demand forecasting.

   I'd like to have cost-analysis support where personnel is concerned. As you have suggested, support in demand forecasting would be extremely helpful. It is very difficult to forecast the personnel which will be needed to keep operations going. A demand forecasting model or reporting system in the area of personnel would help me in my decision making.

Comments

FIGURE 13.3   Continued.

<u>QUESTIONNAIRE</u>

Name   R. Nixon

Department   Accounting

1) Where do you obtain your data and information on which to base your management decisions? Please be specific.

   I obtain my data from reports that come from my department and individual department reports throughout Applia Corporation's Home Appliance Division. I also obtain information by talking to other department managers.

2) Do you find that these sources are sufficient to satisfy your needs for information on which to base decisions? If not, why are these sources insufficient?

   I find that these sources are sufficient to satisfy my basic information needs, but I obtain far too much irrelevant data, and I have to spend much of my time sorting through the data to find information that applies to the decisions I need to make.

3) In what areas are you most interested in having decision making support? For example, product mix and demand forecasting.

   I, of course, would like to have cost-analysis support. As you have suggested, support in demand forecasting would be helpful. It is very difficult to forecast the production demands using manual modeling, and we need that type of information in order to do cost and profit forecasting. A demand forecasting model or reporting system in the area of production would help me in my decision making.

Comments

CARLOS ANDRADE'S MEMO-STYLE PROGRESS REPORT

Carlos Andrade composed this report after the first two weeks on the evaluation project for Evco Electronics. Knowing he had to write a report every two weeks, he kept the format simple and the content requirements flexible.

Format

To minimize ceremony, Carlos Andrade combines two documents, his cover letter and report, into a memo-style report. Like Darlene Deitrick, he uses unambiguous headings to lead the

FIGURE 13.4	A Memo-Format Progress Report.

MEMORANDUM

TO: Franklin Berry, Vice-President
         Evco Electronics
         1100 Computer Way
         Houston, TX 77062

FROM: Carlos Andrade, Training Consultant
         Consultec, Inc.
         1 Main Street
         Amarillo, TX 79153

DATE: November 15, 1988

SUBJECT: Progress Report for November 1 to November 14

The development of your training evaluation is progressing according to the schedule we designed for the timeline in the proposal. No complications or setbacks have occurred. This report outlines the initial developments in two areas: interviewing and collecting data.

Summary
At the end of the second week, the interviews are on schedule, though re-ordered. The preliminary investigation of data collection indicates that the initial phases will be delayed, but with the additional personnel, we do not need to change the closing date of the project. The training evaluation project is on schedule and should conclude within our specified dates.

Interviews
Due to labor-management negotiations, certain upper-level management personnel

FIGURE 13.4   Continued.

have been unavailable for interviews; therefore, we have modified the interview schedule to contact first the managers who are not involved in negotiations.

We interviewed one management director, Alden P. Dayley, Director, Engineering Design.

We have interviewed two members of middle management, Harold Dimplick, Manager, Personnel and Records, and Emilio Martinez, Manager, Payroll.

We have interviewed the following management personnel:
   Ingeborg Hoby, Manager, Tool Distribution and Acquisition
   Thomas Solomon, Manager, Rotary Manufacture
   Kim Kane, Manager, Stamping
   William Werk, Manager, Plating

Data Processing
Research has begun on the financial and manufacturing records for the last five years, the time span covered by the evaluation contract. Marcia Mason, curator of the records library, and Albert Appleby, librarian for the archives, have cooperated in organizing the search for the appropriate historic records. Because all of Evco Electronics' records during the 1981 to 1986 period are stored on paper, the team will need extra time to collect, sort, and process the data for a statistical analysis.

We will need help in the area of data collection, because we must keyboard for computer processing all the information contained on paper in Evco Electronics' archives. I have attached your standard Temporary Personnel Request Form to this report, so we can hire or transfer from another department four data processing specialists. Your prompt review and authorization of the request can help us to keep the training evaluation on schedule.

Attachment: Temporary Personnel Request Form

readers to the specific report sections of interest. He frequently indents both for the start of the paragraphs and for lists to increase the ease of reading. Finally, he writes short, specific paragraphs that are easier to understand and remember.

Content

Carlos Andrade, a more experienced writer than Darlene Deitrick, addresses his reader and creates a shared purpose with Franklin Berry when he refers to "your training evaluation" in line one of the report. He also reassures Mr. Berry, and thus creates interest, by informing him in the first paragraph that no complications have occurred. This paragraph's personal orientation draws the reader into the report. Imitate Carlos Andrade's effort.

In the last paragraph, Carlos Andrade explains very clearly how Mr. Berry can authorize the hiring of the temporary personnel Andrade needs. Carlos makes it easy for the busy executive to cooperate with the evaluation project's goals by attaching the forms Berry must use and by explaining the requirement for extra data processors. Andrade's report focuses on very specific actions, presents specific, individual requests in very logical sequences, and provides the paperwork or forms needed for action, so a busy executive need not address the same problem twice.

Improvements

Although Carlos Andrade's report is succinct, he should consider a separate heading for the last paragraph, his request for extra employees. He might entitle this paragraph "Action Needed," or "Request for Action," or use some other phrase that will alert Franklin Berry to perform an action before he releases the report for distribution and storage. Otherwise, with its narrow focus and its request for a single action, this report communicates its author's message and his needs very specifically and effectively.

FIGURE 13.5   Carlos Andrade's Progress Report Reorganized to Put the Request for a Decision First in the Plan of Organization.

<u>MEMORANDUM</u>

TO: Franklin Berry, Vice-President
         Evco Electronics
         1100 Computer Way
         Houston, TX 77062

FROM: Carlos Andrade, Training Consultant
         Consultec, Inc.
         1 Main Street
         Amarillo, TX 79153

DATE: November 15, 1988

SUBJECT: Progress Report for November 1 to November 14

The development of your training evaluation is progressing according to the schedule we designed for the timeline in the proposal. No complications or set-

FIGURE
13.5

Continued.

backs have occurred. This report outlines the initial developments in two areas: interviewing and collecting data.

Decision Needed
We will need help in the area of data collection, because we must keyboard for computer processing all the information contained on paper in Evco Electronics' archives. I have attached your standard Temporary Personnel Request Form to this report, so we can hire or transfer from another department four data processing specialists. Your prompt review and authorization of the request can help us to keep the training evaluation on schedule.

Summary
At the end of the second week, the interviews are on schedule, though re-ordered. The preliminary investigation of data collection indicates that the initial phases will be delayed, but with the additional personnel, we do not need to change the closing date of the project. The training evaluation project is on schedule and should conclude within our specified dates.

Interviews
Due to labor-management negotiations, certain upper-level management personnel have been unavailable for interviews; therefore, we have modified the interview schedule to contact first the managers who are not involved in negotiations.

We interviewed one management director, Alden P. Dayley, Director, Engineering Design.

We have interviewed two members of middle management, Harold Dimplick, Manager, Personnel and Records, and Emilio Martinez, Manager, Payroll.

We have interviewed the following management personnel:
    Ingeborg Hoby, Manager, Tool Distribution and Acquisition
    Thomas Solomon, Manager, Rotary Manufacture
    Kim Kane, Manager, Stamping
    William Werk, Manager, Plating

Data Processing
Research has begun on the financial and manufacturing records for the last five years, the time span covered by the evaluation contract. Marcia Mason, curator of the records library, and Albert Appleby, librarian for the archives, have cooperated in organizing the search for the appropriate historic records. Because all of Evco Electronics' records during the 1981 to 1986 period are stored on paper, the team will need extra time to collect, sort, and process the data for a statistical analysis.

Attachment: Temporary Personnel Request Form

RUTH SILK'S SUMMARY REPORT

Ruth Silk wrote this report to compile in one quick reference report a summary of volumes of information about soil conservation practices. Unreferenced as it is, the report is not intended as primary research. Her report, which does not contain new information for anyone familiar with farming, collects the main points of current knowledge about conservation tillage. Ideally, it would save time for nonfarmers, agricultural hobbyists, and local gardeners who want to learn about the importance of soil conservation.

FIGURE 13.6   A Sample Summary Report.

PREVENTING CROPLAND SOIL EROSION

Cropland soil erosion is a serious problem because it threatens to ruin our topsoil. Currently the United States loses 3 billion tons of valuable soil each year because of erosion. This report describes the current erosion problem and outlines conservation tillage practices that can save and replenish the soil.

Factual Summary

Soil erosion results when particles of soil break off and are either blown away by wind or washed away by water. Soil erosion first became a topic of public concern in the 1920s and 1930s when the "dust bowl" that resulted from depleted farmland laid waste the family farms of the Midwestern and Southern United States. The call for land reform brought on by the dust bowl resulted in the formation of the Soil Conservation Service (SCS) in 1935 to monitor the erosion rate and instruct farmers in conservation techniques.

A 1977 SCS survey estimated that over 2.8 billion tons of topsoil were eroded in the United States. The survey also showed that the total erosion rate for 1977 was 6.8 tons per acre. This rate compared unfavorably to the set acceptable erosion value or T-Value of 5 tons per year. Because much of the nation's farmland is not the best land for cultivation, that land is at risk of erosion with serious consequences for the food production and agricultural economy of the United States.

Recommendations

Six techniques can be used to prevent erosion.

1. Use chemical additives and fertilizers carefully to improve soil composition, thus fighting the loss of nutrients due to soil exposure.
2. Plant crops on erosion-resistant soil only. Let poor quality or erosion-prone soil be fallow.
3. Use crop rotation to alternate row crops with cover crops. In Northwest Ohio, farmers usually plant corn one year and wheat or alfalfa the next, allowing the fields to rest every other year.

**FIGURE
13.6**

Continued.

4. Use strip cropping to slow runoff. Plant a combination of corn and alfalfa in strips. The alfalfa serves as a buffer from wind and catches sediment or eroded land from runoff.
5. Retain crop residue. Leave residue to protect the field from wind and rain during the winter.
6. Reduce wind erosion by planting trees and shrubs at right angles to the wind. This form of conservation can reduce erosion by 5%. Also, cover crops such as wheat and alfalfa can reduce wind erosion almost 100%.

Finally, practice conservation tillage to limit the soil's exposure to erosion. Five practices are beneficial.

1. Contour plowing entails plowing around a hill or slope rather than plowing up and down the hill. This cuts the amount of runoff and sedimentation.
2. No till and minimum till, practices in which residue is kept from the previous crop, require little or no tillage. The new crop is planted through the residue in the spring, and the residue acts as a cover to prevent wind and soil erosion as the new crop takes hold.
3. Ridge planting requires that ridges are developed and seeds are planted in the ridges. The rows between the ridges catch the runoff.
4. Plow planting eliminates spring field preparation and reduces the time during which a field is exposed to erosion.
5. Chisel plowing is useful for areas that have become compacted from heavy machinery. The chisel plow loosens soil without inverting it and retains some of the residue.

Cropland soil erosion is a problem that can be prevented and stopped through conservation efforts. Conservation techniques range from simple to complex and from inexpensive to costly. The individual farmer should consult the county Agricultural Extension Service for advice about the best methods of conservation for each region.

Format

The report relies on generic section headings: factual summary, recommendations, conclusions, and definitions. These straightforward headings enable readers to select the sections they wish to read and to skip over what they may feel is less important. Like the Andrade report, this report relies on enumeration to isolate and call attention to its central concerns, its recommendations, and its calls for action. The enumerative technique also enables the author to "chunk" the text into short blocks for easy reading and retention.

Content

To give the call for action an appropriate perspective, the author writes a brief history of soil erosion problems in the United States. This orientation helps amateur farmers understand why their own efforts are necessary to halt erosion. Throughout the report, the author uses action words and a good ratio of active to passive voice to retain the ease of reading so important in reports for the non-technical reader. Verbs such as "adopt," "enhance," and "control" help to make the topic specific and easy to understand. Specific nouns like "crop rotation," "crop residue," and "contour plowing" help the author present concepts and processes succinctly.

Improvements

A report this general in scope should contain a general reference or reading list at the end, so the readers can examine the author's sources and study books and articles that discuss the topic more specifically. In the absence of footnotes, which increase reading difficulty, such a list lends authority to the author's viewpoint. The author should also include a glossary of farming terms that may be new to readers. For example, "ridge planting" and "chisel plowing" are terms only people in agricultural industries know.

DENISE WOODROW'S RECOMMENDATION REPORT

Managers frequently write recommendations, documents that make a specific request, or call for a particular course of action.

Format

Although most reports call for decisions or actions, recommendations and justification reports use a less conservative plan or organization, and concentrate on the request for action rather than its rationale. Thus, as with the reorganized version of Carlos Andrade's report, the recommendation report uses a reverse-order outline.

Thesis:	We write to inform you of a decision.
Recommendation:	Purchase the site in West Caldwell.
Rationale:	We used six criteria.
History:	The bank had decided to open in New Jersey.

This is a very emphatic plan or organization, one designed to waste as little of the reader's time as possible. Note that this report is in memo-form. Since recommendation or justification reports are typically written for internal distribution, their form is more direct.

Content

The recommendation report content is typically divided into three areas: the recommendation, rationale, and background. Since the writer recommends a decision, the facts come first. The rest of the report explains or "sells" the manager on the wisdom of the choice. The background section also reminds him that there is a basis for the decision.

Improvements

This report has a crisp, direct style that sounds efficient. Denise Woodrow has used "chunking" to her advantage to write lists of facts and decisions that are easy to follow. She should be cautious, though. The crisp style in this short, recommendation memo might be offensive in a longer report, for example, a feasibility study. Thus, she should attend to her style by using well-placed transitional phrases and adverbs to ease the reader to the next idea.

FIGURE 13.7

A Recommendation Report (Memo-Form, Internal Distribution).

TO: David Rhoades

FROM: Denise Woodrow

DATE: August 11, 1988

SUBJECT: Recommended Location for Branch Bank

Since May, we have been studying the West Essex County area for a suitable location for a branch office. We have located a site, and I write to inform you of our choice.

Recommendation: A lot is available just west of the Rte. 46 Fairfield Ave. intersection. The site meets the six criteria we are interested in; it has commercial zoning, and it offers access to the two busiest streets in West Caldwell. Since land is at a premium in this area, we recommend that the board approve an offer of $489,000 for lot #214, Fairfield Avenue, West Caldwell, New Jersey.

For further information, please call because I have the realtor's material, the survey information, the pertinent ordinances, a contact list and other site information available for you at the office.

How We Selected West Caldwell: We divided Essex County into 10 zones for possible drive-in and ATM sites. We then tested each available site for the six criteria selected at the May 2 meeting. Those criteria were:

SUMMARY

In this chapter, you learned to compose progress and summary reports. You learned that short reports of five or less pages offer special advantages for specific audiences. After reading this chapter, you should be able to:

- Use a checksheet to examine a report you read, write, or edit to ascertain that the report contains all the required elements.
- Modify a checksheet to meet the requirements of your own reports.
- Use a generic outline to compose a short report.
- Modify a generic outline to meet the organizational requirements of special types of short reports.
- Choose when to write a report that contains a cover memo or one that combines a memo and report.

FIGURE 13.7   Continued.

* Ease of access by major thoroughfares
* Cost of demolition/construction or land preparation
* Land cost factors and commercial tax
* Presence of "feeder" businesses
* Growth potential—commercial and private accounts
* Prestige, image, and corporate access

In the first three categories, West Caldwell tested best. Also, we actually found a vacant lot requiring minimal preparation for construction. It is unusual to have such an opportunity in the West Essex area. Finally, the owner, Commercial Properties, Newark, agrees to meet our terms for the contract including surveys, sewage, and utilities.

Why We Need A West Essex Location: At our April 4 meeting, George Leone discussed our plans for Northern New Jersey. The Essex County area is currently the number two growth area for commercial accounts after Hartford. Leone pointed to:

* The presence of our two competitors, Consolidated and Federal
* The saturation of the banking market in our number three area
* The steady growth of residential accounts in West Essex
* The strong growth in commercial accounts throughout Northern New Jersey.

We have thirty days in which to make our offer. Your prompt response can secure our first-choice site and our continued growth.

■ Discuss the uses of summary reports and determine when a summary report is an appropriate means of communication.

ASSIGNMENTS

1 Write a memo report to your instructor about any lengthy assignment you are now already preparing. Consider this assignment a progress report. Explain: what the assignment was, what you have accomplished, how you will proceed, and when you will finish. If you need assistance, ask your instructor for some specific help with the report, for example, a review.

2 Find and study two short reports. Be prepared to discuss them for the class. One report should be a "good" report, that is, a document that reads well, has an appropriate format, and accomplishes the objectives this chapter outlines. The other report should be one that has room for improvement. Prepare to explain to the class how you would edit and improve the document.

3 Copy the "bad" report you discussed in class. Then edit the report to improve its content and format. Submit both the original and the revised versions to your instructor for comments and a possible grade.

4 Prepare a short summary report about a pressing professional issue in your chosen occupation. Study our summary report on soil conservation as an example of such a report. Assume that because your readers are vaguely familiar with your field, you must briefly explain the issues and the technical topics. For example, if you are in medical records management, you may wish to report on the costs and complications of billings. As a construction technology major, you may want to discuss the hazards presented by new building materials. Or as an engineer, you may want to discuss the implications of engineering salaries to your profession. Every field presents issues; discuss one that is critical in your profession.

5 Practice up on your abilities as a writer of self-reports. Compose a memo report to your technical communication instructor discussing your progress in the technical communication class. What were your goals at enrollment? What grades have you earned? What has the instructor said about your work? Have any problems occurred that your instructor should know about? Do you need assistance? What will the outcome be from this vantage point?

CHAPTER 14

Closeout, Investigative, and Feasibility Reports

This chapter introduces you to the organization and composition of various kinds of long reports. After reading this chapter, you will be able to:

- Explain how a technical report differs from a college term or research paper.
- Review a generic outline for a long report and modify it to fit your requirements.
- Explain the purposes of closeout, investigative, and feasibility reports.
- Describe the similarities of the various types of long reports.
- Describe the differences of the various types of long reports.
- Compose, edit, and produce the correct type of long report for the requirements of your audience.
- Evaluate the lengthy reports you receive for their accuracy, appropriateness, and thoroughness.

CASE STUDY   A Consultant Writes a Closeout Report

Consultec, Inc., a consulting firm with a specialty in training and evaluation, has completed its lengthy evaluation of Evco Electronics' Technical Modernization program. During the three-month study, Carlos Andrade, Consultec's evaluator, has gathered two volumes of information.

Although Carlos intends to leave his two-volume study with the Evco managers, he knows they will not read a lengthy report. Instead, he composes for the manager a 10- to 15-page overview of the evaluation process and its results. This closeout report serves as the basis of an oral presentation Carlos will conduct to explain his investigation to the managers, and the report refers the managers, as well, to the two-volume study for all the details.

This chapter examines Carlos's closeout report because this type of report is usually a requirement at the conclusion of any major corporate or institutional effort. We will also examine an investigative report and a feasibility study.

LONG REPORTS—SOME DEFINITIONS

In view of the assortment of needs and audiences for reports, some definitions are useful because many different types of reports can be discussed (and new categories for reports develop as different occupations and needs emerge). Specifically, we classify all reports as either short or long, because length determines content, format, method of distribution, and many other factors as well. Thus, our chapter on short reports—reports around five pages long—discussed formal and informal progress reports, and recommendations and summary reports. These short reports share certain common features in their organization and narrative style. The same is true of the three types of long reports defined here.

1. Closeout reports (also called summary reports) tell the reader a project has ended. They are a complete record of the activities just concluded, including such conventional information as the purpose of the effort, the method of evaluating the effort, the results, and the conclusions or recommendations (but usually not both). Figure 14.3, Carlos Andrade's report, is a closeout report.
2. Feasibility reports define a problem, describe the available solutions, and discuss in detail the preferred solution. Figure 14.5, Bill Dorsey's report to the Miskas, is a feasibility study that examines several possible solutions and makes multiple recommendations about the question, "Should the Miskas move into the insulation business?"
3. Investigative reports define a problem, describe the method of analysis used for the problem, and present the factual evidence the investigators gathered. Conclusions and recommendations may be excluded, depending upon the requirements of the audience and the subject. Figure 14.4—a report on controlling noise in the Tiffin factory—is such a report even though the authors make recommendations.

To some degree, the types of reports this chapter discusses have developed from the traditional scientific report and its five-part method: statement of problem (hypothesis), survey of pertinent literature, statement of method, presentation of results, and discussion of conclusions. We do not discuss the scientific method in detail here because even as the various types of technical reports are derived from scientific method, the actual reports have become quite different from the investigative reports of scientists.

For example, as you read the reports in this chapter, you can see that the categories of reports are not mutually exclusive. A progress report may at the same time be an investigative report since the writer outlines the events that occurred during a given week of an eight-week investigation. A feasibility study may use the techniques of an investigative report to examine problem solutions. In actual practice, the categories begin to merge depending on a given audience's needs.

To determine the type of report to write, define your document's audience and objective. If, for example, your analysis indicates that you, like Carlos Andrade, must write a program evaluation, you may combine elements of the investigative report and the feasibility study as you address your topic. For this reason, we adhere to generic outlines you can modify to suit your requirements.

RESEARCH AND YOUR REPORT

All writers, including creative writers, perform research to inform themselves about their subjects. "Research" refers to the reading, note-taking, writing of survey instruments, interviewing, collecting of data, and a range of activities performed to learn a subject well enough to write about it. This chapter's long reports result from their authors' traditional library research, as the appendixes of the noise pollution reports show.

But these examples of job-related reports also show the importance of what is loosely termed "action research." Action research refers to the investigation of nonlibrary sources that appear as factual evidence in your report. For example, Carlos Andrade uses some traditional, statistical research measurements in his study, but he also visits the factory, interviews managers and supervisors, and observes the production process. Similarly, Kim Ash and Mark Graves, who investigate noise containment, visit the factory and show a floor diagram as part of their factual evidence. These authors mix their traditional library research, which we discuss in Chapter 4, with the elements of practical investigation required for job-related reporting.

Do not confuse these technical reports with the traditional "term" or "research" papers prepared for other classes. Our concern in this chapter is the technical reporting of practical, job-related findings in appropriate formats. Our visual aid helps you to see the difference between traditional "term" papers and the applied research that you may perform to prepare technical reports for business.

THE USES OF LONG REPORTS

In the age of instant feedback, electronic mail, and overnight deliveries, why bother with a long, time-consuming report? Is there a simpler way for a business to compile information?

FIGURE 14.1

A List Contrasting Term Papers and Technical Reports for a Specific Business Audience.

HOW TERM PAPERS AND TECHNICAL REPORTS DIFFER

Term papers and technical reports differ in many ways, including their readers, plans of organization, approaches to the topic, methods of citation, and approaches to the research problems. We have grouped some of the general differences into the three areas shown here. As you compare your term papers with your technical reports, what other differences do you find?

Contents

Term Papers	Technical Reports
Original text based on:	Original text based on:
Books	Books
Journal articles	Journal articles
Abstracts	Abstracts
Government documents	Government documents
Other primary research sources from libraries	Other primary research sources from libraries
	Secondary sources from libraries
	Unpublished material from the company's library
	Existing reports from the company's library and files
	Action research materials including:
	Site visits
	Measurements
	Interviews
	Observations
	Other data and analyses

Format

Term Papers	Technical Reports
Typical five-part plan of organization for science paper:	Typical plan of organization of a business report:

FIGURE 14.1	Continued.

Statement of the problem
Review of the literature
Discussion of the method
Presentation of the results
Conclusion

Typical three-part plan of organization
for humanities paper:
   Statement of the problem
   Review and analysis of the
    literature
   Conclusions

Science- or humanities-style reference
  list

Selected bibliography

Cover letter
Front matter
Summary
Body—often containing:
   Methods
   Results
   Conclusions or recommendations
Copious illustrations (throughout
  the text or in appendixes)
Appendixes
Conclusions

Variation in the documentation of
  sources including the use of:
   Appendixes as sources
   Bibliographies
   Footnotes
   References to sources in the text only

Tone

Term Papers	Technical Reports

Third-person narrative
Some distance between writer and
  reader
Contemplative, thoughtful approaches
  to the topic
Statements carefully qualified by footnotes from library research

First- and second-person narrative
  used often
Distance between writer and reader
  varies with the subject
Less contemplative approach to the
  topic
Calls for action in a recommendations
  section
Statements carefully qualified by
  evidence from various kinds of
  research and investigations

FIGURE 14.2

A "Generic" Outline for Long Reports.

A SAMPLE OUTLINE FOR A LONG REPORT

All long reports share certain common elements of organization. For example, all long reports contain specific types of front matter, and all long reports should begin with some type of executive summary. This outline stresses the elements that the various types of 10- to 30-page reports have in common. Experiment with modifying the outline to meet your requirements.

Front matter contains all the required elements.
    Cover letter identifies the report, its origin, the sender and the receiver.
    Title page states the correct title of the report and lists the writer and the chief recipient.
    Abstract lists the report title, author, and all the other important identification numbers or categories for filing or invoicing purposes.
    Table of contents lists each item in the report, names the report's major sections correctly, and has correct page numbers.
    List of illustrations has a reference to every visual aid.

Summary (one or two pages) reads easily for a busy executive.
    First paragraph states the reason for the report.
      Subsequent paragraphs state the results, conclusions, and recommendations for an executive to scan.

Body (4 to 20 pages) explains the details of the investigation.
    Methods statement (one to four pages) describes your approach.
    First sections describe how the research or investigation proceeded, for example, by interviews, surveys, or on-site examination.
    Subsequent sections describe the equipment and materials used.

Results (one to four pages) state the findings and nothing else.
    Factual evidence of the investigation appears in enumerative organization pattern.
    Equations, financial abbreviations, or other symbols appear as needed to save space.

Conclusions or recommendations (2 to 15 pages) define a plan of action or suggest a basis for further study.
    Conclusions equal results, one conclusion for each result.
    Recommendations appear to result logically from the results and conclusions and are generally suggested actions that the recipient of the report should undertake.

End matter (four to any number of extra pages) gives the readers all the relevant information that didn't belong in the text of the report.
    Glossary lists and defines any special language.
    List of symbols explains all abbreviated notations.
    Footnotes when needed are easy to read and offer adequate detail.
    Reference list provides basis for further study.
    Appendixes of various kinds amplify the details of the method of investigation, the materials used, the conclusions, and the results.

Although these questions occur to anyone working overtime to meet a report deadline, businesses and institutions still use long reports for several purposes.

Records

Companies require various kinds of records just to stay in business. For example, a company must document its compliance with the Occupational Health and Safety Agency's guidelines for workers' safety in order to avoid lawsuits which claim negligence as a cause. Also, a company requires records of a product's or process's development, so the managers can analyze the costs of production, labor, inventory, and other functions. For example, Carlos Andrade's report is based upon a similar study performed by a consultant, so that a corporate contractor could cut its training costs as new machines were installed.

Histories

Corporations have realized recently that they gain substantial benefits by compiling the histories of their corporate franchises, their products, and their various campaigns. Thus, as a major manufacturer develops a new advance in avionics, a record may prove useful. Thus, each product, program, and phase of corporate development should be reported, so records can be compiled in the growing field of corporate history. These records also allow the firm to study, after the fact, how their successful products and technologies developed, so that their managers can engineer a greater number of successes.

Research

Research is the source of new products and sales for companies as diverse as pharmaceuticals corporations and electronics manufacturers. The specialists who develop new products require primary research on consumers' behavior, social trends, buying patterns, regional preferences, trends and emerging needs, and many other areas, depending upon the firm's products and its plan for growth. Research appears in the form of a report, because the report is still the most cost-effective method for a company to collect and store information about its specific interests.

THE ADVANTAGES OF LONG REPORTS

The full-length technical report offers three distinct advantages that help to offset the time spent on its creation and production.

Creation of a Product

The long report can show the scope of a project by allowing its authors to explain their work and display examples of the materials and techniques that contribute to a new process's development. Thus, the report itself becomes a product. For example, the long report makes it possible for Bill Dorsey to give the Miskas, his clients, a comprehensive collection of information about the impact of noise pollution on the residential construction market. In the same manner, Carlos Andrade supplies a product, a long report, as a tangible proof of his authority as an evaluator of training programs.

Development of a Record

Some companies require their employees to write closeout reports, so a record of a product's research and development will exist. The government and various state agencies may also require that the research and development personnel assigned to a product keep detailed records of their compliance with the regulations for safety, uniformity, and various other industry requirements.

Evidence for Evaluation

Reports help to explain a project and to educate unfamiliar corporate personnel about its merit. For example, Carlos Andrade's report alerts Evco's managers to several production problems. If the company's executive directors have hesitated to correct these problems, Andrade's evidence can help the managers in their campaigns for changes. A lengthy report can also explain how a project benefits a company, and in doing so, educate the executives about the project's importance. For example, a report that shows how an outplacement program builds goodwill for a company can convince its managers to continue that program despite its costs.

THE SUMMARY REPORT

Often in business, the summary report is composed, among other reasons, to signal the end of a project, to describe how employees have spent their time and the firm's budget, and to secure more funds.

Audience. Usually summary reports travel "up" the corporate ladder; that is, subordinates draft these reports to inform their managers of a project's outcomes. The readers of a summary report want to know:

- ☐ Did the project end on time?
- ☐ Were the managers within, below, or over budget?
- ☐ What were the objectives?
- ☐ Were the objectives met?
- ☐ How did the employees meet the objectives?
- ☐ How were accomplishments measured?
- ☐ How did the project affect the firm's image?
- ☐ Are there implications for the area I manage? (If the managers don't already know.)

Organization. A closeout report should follow a conservative plan of organization, a plan much like the generic outline shown in Figure 14.2. A list of elements might include:

- ☐ Front matter—every item listed under front matter in the generic outline.
- ☐ Executive summary—should read as a mini-report about 10 to 20 percent of the length of the entire report.
- ☐ Methods—should answer three questions:
    What were the objectives?

What were the desired outcomes?

How did you measure progress?

☐ Results—preferably in quantitative terms:

When you measured the project's progress, what results did you obtain?

☐ Conclusions (optional)—a narrative that explains how the manager should think and feel about the results.

☐ Recommendations (optional)—a call for decisions or actions that will benefit the institution.

☐ End matter—all the attachments needed and listed in the generic outline.

Resources. Compose a summary report from your observations, measurements, and records of a project. The techniques of action research rather than library research are most appropriate for this type of report.

Special Concerns. This report is a permanent record of your accomplishments or those of your group. Therefore, it should have a simple, consistent design including a cover, binding when appropriate, and all the visual "extras" that suggest the project's or firm's prestige. Although the text may be densely numeric, like Carlos Andrade's, it should be woven together with a traditional approach and the traditional prose forms of the sentence and paragraph. Use headings liberally, but they should indicate restraint, not humor.

CARLOS ANDRADE'S CLOSEOUT REPORT

As you study the format and content of Carlos Andrade's report, remember his audience and purpose. Most members of his audience are electrical engineers with a strong background in computer programming, because Evco hires and promotes engineers to management. Thus, this audience is comfortable with tables, abbreviations, and succinct explanations of mathematical processes that might intimidate other audiences. For a lay audience, Carlos Andrade's report, peppered with the names of variables he measured and composed in a very restrained style, may be difficult reading. But Andrade's report shows the level of detail and the amount of tabular matter that should reinforce a truly technical report for an expert audience.

Furthermore, Andrade's report shows how an experienced corporate writer presents a "politically touchy" subject. Usually, students don't have the opportunity to read a "bad news" report. As you read Andrade's report, you learn that he discovered Evco Electronics is badly managed. He has, therefore, unpleasant news to deliver, so he must establish a basis in fact for his conclusions and recommendations. Without misleading his client, Carlos Andrade must "soften" his news, so the managers will follow his recommendations instead of neglecting his advice.

Format

The study was bound in a blue folder with the Consultec logo and design featured on the cover. Along with the study, Andrade handed in his two volumes of comprehensive information bound in similar blue folders to create an impression of continuity and professionalism.

(Text continues on p. 352.)

FIGURE 14.3   A Sample Closeout Report Showing the Length and Technical Complexity of a Long Report Prepared for a Specific Audience.

Consultec, Inc.
1 Main Street
Amarillo, TX 79153

September 15, 1988

Franklin Berry, Vice-President
Evco Electronics
1100 Computer Way
Houston, TX 77062

Dear Mr. Berry:

Here is the closeout report that we discussed in April. I have translated the two-volume study into a brief discussion with abbreviated results and discussion sections. You will notice, too, a recommendations section at the end.

Because of the recent changes in personnel and management at Evco Electronics, I look forward to reviewing our discussion of this evaluation process. We may wish to review any materials relevant to your request that quantify the attitudes of the hourly personnel toward the training program. Also, I still plan to prepare the materials that can assist you in identifying specific costs and their relationships with training, education, and other plant activities. Since the previous study did not focus on these issues, I think we should discuss the information I will need to complete this phase of the project.

The Closeout Report offers some specific recommendations for your managers and for your examination and analysis. I hope you find the study useful, and I welcome any calls you have for further information.

Cordially,

Carlos Andrade
Training Consultant

**FIGURE
14.3**

Continued.

EVCO ELECTRONICS EVALUATION

Houston Manufacturing Center
Houston, Texas

Closeout Report

Carlos Andrade, Training Consultant
Consultec, Inc.
Amarillo, Texas

September 15, 1988

FIGURE 14.3   Continued.

Contents

List of Tables

**FIGURE
14.3**   Continued.

EVCO ELECTRONICS EVALUATION

Closeout Report

I was commissioned by the Division of Training and Human Resources to perform a program evaluation of the Technical Modernization Program at Evco Electronics. This discussion is taken from the detailed statistical report that I submitted in August 1988. Throughout this document, I refer to the two volume study as the "Original Study."

The managers at Evco Electronics actually helped to develop the questions answered in this report, as they assisted with advice regarding my collection of the data analyzed in both studies. The selections of questions and data were made during a series of discussions organized for the specific purpose of shaping the original study. From these meetings, the managers and I arrived at a consensus for the purpose of the original study—to investigate the effects of the training and educational programs on the various activities at the Houston plant.

The abbreviated discussion in this document will describe briefly the activity of the original study. Portions of the Results and Discussion sections have been simplified and presented here. For example, I have rewritten portions of both sections to comment on the general patterns of the evaluation within a limited text space. Similarly, I present the data in general terms to show trends rather than finite details.

I have used the evaluator's term, "R^2," to explain the variables that demonstrated the ability to predict reliability or validity. To clarify this concept, I use the analogy of the notion of percent; this well-known term suggests the direction of the movement shown by the variance in the selected data. Finally, the original study was not financial in nature and not detailed in the selection of data to provide specific functional relationships—although a manager may wish to make decisions that have financial impacts resulting from the evaluation of the Technical Modernization Program.

VARIABLES

Fifty-eight variables were identified by management as necessary to execute the present study. They appear in Table 1.

CORRELATIONS

A correlation matrix of 60 by 60 variables was provided as a reference for all subsequent statistics. (See Chapter 3, Results, Statistical Interpretation, Correlations, p. 31, Original Study.)

1

**FIGURE
14.3**   Continued.

TABLE 1

VARIABLES

1. CASEID	21. SCRAP	41. TMFMSDDP
2. CARDNUM	22. REWORK	42. SITMFMDD
3. YR1982	23. YIELD	43. S2TMFDP
4. YR1983	24. RETVEND	44. S3TMFDP
5. YR1984	25. RETREWK	45. TMFMIMP
6. YR1985	26. RETSCRAP	46. S2TMFIMP
7. YR1986	27. RESPVEND	47. S3TMFIMP
8. JAN	28. RESPURCH	48. S1TMFAS
9. FEB	29. RESPENGI	49. S2TMFAS
10. MAR	30. WARREXP	50. TMMFCON
11. APR	31. RECURSAL	51. TMSPWIMA
12. MAY	32. NETALOHR	52. IMQUASUR
13. JUN	33. OVRTIMHR	53. TMSTELAS
14. JUL	34. OUTIMCAS	54. TMCNCMC
15. AUG	35. INVENTRY	55. TMCADCAM
16. SEP	36. RESPQUCO	56. ENGCACAM
17. OCT	37. RESMFG	57. ETHEANET
18. NOV	38. RESPPRPL	58. TMSHMECO
19. DEC	39. FOTNGCOS	59. COPICS
20. PROTIVTY	40. TNGCOS	60. PETNGARS

This list was created after several discussions individually and in group meetings with management personnel.

FACTOR ANALYSIS

Factor Analysis is a statistical procedure suitable for a complex data field where general structures are under investigation. It allows the investigator to identify the unique contributions of individual variables to a larger group of variables—called "factors." The "factor" is a collection of these statistical contributions that must be interpreted to provide meaning.

Therefore a factor is a mathematical "picture" of the correlative relationships among all of the variables. The first "Principal Axis Factor" is simply the factor that collected the largest correlations. The second factor—in the present VERA-MAX study—is mathematically unrelated to the first factor. Therefore the correlations on the second factor are not related to the correlations on the first factor.

Each subsequent factor is mathematically unrelated to the previous factors. Less and less variance occurs with each succeeding factor. The evaluator decides

FIGURE 14.3   Continued.

which factors represent useful variance. In the present case, based on the eigen-value, the first four factors appeared to contain significant amounts of informa-tion. The rest of the factors represent mathematical quantities too small to con-sider in the present study.

QUESTION ONE

The first question defined by management was: "What factors appear in the present data over the past six-year period?" When the data were exposed to Fac-tor Analysis (SPSS, 1985), four Principal Axis Factors with interpretable individ-ual variable loadings resulted. (See Factor Matrix: Six Years of Data, Original Study, p. 89.)

The first Principal Axis Factor is comprised primarily of Technical Modern-ization variable loadings. Given the extent of changes in the Evco Electronics plant that are the result of the Technical Modernization program, this statistical result is expected in the first two or three factors. It is within the theoretical set to name the first Principal Factor the Technical Modernization "A" factor.

The second Principal Axis Factor is comprised primarily of Technical Mod-ernization variable loadings with two additional variables, YIELD and PROTIVTY loading heavily on the factor. This is within the theoretical set. YIELD and PRO-TIVTY are mathematically related and both can be presumed to be heavily cor-related with the Technical Modernization program as variations in production are directly tied to the Technical Modernization processes. It is within the theoretical set to name the second Principal Factor Technical Modernization "B" factor.

The third Principal Axis Factor is comprised of NETALOHR from which the TNGCOS and PROTIVTY factors seem to draw their primary mathematical "strengths." Further, TNGOOS and PROTIVTY as well as NETALOHR have been shown to be highly correlated. It is within the theoretical set to name the third Principal Factor "Production."

The fourth Principal Axis Factor is comprised of YIELD, RETSCRAP, RESP-VEND, WARREXP, S2TMFAS, and S2TMFAS. This factor is not clear. YIELD is a mathematically unstable variable as it is an artificial ratio, previously weighted for managerial purposes, clouding its interpretability. It is difficult to show a theoretical relationship between RETSCRAP and S2TMFAS and S3TMFAS. RETSCRAP and WARREXP could be related but the relationship is not clear. This Principal Axis is the last in this investigation subjected to theoretical explana-tion. Factor 4 is not clear within the empirical bounds and will remain simply "Factor 4."

QUESTION TWO

The second question defined by management was: "What factors appear in the present data over the past two year period?" When the data were exposed to

FIGURE 14.3   Continued.

Factor Analysis (SPSS, 1985), four Principal Axis Factors with interpretable individual variable loadings emerged. (See Factor Matrix: Six Years of Data, Original Study, p. 89).

The mathematical structures of the six-year Factor Analysis and the two-year Factor Analysis are essentially isomorphic. This is due primarily to the measurement procedures used in collecting the variables as a result of the Elapsed Time Model characteristic of the entire study. The six-year Factor Analysis is, due to the data collection procedure and the time-bound character of the program under investigation, mathematically identical to the two-year Factor Analysis. This was an expected event. The second, e.g., two-year analysis was executed primarily as an exercise to demonstrate the validity of the six-year analysis.

Summary

Four strong Factors appeared in the data. Each Factor is constructed of variables, represented by correlations that contribute to the Factor. The Factor description and interpretation are based on the numeric correlation and a knowledge of what real events the variables represent. The differences between the six-year Factor Analysis and the two-year Factor Analysis are small and the Factors are remarkably stable.

The four Factors that appear in the Evco Electronics study, derived from the present variables and in descending order of contribution, but each contributing significantly to the picture of events, are listed here:

Factor One—A Technical Modernization Factor "A."   The most significant mathematical factor, it contributes the most variance to the Factor Analysis heavily correlated with the Technical Modernization variables.

Factor Two—A Technical Modernization Factor "B."   The second most significant mathematical factor, it contributes the second largest amount of variance to the Factor Analysis of the data. The difference is that "B" includes loading Productivity and S1TMFMOD ($r = .40$ .50) and Yield ($r = .30$ .40).

Factor Three—A Training and Production Factor.   This is the third most significant mathematical factor. It contributes the third largest amount of variance to the Factor Analysis of the present data. Production, Training (Cost), and Net A Lot Hours are the major contributors to this factor. This history of training literature predicts this factor and its structure.

Factor Four.   The fourth Principal Axis Factor is too generalized and, therefore, not totally clear. Its principal variable loading is Yield. Yield is an artificial variable created by ratio and weighting for management purposes. As such, it does not represent real events in the plant and is uninterpretable.

4

**FIGURE
14.3**   Continued.

Discussion

The first four Principal Axis Factors are mathematically strong. The first three Factors can be interpreted within theoretical expectations. The Fourth Principal Axis Factor is not interpretable as it is mathematically artificial.

The First and Second Principal Axis Factors are two mathematically unrelated sets of Technical Modernization data. The Second Principal Axis Factor includes Yield, Productivity, and SlTMFM with the same set that appeared in strong relationship with the First Principal Axis Factor. The broad interpretation of these two factors is that they represent separate characteristics of the Technical Modernization Program. Specific interpretations should be made by individuals with detailed knowledge of the program.

The Third Principal Axis Factor is a Training and Production Factor. Training (Cost), Production, and Net A Lot Hours combine in a strong set. It is an expected outcome predicted by both literature and experience in Training Programs in the present setting. The strong relationships between Training figures, Production figures, and Net A Lot Hours figures are expected and predicted by the history of experience in training programs and reflected in the literature describing the history of the effect of training programs.

The three areas of the study that will be of interest to management may be condensed as follows.

Table 2

Training

Variables

Page	Criterion	Predictor	Direction	Source	R^2
40	PROTIVTY	FOTNGOOS	Apart	Professional	− .02440
46	SCRAP	FOTNGOOS	Apart	Professional	− .01699
74	NETALOHR	FOTNGOOS	Apart	Professional	− .02808
56	REWORK	PETNGHRS	Apart	Hourly	.00921
42	PROTIVTY	TNGOOS	Together	Hourly	.08699
66	REWORK	TNGOOS	Together	Hourly	.06298
72	RETSCRAP	TNGOOS	Together	Hourly	.03639
60	NETALOHR	TNGOOS	Together	Hourly	.13604
58	RETVEND	FOTNGOOS	Together	Professional	− .01441
70	RETSCRAP	FOTNGOOS	Together	Professional	− .02120
74	WARREXP	FOTNGOOS	Together	Professional	− .02679
82	OVERTMHR	FOTNGOOS	Together	Hourly	− .14138
84	OVERTMHR	PETNGHRS	Together	Professional	− .00921

5

FIGURE 14.3   Continued.

In the table "Training" above (Original Study, p. 143), there are two training figures. Formal Training Cost (FOTNGCOS), produced primarily by professional staff, and Program Training (PETNGHRS), produced by hourly personnel. As PETNGHRS rises, REWORK falls, which is the expected outcome of a good training program.

Further, as FOTNGCOS increases, SCRAP figures fall. This is also within expectation of a training program. However, as FOTNGCOS increases, PROTIVTY and NET A LOT HOURS fall. This result is not in keeping with the training literature; however the majority of FOTNGCOS is for the education of the professional staff, and the training literature does not address the education of professionals. A detailed knowledge of the activities of the professional staff is necessary to interpret this information. In any case, increases in FOTNGCOS predict decreases in NET A LOT HOURS and PROTIVTY (Original Study, p. 146).

It is difficult to interpret NET A LOT HOURS. Like YIELD, it is an artificial number created by weighting and ratio. It does not reflect real activities in the plant. While such a construct may be useful as a management tool, it is not useful in an empirical investigation.

In summary, there are two sets of relationships reviewed in the linear section, the predicton of <u>increase to decrease</u> and <u>increase to increase.</u>

<u>INCREASE TO DECREASE</u>

Increase in TNGCOS predicts decrease in PROTIVTY, REWORK, RETSCRAP, and NETALOTHRS.

Table 3

Factor Analysis

Factors	Name	Source of Variance
Factor 1	Tech Mod "A"	Tech Mod Variables
Factor 2	Tech Mod "B"	Tech Mod Variables
		YIELD
		PROTIVTY
Factor 3	Production	NETALOHR
		TNGOOS
		PROTIVTY
Factor 4	Noninterpretable	

| FIGURE 14.3 | Continued. |

INCREASE TO INCREASE

Increase in FOTNGCOS predicts increase in RETVEND, RETSCRAP, WARREXP, and OVERTIMHR.

Increases in PETNGHRS predict increases in OVERTIMHR and REWORK.

The R^2 of the variable may be viewed as a figure similar to a percent. It is arrived at by "subtraction." A mathematical model with all the variables is "run" to establish Full Model R^2. This is "100%" of the possible predictive "power" of the variables. Then the variable of interest, say, FOTNGCOS, is removed, and the

Table 4

Technical Modernization

Variables

Criterion	Predictor	Direction	Source
PROTIVTY	TMMFOON	Apart	Tech Mod
	IMQUASUR	Apart	Tech Mod
SCRAP	TMFMSDD	Apart	Tech Mod
	S3TMFMDD	Apart	Tech Mod
	TMFIMP	Apart	Tech Mod
	TMFOON	Apart	Tech Mod
	TMSPWIMA	Apart	Tech Mod
	TMSTELAS	Apart	Tech Mod
	TMCNCMC	Apart	Tech Mod
	TMCADCAM	Apart	Tech Mod
	ENGCACAM	Apart	Tech Mod
REWORK	IMQUASUR	Apart	Tech Mod
PROTIVTY	S1TMFMDD	Together	Tech Mod
	TMSTELAS	Together	Tech Mod
	TMCNOMC	Together	Tech Mod
	ETHERNET	Together	Tech Mod
REWORK	S1TMFMDD	Together	Tech Mod
	S2TMFMDP	Together	Tech Mod
	TMSTELAS	Together	Tech Mod
	TMCNOMC	Together	Tech Mod
	TMCADCAM	Together	Tech Mod
	ETHERNET	Together	Tech Mod

7

**FIGURE
14.3**   Continued.

model is run again. This is the Restricted Variance, restricted by the loss of FOTNGCOS. The difference is the unique R^2 variance attributable to the restricted variable above and beyond all other variables.

If the criterion, PROTIVTY, falls, and FOTNGCOS constantly predicts that fall by rising, the R^2 represents the amount of the decline in PROTIVTY that is predicted by increases in FOTNGCOS. If the R^2 is .02440, it means (roughly) that changes in FOTNGCOS predict about .024% of changes in PROTIVTY of the 100% of predictive power of the Full Model including FOTNGCOS. Ths is the method by which a percent notion can be attached to R^2. The R^2 is not a percent, and it is not derived from the arithmetic of percentages, but the analogy is sound.

It is not possible to translate these figures into representative dollars and cents and to relate money and specific activities. The data that were made available were too general and the time intervals were too great for such estimates. The routine manner in which Evco Electronics collects and stores data prevents a more intimate exploration of the data.

Further this is a condensed discussion taken from a study requested by the Division of Training and Human Resources. The variables recommended by management as well as the questions developed by management do not represent detailed financial concepts. The commissioned study was not directed toward financial predictions by management directors.

DISCUSSION OF THE EVALUATION PROJECT

The Evco Electronics evaluation evolved into four major activities: the collection of history, the collection of data, the statistical evaluation, and the interpretation of the statistics. The process of the four major activities collectively comprises a fifth activity—observation.

Observation may be the oldest and perhaps the best technique for collecting information, a method in the tradition of Western hypothetico-deductive analysis. Observations contribute an important aspect of an evaluation. The trained and skilled observer is as useful a tool to management as the computer, the engineer, and the hourly worker.

Technical Modernizaton Program

The present Technical Modernization program assumes both hourly and professional staff will retrain to use new Computer Aided Design and Computer Aided Manufacturing techniques. The involvement of Evco in the Japanese-led Quality Circle program depends upon the concept of training in the workplace.

The Evco Electronics plant is a setting known, accurately or inaccurately, for rather poor labor-management relations. Fortunately, the relationship between management and labor in the Evco Electronics plant has no effect on productivity

| **FIGURE 14.3** | Continued. |

or profit because production is controlled by contract quota. The sales totals control the contracts. Fixed numbers of products are to be delivered at fixed times. To manufacture more would waste time and money. To manufacture less would take a concerted effort as the actual numbers of products produced are small.

Training

The present work force is experienced and produces the product with little or no detailed supervision. The product lines are fixed by contract and by the very nature of the products, which change very little. Changes in manufacturing technique and changes in quality assurance technique occur regularly. These do not change the product lines, and training programs provide for smooth transitions in management, professional, and hourly personnel.

Professional Personnel

The Evco Electronics plant has approximately 400 hourly workers on the floor. The Evco Electronics plant employs approximately 400 additional professional personnel. If any sensitive projects exist, they must involve no more than a small portion of the entire professional staff. It must be concluded on prima facie evidence that the ratio of overhead to production is excessive. I fully realize that this observation is made without access to closed information that could justify the maintenance of so large a professional staff. This evaluation would be less than candid if the observation—your oversupply of engineers—were not called to your attention.

Manufacturing

The manufacturing and testing processes are thorough and rigorous. Some exotic materials are used in special applications for thermal and vibrational stress, but no manufacturing or testing process is unusual. Products are manufactured and tested by workers, not by machines, essentially as a quality assurance procedure. Techniques are simple and basic but must be done with great care and constant testing. The success of each step depends upon training and development in the hourly personnel. The present setting and the company's growth require a well-trained and aware hourly employee.

Training and Development

Mass production techniques are of questionable value in this setting. Except for the punching of laminates for rotating assemblies, the plant reflects this fact in equipment and personnel. Because of the Technical Modernization program, the plant offers numerically controlled manufacturing. This Technical Modernization program will improve some aspects of the production of rotary shafts and assembly housings. However, since volume does not affect the financial success of the plant, the machines will operate infrequently.

9

FIGURE 14.3

Continued.

Thus, the Technical Modernization program depends upon the presence of skilled personnel in professional and hourly positions. This translates to new training programs or the hiring of newly-trained personnel in the necessary positions. Since the plant already has an internal training program, the managers may consider the development of appropriate in-house training.

Middle Management

The role of middle management eludes definition. Line supervisors and hourly workers are involved with the product directly in parts cueing and immediate manufacturing problems of a practical nature. Middle managers, on the other hand, participate in a more ceremonial series of activities at Evco Electronics than in most manufacturing settings. In this capacity, middle management has moved physically from the manufacturing setting to a white collar environment. As a consequence, the older, experienced workforce has become self-directing and appears to require little managing.

RECOMMENDATIONS

Changes should occur in inventory, management, and labor.

Inventory

Inventory appears to be excessive, and the processes by which the majority of the raw materials are delivered to work stations is primitive. Although the Technical Modernization program attends to part of this problem, managers could improve the entire process with more attention to decreasing the inventories and increasing the movement of parts through the plant and to the workstations.

For example, management can control the inventory precisely because in rotary assemblies, and in most of the large assembly areas, the production level is contractual. Contract sales determine the number of products to be manufactured. Thus, the numbers of items to be inventoried are clear, including estimates of faulty products delivered, estimates of numbers of products that fail to test properly, and numbers of items manufactured incorrectly.

Inventory cueing, a nuts-and-bolts activity and function of middle management in the manufacturing process, does not seem to be exercised vigorously at Evco Electronics. The process seems to have evolved on its own until a high technology solution appeared feasible in the form of the Technical Modernization program.

Labor

The workforce comprises only 10% to 11% of the cost of manufacturing. Thus, a 10% change in labor cost represents a less than 1% change in total cost. Although a change of this magnitude is important, the time and energy required to make useful increments are not justified. Further, changes in this cost are

**FIGURE
14.3**

Continued.

limited to the contract negotiation period. The traditional areas of managerial
concern—labor and production—should be of relatively low priority at Evco
Electronics.

<u>Management</u>
    Your Management Directors need the tool of empirical evaluation. Since a
Management Information Systems group already exists, you could easily create
the priority for developing a functional data bank. The structure and function of
the data bank is essentially standard and routine in most business and manufacturing settings. Currently, your divisions collect data separately causing duplication and wasted time. More important, the Management Directors have a difficult
time obtaining accurate information.
    Evco Electronics needs a free-standing Management Information System.
Such a group should provide Management Directors and other divisions with raw
information, reports and evaluations of any activity in the plant. Both routine
and requested reports could be provided quickly. A functional Management Information Systems group would free other divisions to execute their primary activities, to increase their effectiveness, and to monitor Evco's inventory.

The two backup volumes had information produced on daisy wheel and dot matrix printers; the closeout report itself was laser-printed with a typeset cover.

Throughout the report, Andrade relied on two levels of headings to keep his facts easy to find. For the same reason, he relied on simple, direct headings such as "Question One" and "Technical Modernization Program." The table of contents presents no ambiguity, and it suggests a short, readable analysis.

Several decisions about the organization of the text show in Andrade's format. Note the absence of an executive summary. Instead, he starts the report with four paragraphs that explain his scope and restrictions as an investigator. At the start, he wants the managers to understand his report's limitations. He then moves directly to a table listing his variables. Note that Andrade does not explain statistics to his readers. He assumes the engineers can understand his mathematical model and tables. He writes quite crisply about his mathematical procedures. Thus, the tables seem to interrupt the text to show many unexplained results in one snapshot. Andrade avoids footnotes and reference lists, preferring instead to refer to his sources for the text's statistics.

Content

Although this is a "bad news" message, Andrade softens his news by avoiding shocking or "red flag" labels. Thus, as he turns to his less numeric observations, he labels the section "Discussion of the Evaluation Project," so the engineers will continue to read the report. He also divides the discussion into five distinct parts so the managers can read first about their own divisions. Finally, his call for improvements occurs in a low-key section labeled simply, "Recommendations," so the traditional format of the report can reassure the readers.

Note that Andrade uses his language very carefully in his discussion and recommendations sections. He does not want to make accusations or to create hostility that could deflect his report's impact. He uses the passive voice frequently to create a contemplative tone, and he repeatedly uses the conditional tenses by inserting the verbs "could" and "can" to allow his readers to see possibilities rather than accusations. For example, he writes ". . . management can control the inventory . . ." instead of "management should control the inventory." The more conversational style of these recommendations helps to diffuse their unpleasant message that the overly managed Evco Electronics is losing profits.

Improvements

Despite the crisp, concise style that is so appropriate for this report, Andrade should have considered an executive summary. Since his audience has a rather traditional orientation toward language and many have not formally studied writing since college, a traditional orientation might reassure the readers. For the same reason, Andrade might experiment with more descriptive titles. Instead of entitling a section "Question Two," a heading used often in research writing, he might write a heading which states question two. For example, he might write the headline "Four Factors Created the Technical Modernization Program." Or, he might express an opinion in his headline by writing "Factor A, Manufacturing, Affected the Technical Modernization Program Positively." As you compare this closeout report with your generic outline, what other recommendations would you make to Andrade?

THE INVESTIGATIVE REPORT

Investigative reports are very common in business, because professionals and technicians must both examine methods of solving problems and recommend improvements. Investigative reports could result from any of these activities:

A training manager visits a local plant to survey employees about their perceptions of their training needs.

An insurance investigator visits the site of a building's collapse.

An employee relations panel at a public agency studies six insurance companies to determine the best medical benefits package.

A corporate accountant determines whether a firm should pay for rental cars or invest in a fleet for its sales representatives.

Audience. For work, the typical audience of an investigative report is a manager who needs information for a decision. Perhaps like the factory manager in Tiffin, the manager must regain the approval of a public agency in order to continue business. Often an investigation is for cost reduction. At other times, a manager may conduct an investigation for supervisory reasons.

Organization. Investigative reports are formal or informal. If the investigator is a firm member, the report tends to be informal. If the investigator is an outside consultant, the report is more formal.

☐ Front matter — if the report is formal, it should include all the items in the generic outline; if informal, a cover letter and title page may be adequate.

☐ Summary — 10 percent of the length of the report, the summary is optional in an informal investigative report.

☐ Method — explains methods of action research used, such as site visits, interviews, or examinations of records. In a formal scientific report, use the procedure's proper name.

☐ Results — quantitative results are presented here in tables, columns, or other forms appropriate to the formal report. In the informal report, a narrative description of the outcomes is adequate.

☐ Conclusions — in both formal and informal reports, a narrative section explains to the reader how to interpret the findings. This section usually concludes the report; no recommendations need to be stated, although Mark Graves' and Kim Ash's does.

☐ End matter — as needed for the subject and audience.

Resources. The library is often students' best resource. In the case of the Tiffin factory report, Kim Ash and Mark Graves researched noise-related topics in the library, made repeated visits to the factory, interviewed employees, and made measurements as well as cost calculations.

Special Concerns. Remember to maintain an objective tone in an investigative report. You may find it difficult to be objective, especially if you are required to investigate something unpleasant such as thefts at your place of work or complaints about a discourteous worker. Nonetheless, you have an obligation to be truthful and to use descriptive rather than emotive language. For example, the noise in the Tiffin factory must have been distressing for the workers there, and Kim Ash and Mark Graves would have known that fact. Yet, their report assumes a professional, nonjudgmental tone toward the plant's manager.

KIM ASH'S AND MARK GRAVES' INVESTIGATIVE REPORT

Because both writers worked in a noisy factory in Tiffin, Ohio, the topic of noise pollution in Chapter 4 caught their interest. They investigated their factory for a redesign that could contain noise and create a safer workplace. Thus, the report shows the various aspects of a writer's research: action research, including measurements within the factory; library research, along with all the basic reading about the definition of noise pollution; and the combination of the two in the conclusions and recommendations.

Format

The authors used a traditional plan of organization to create readability. This plan, with its historical overview and executive summary at the front of the report, enables a busy executive to read a capsule report of the investigation in five pages. A more detailed discussion then follows in the remainder of the report. Although this method can create some problems of continuity, it does help the readers to quickly find the sections that pertain to them.

Two other features of the format also promote readability. The authors integrated the illustrations into their text instead of placing them in appendixes. This feature creates a more informal report, but it helps to create continuity and unity in the text. Ash and Graves also used the "chunking" technique throughout; each page contains headings, lists, italics, and other features to facilitate reader comprehension.

Content

Ash and Graves chose to speak directly to their employer with a personal orientation throughout. Because they have known their employer for several years, this direct approach is appropriate and anything more formal would seem awkward. To preserve the content's readability, Graves and Ash mixed the active and passive voices in good proportion, too. Active voice, descriptive verbs such as "cited," "endorsed," and "reveals," create an appropriate level of detail throughout the text.

Study this report's content for its mixture of action research with library sources. Confident, well-informed writers mix their knowledge of the subject with the facts of their investigation to create a unified report. Thus, Ash and Graves, who have studied the Occupational Safety and Health Administration's (OSHA) requirements regarding noise, can comment on this information as they describe the Tiffin factory's particular problems. Cumbersome footnotes do not clog the text; instead, a selected bibliography appears at the end of the report. Try to obtain a similar mixture of reading knowledge and on-site experience when writing investigative reports.

(Text continues on p. 368.)

FIGURE 14.4
A Sample Investigative Report.

Title Page

<u>CONTROLLING CENTRIFUGAL FAN NOISE</u>

<u>A Feasibility Study for Reynolds Industry</u>

by
(Mark Graves & Kimberlyn Ash)
Miska Consulting

FIGURE 14.4

Table of Contents

Continued.

TABLE OF CONTENTS

List of Illustrations

LIST OF ILLUSTRATIONS

FIGURE 14.4   Continued.

Background Statement

INTRODUCTION

A recent inspection by OSHA reveals that a centrifugal radial-blade fan, in your Tiffin, Ohio plant, emits noise levels above the 90 dB/eight-hour work period standard as determined by OSHA. If your plant does not improve in the areas cited by OSHA, a decrease in production due to the physiological effects of noise on employees and a possible shutdown in production may occur. This report presents the results of our feasibility study in which we establish the best method to control your centrifugal fan noise pollution. Such information should allow you to take quick action to control centrifugal fan noise.

Background or Historical Summary

After the installation of the centrifugal fan ventilation system in January, OSHA cited you for exceeding the legal noise level standards. They gave you a 90-day grace period in which to correct the problem or else face a shutdown. In addition to your concern for production loss, employee safety remained a primary consideration as well. As a result, you spoke to the news media in an effort to warn your employees and the surrounding community about this health hazard. The subsequent article prompted Miska Consulting to submit a proposal.

In this proposal, we suggested you commission us to conduct a feasibility study. After you reviewed our proposal in March, you hired us to conduct a feasibility study. Over the past month, we have conducted that study, and this report presents the results. In the preliminary examination, we endorsed OSHA's determination of the source of the noise problem. And after an analysis of the citation, we narrowed down all the possible methods of alleviating the problem to three, according to your needs. The three methods we feel meet the immediate needs of your company include the following:

1. Return to prior method of ventilation.
2. Muffle the fan with a shield.
3. Build a concrete encasement for the fan.

Admittedly, a number of methods could solve your problem, but after an analysis of your immediate needs and the OSHA citation, we feel three would best eliminate your problem.

In addition, we believe these three best fit the criteria we established during the course of study. Those criteria are as follows:

1. Cost-effectiveness.
2. OSHA standards
3. Timely installation.
4. Overall benefits to plant operation and employee safety.

1

**FIGURE
14.4**

Continued.

We arrived at these criteria after our preliminary investigation. After a comparison of methods to criteria, we feel these factors best fit your needs.

Statement
of Method

Plan of Action
    As outlined in the proposal, we studied your noise control problem according to a specified plan of work. We broke the study into the following four tasks:

1. An analysis of the OSHA citation.
2. A determination of plant deficiencies in the area of noise management.
3. The establishment of criteria for meeting the regulations and improving working conditions.
4. A selection of the best method for diminishing the fan noise in a timely and cost-efficient manner.

    We narrowed down the possible solutions for solving your noise control problem to the three we feel will best meet your needs. Those three methods include the following:

1. Reinstall the old system of ventilation.
2. Muffle the fan with a shield.
3. Build a concrete, lead-lined enclosure for the fan.

    We compared those methods to the following criteria, which we feel best take into account your needs:

1. Cost-effectiveness.
2. OSHA standards.
3. Timely installation.
4. Overall benefits to plant operation and employee safety.

    On the basis of our criteria, we conclude that building a 6-inch thick, lead-lined encasement will best meet your needs. This method meets your budget limitations, costing you approximately $1850.00. This method complies with OSHA standards by absorbing the excessive noise. A timely 3-day installation assures that you will meet the 90-day OSHA grace period. Moreover, by taking this step to eliminate excessive fan noise, you will ensure employee safety and eliminate noise interference in the work place.

**FIGURE
14.4**   Continued.

Besides the benefits, you should consider fire safety standards as well.
Because we advocate the enclosure of a heat-generating mechanism (the fan
motor) in the vicinity of flammable materials (sawdust), fire safety becomes an
additional concern. A consultation with the local fire inspector would easily
remove all doubt from your mind in regard to fire safety (See Figure 1, Tiffin
Plant Floor Plan).

Figure 1. Tiffin Plant Floor Plan.

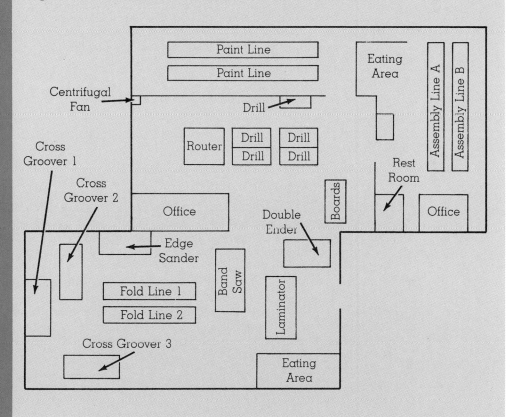

FIGURE 14.4

Continued.

Results

CONCLUSIONS

Our conclusions fall into two categories, with two conclusions under each subdivision. As a result of the comparison of methods to criteria, we can draw the following conclusions:

Related to the Problem
1. Your company cannot continue operating at current noise levels caused by the centrifugal fan.
2. Your company faces a shutdown by OSHA.

Related to the Solution
1. Your company cannot operate without the ventilation provided by the centrifugal fan.
2. Your company needs to install a concrete enclosure around the centrifugal fan.

Optional Recommendations

RECOMMENDATIONS

Derived from the conclusions, we recommend you take the following steps to solve your noise control problem:

1. We recommend you take action by encasing the centrifugal fan in a lead-lined, concrete enclosure. The enclosure we propose should be approximately 6 inches thick, and should encase the fan on three sides around the motor.
2. We recommend you contact the necessary companies and contractors, taking out bids to complete the work. Time is important. With warm weather approaching, contractors will be occupied with reservations for outdoor construction.
3. We recommend you contact OSHA for a noise re-inspection. As a government agency, OSHA must strictly adhere to government regulations regarding time allowances. Also, as a government agency, they follow a strict schedule.
4. We recommend you contact fire officials to inspect our method for safety. Because we recommend the encasement of a potentially hazardous mechanical apparatus (the motor) in the area of flammable materials (sawdust), fire regulations become a concern.
5. We recommend you periodically check noise levels to ensure that our method works effectively over time. This preventive measure will allow you to recognize any additional concerns before they become a problem.
6. We recommend you commission another study if you do not choose our method. The problem requires an immediate solution, so time is of the essence in resolving this situation.

4

**FIGURE
14.4**

Continued.

DISCUSSION
Centrifugal fans "move air by centrifual action" (McDermott, 264). This means that the fan rotates on its axis, causing air to flow out from the center. Centrifugal fans come in three different types, depending on the blade shape. These fan types include forward-curve fan, backward-curve fan, and the radial-blade fan.

The centrifugal radial-blade fan "resembles in its characteristics the older paddle wheel impellers . . ." (Strock, 2–63). The curve of the blade at its outer tip meets the axis (see Figure 2, Radial Blade Fan). Because of this fan's blade shape, the centrifugal radial-blade fan does not collect dust and grease on its blade. Therefore, it efficiently removes dirt and grease from the air, thus making it an overall effective fan design for your ventilation needs.

Although the centrifugal fan meets your ventilation needs, it emits excessive noise which raises your plant's overall noise level above OSHA's standard of 90 dB's/8-hour work day. A comparison of methods to minimize or eliminate this problem by established criteria aids in a better understanding of our conclusions and recommendations.

Review of Methods
As previously mentioned, we narrowed down all possible alternatives to the following three methods, according to your immediate needs.
1. Return to prior method of ventilation. This first method of solving your problem entails the removal of the current ventilation system and the reinstallation of the old ventilation system.

Figure 2.   Centrifugal Radial-Blade Fan.

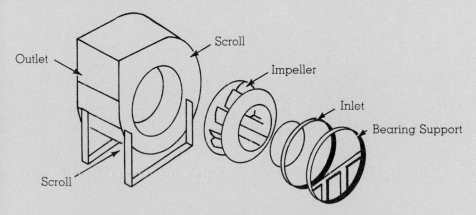

5

**FIGURE
14.4**

Continued.

2. Install a shield (also known as a muffler) around the motor. A second method of controlling your noise problem includes the installation of a shield muffler around the fan motor.
3. Install a barrier around the motor. The third method of reducing your noise to an acceptable level consists of installing a lead-lined, concrete barrier around the motor. This 6-inch thick barrier would encompass three sides of the motor, leaving the fourth side open for heat escape (See Figure 3).

<u>Review of the Criteria</u>

We establish four criteria unique to your individual problems, which acted as a guideline for our selection of the most feasible method. The selection of these criteria followed extensive research into your plant's facilities, needs, problems, etc. and of OSHA's citation. The criteria we selected include the following:

1. Cost-effectiveness. We recognize that because of the size of your industry, cost becomes an important consideration. Therefore, we consider cost a very important concern.
2. OSHA standards. Because we realize the importance of your meeting OSHA standards, we view compliance with OSHA standards as a crucial criterion, as well.
3. Timely installation. As a part of the citation OSHA granted you a grace period of 90 days. Therefore, finding a solution and implementing it within a limited amount of time becomes vital.

Figure 3   A Cutaway of the Barrier.

**FIGURE
14.4**   Continued.

4. Overall benefits to production and employee safety. Your company needs a
solution that will enhance your already high production rate. In addition,
your company, concerned about your employees, wishes to ensure their
safety. These two concerns justify these factors as a fourth criteria.

A comparison of the methods to the criteria provides a basis for our conclusions and subsequent recommendations.

Cost-Effectiveness
Your company needs a method of controlling your noise problem that costs
a reasonable amount of money. A comparison of each method's cost allows us
to conclude that installing a barrier around the motor would be the most cost-
effective means of solving your noise problem (See Figure 5, A Comparison of
Methods to Criteria).
The cost for returning to the old method of installation includes the cost for
removing the centrifugal fan, in addition to the cost for reinstalling the old
method of ventilation, the cost for maintenance on the old system, and the cost
of the centrifugal fan that has been removed (because you cannot return it or get
the money back). The costs of these items are as follows:

$ 500 (labor to remove centrifugal fan)
500 (labor to re-install old system)
500 (maintenance on old system)
$1500 (plus the hidden cost of the
centrifugal fan)

The cost for installing a shield around the motor to muffle the noise includes
the cost for the shield and an extra cost for maintenance (because the shield is a
sensitive instrument that needs much repairing). A breakdown of the costs to install a shield follows:

$2000 (price of a muffler system)
500 (maintenance on the muffler system)
$2500

The cost for installing a lead-lined, six-inch concrete barrier around three
sides of the motor includes the cost for the concrete and lead lining. The costs of
these products are as follows:

$1100 (6-inch concrete barrier)
750 (lead lining)
$1850

**FIGURE
14.4**

Continued.

By looking at Figure 4, A Cost Comparison, you see that returning to the old method of ventilation seems to be the most cost-effective means of controlling your noise problem. However, after considering the hidden cost of the centrifugal radial-blade fan, this method becomes extremely costly. Thus, the installation of a lead-lined, concrete barrier around the motor still remains the most cost-effective solution.

OSHA Standards

According to the results of our research, all three methods will lower your noise to the acceptable level of 90 dB/eight-hour work day. As you can see in Figure 5, the Comparison of Methods to Criteria, we found that all methods would meet OSHA standards.

Timely Installation

As you enter the 58th day of the 90-day grace period granted by OSHA to correct your noise problem, a timely installation becomes more crucial. Therefore, you need a method that will meet your time limitations. A comparison of the three methods according to their installation time indicates that constructing a lead-lined, concrete barrier would best meet your time constraints.

Figure 4.   A Cost Comparison.

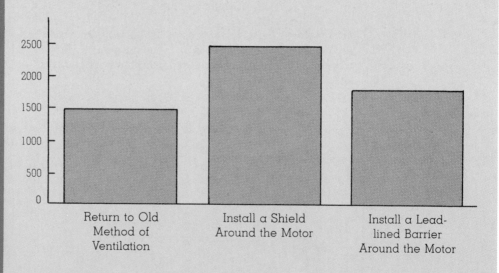

**FIGURE
14.4**

Continued.

Replacing the new centrifugal fan with your prior method of ventilation would take approximately four days. Removing the centrifugal fan would require one to two days, while reinstalling the previous ventilation system would take an additional one to two days. In addition to this four-day installation period, you must consider the additional time needed for the maintenance and supervision of the system to ensure its proper operation. Therefore, with installation and maintenance time, a month could easily pass before the system operates efficiently.

Installing a shield around your motor will take approximately one day. Unfortunately, in addition to installation time, you must allot time for ordering and shipping the muffler. These delays take up to six weeks. Therefore, the time required to implement this method exceeds your 90-day grace period.

The construction of a lead-lined, concrete barrier around the motor would take approximately three days of labor, depending on how soon you contract a company to install the barrier. This construction remains the most efficient manner in the area of time. Thus, contacting the necessary companies to construct the barrier should become a priority in your time plan.

In regard to a timely installation, our research reveals that the lead-lined, concrete barrier remains the most time-efficient method of controlling your noise problem. Whereas returning to the old method of ventilation and installing a muffler could take months to fully implement, the construction of a barrier requires only three days of labor, provided that you begin to contact companies to install the barrier (See Figure 5, A Comparison of Methods to Criteria).

Figure 5.   A Comparison of Methods to Criteria.

The numbers represent a ranking of the methods in comparison with one another.

1 = Most Successful   2 = Moderately Successful   3 = Least Successful

	Return to Old Method Ventilation	Install a Shield Around the Motor	Install a Lead-Lined Barrier Around the Motor
Cost-Effectiveness	3	2	1
OSHA Standards	1	1	1
Timely Installation	2	3	1
Overall Benefits	3	2	1

**FIGURE
14.4**

Continued.

Overall Benefits to Employee Safety and Plant Operation

The method used to correct your noise problems should also ensure employee safety and enhance your plant's operation. Our research indicates that installing a lead-lined, concrete barrier not only complies with our other criteria, but will also increase employee safety and plant operation.

The reason for the installation of a centrifugal radial-blade fan included a concern for employee safety. The previous method of ventilation failed to adequately remove the dust that accompanies a growing industrial plant such as yours. Therefore, if you returned to the old method of ventilation, you would defeat your purpose for installing the centrifugal fan. In addition, the high content of dust in the air may harm your employees' lungs.

But the physical problems that would occur with the removal of the centrifugal radial-blade fan, combined with the constant repairs and maintenance that the previous ventilation system would require, could force your company to focus attention on the ventilation system instead of more profitable areas of your plant. Therefore, returning to the old method of ventilation not only inadequately meets your employees' health needs, but could reduce your management's effectiveness.

Installing a shield around the motor will enhance employee safety but will not improve plant operations. Moreover, the installation of a muffler around the motor will absorb the sound produced and will allow the fan to operate efficiently, thus providing adequate ventilation for employees. Unfortunately, the shield is a very sensitive device. As a result, frequent maintenance often follows the installation of a shield. This inconvenience would become a burden to your maintenance personnel, taking their attention away from other maintenance problems. In addition, excessive maintenance concerns could become an additional problem for management, drawing their focus away from company productivity.

Constructing a lead-lined barrier around the motor would reduce noise, provide adequate ventilation, and enhance plant operations. This method of managing your noise problem controls the noise by acting as a barrier between the noise source (the fan motor) and employees. In addition, the barrier allows the fan to operate efficiently and ventilate the area. Finally, the barrier produces few maintenance concerns. Once constructed, a barrier requires only preventive maintenance. Thus, the worries of maintenance and constant care diminish. This helps your maintenance staff (they do not have to struggle with an additional maintenance problem), your employees (they have a pleasant noise- and dust-free environment), and your management (they do not need to concern themselves with an additional burden). Please refer to Figure 5, A Comparison of Methods to Criteria.

Therefore, on the basis of this comparison of the methods to the criteria, we recommend you thoroughly consider the possibility of constructing a lead-lined,

| FIGURE 14.4 | Continued. |

concrete barrier to solve your noise problems. The method provides the following advantages:

1. Cost-Effectiveness. This method costs only $1850.00.
2. OSHA Standards. This method reduces noise to a level that adheres to OSHA's 90 dB/eight-hour work day.
3. Timely Installation. This method would take approximately three days to construct, providing that you contact necessary companies to do the work.
4. Overall Benefits to Plant Operations and Employee Safety. This method reduces noise and dust in the work area without becoming an unnecessary burden to maintenance or management.

BIBLIOGRAPHY

Baturin, V.V. Fundamentals of Industrial Ventilation. New York: Pergamon Press, 1972.

Carpenter, Rolla C. Heating and Ventilating Buildings. New York: John Wiley & Sons, 1915.

Cheremisinoff, Paul N. and Peter P. Cheremisinoff. Industrial Noise Control Handbook. Ann Arbor: Ann Arbor Science, 1977.

Dorman, R.G. Dust Control & Air Cleaning. New York: Pergamon Press, 1974.

Feldman, Alan S. and Charles T. Grimes. Hearing Conservation in Industry. Baltimore: Williams and Wilkins, 1985.

Hayoshi, Taro, Ronald H. Howell, Nasuree Shibata, and Katsuhiko Tsuji. Industrial Ventilation and Air Conditioning. Boca Raton: CRC Press, 1985.

Khashab, A M. Heating, Ventilating, and Air Conditioning Systems Estimating Manual. U.S.: McGraw-Hill Inc., 1984.

McDermott, Henry J. Handbook of Ventilation for Contaminant Control. Boston: Butterworth Publishers, 1985.

McQuiston and Jerald D. Parker. Heating, Ventilating, and Air Conditioning: Analysis and Design. New York: John Wiley & Sons, 1982.

Miller, Richard K. Handbook of Industrial Noise Management. Atlanta: Fairmont Press, 1976.

Petrusewicz, S.A. and D.K. Longmore, ed. Noise and Vibration Control for Industrialists. New York: American Elsevier, 1974.

Strock, Clifford, ed. Handbook of Air Conditioning, Heating, and Ventilating. New York: The Industrial Press, 1959.

Thuman, Albert and Richard K. Miller. Secrets of Noise Control. Atlanta: Fairmont Press, 1974.

Wallis, R. Allan. Axial Flow Fans and Ducts. U.S.: John Wiley & Sons, 1983.

11

Improvements

This draft, though good, could be improved with more illustrations of the problem and its solutions. Graves and Ash should illustrate the idea of noise pollution so their primary reader, the factory manager, can understand how painful and damaging excessive noise is for his employees. They might borrow some ideas from Bill Dorsey's final report for the Miskas, so they can make the problem more realistic for their practical, cost-minded manager. For the same reason, they should include a more detailed cost analysis that shows the reader the improvements are affordable and that the costs can be absorbed through any tax write-offs that may apply.

Finally, Graves and Ash could improve their style dramatically by eliminating the gerund phrases that can clog up the beginning of a sentence. For example, they write, "By looking at Figure 4, . . . you see that returning to the old method of ventilation seems to be the most cost-effective means of controlling your noise problem." What is the subject of the sentence? The sentence confuses the reader with the repeated use of "ing" words instead of simple nouns. It might read, "Figure 4 shows that your old method of ventilation can contain costs best." Simple sentences greatly improve ease of reading complicated materials.

THE FEASIBILITY STUDY

Feasibility studies are challenging to write because, typically, they have a fairly broad focus that the writer must narrow with care. For example, Bill Dorsey had to consider the entire problem of noise pollution, so that he could analyze the areas in which Steve and Elaine Miska would find opportunities to work.

Audience. A feasibility study usually has a very narrow audience: Steve and Elaine Miska and others like them, the board of directors of a company, the local school board, the members of the county energy commission. Thus, the audience tends to have more in common than any of the other report audiences you've studied.

Organization. The generic outline is again a good place to start. A feasibility study is almost always a formal report. It should contain:

☐ Front matter — everything listed in the generic outline.

☐ Summary — a traditional executive summary that is 20 percent of the length of the report. Again, think of this as a mini-report.

☐ Methods Statement — in two parts, first, a history of the problem including its scope and any issues related to the problem such as costs, politics, or time constraints; second, a discussion of how you performed the investigation.

☐ Results — a narrative and graphic presentation of the information you obtained. Note Bill Dorsey's explanation of how sound buffer systems operate as part of his results section.

☐ Recommendations — is there a requirement for action? Should Steve and Elaine go into insulation? Should the company apply for a patent on this product? Should we investigate further? In the feasibility study, specific and general recommendations are called for; they are much more important than conclusions.

☐ End matter as needed for the subject.

Resources. Here too, action research is necessary. Bill Dorsey worked in the library with books, periodicals, and government documents. He visited sound buffer walls along U.S. Highway 275 south of Detroit, and he interviewed experts in the Toledo area.

Special Concerns. Remember that the feasibility study is often an educational document. Specifically, the writer must often explain new terms, new theories, and complicated research to the audience just to help them to understand the proposed course of actions or solutions. In feasibility studies, allow space for the explanations of background information for the persuasive writing you must do. For example, Bill Dorsey allows space for a discussion of the impact the new sound control legislation may have. By doing so, he prepares his audience for the course of action they should follow.

BILL DORSEY'S FEASIBILITY STUDY

Remember Bill Dorsey? In our proposals chapter, he wrote the proposal to Miska Construction for a feasibility study of noise control devices. After sending out his proposal, he composed a feasibility study that his firm can tailor to the needs of various contractors and municipal ordinances.

Format

Bill Dorsey uses a fairly traditional format that he can modify to accommodate his various clients. The table of contents shows a mix of generic headings, such as "Part Two," and specific headings, such as "Using Buffer Walls to Keep Sound Where It Belongs." This mixture creates a report that is at once traditional in structure but specific and easy to read. Bill Dorsey omits an executive summary, but he includes an abstract which gives an overview of the report.

Content

This text, like Kim Ash's and Mark Graves', contains a healthy mixture of text that is "chunked" and enumerated for ease of comprehension. The report is also amply illustrated with line drawings and tables. Headings interrupt each page to stimulate the reader's interest in the specific sections. Even though this report is lengthy, it is unintimidating because Dorsey incorporates current research about page layout into his composition. Consider how different this report looks from a term paper.

Improvements

Like Kim Ash and Mark Graves, Bill Dorsey has a really interesting draft here. The draft is worthy of improvement. For example, Dorsey should make a decision about his heading format, so the level two headings will be grammatically parallel. The parallelism can help to create continuity among the report's sections. Also, to increase readability, he should type his

(Text continues on p. 387.)

**FIGURE
14.5**

A Sample Feasibility Report.

Front
Matter

CONSTRUCTION RAMIFICATIONS OF
PROPOSED COUNTY ORDINANCE
PL G101.7
A REVISED FEASIBILITY STUDY

Prepared for
Miska Insulation Company

by
William Dorsey

1 May 1988

**FIGURE
14.5**   Continued.

TABLE OF CONTENTS

**FIGURE
14.5**

Continued.

Abstract

ABSTRACT

Despite setbacks in recent years, local governments continue to press for
regulations regarding noise control. Proposed County Ordinance PL G101.7 is an
example of this effort. Its passage will be aided by heightened public awareness
and support. With its passage, new construction markets will be born. Prepared
contractors will find applications for insulation, the interior wall, and sound buffer walls in residential, industrial, and transportation facilities of this county. It
will be important to keep abreast of technological advancements as well as regulatory refinements.

FIGURE 14.5

Continued.

PART ONE

SOLVING THE PROBLEMS OF NOISE AND SOUNDPROOFING

Presently, a proposed county ordinance exists that would attempt to deal with the growing problem of noise pollution. Ordinance PL G101.7 would introduce measures to control noise through the regulation of residential, industrial, and transportation facilities. This ordinance would bring about vast, untapped wells for construction companies.

The feasibility study that follows will provide data to assist Miska Insulation Company in tapping these wells and placing your company on the forefront of noise control contracting.

We will begin with a historical perspective of noise pollution, and a background of noise control. Next, we will demonstrate the need for noise control in our community. Then, we will reach into the specifics of the ordinance and the associated markets, and also examine possible future regulations. We will explore some design considerations for the control devices and conclude with recommendations that Miska may find useful.

HOW THE WORLD BECAME SUCH A NOISY PLACE

Perhaps the roots of modern noise concern found their hold in the industrial revolution. Certainly, humanity was faced with new challenges in industrial pollution. Journals began to record the problems of noise in the late 1800s. By 1940, concern about noise control had grown to such an extent that the Noise Abatement Council in New York declared October 21–26 National Noise Abatement Week. Heightened awareness came in the 1950s with the advent of the supersonic jet. As technology advances, it seems, so does the noise pollution problem. Figure 1 demonstrates how some of today's familiar sounds affect our ears.

THE GOVERNMENT SPEAKS UP ABOUT NOISE

In 1972 the federal government responded to growing concern. The 97th Congress authorized the government to adopt a comprehensive regulation for noise control. The Noise Control Act was aimed at products in general and interstate motor carriers specifically. The Act states, "The Congress finds that inadequately controlled noise presents a growing danger to the health and welfare of the Nation's population . . ." The Environmental Protection Agency (EPA) was given the power to administer the Act.

A 1978 amendment, known as the Quiet Communities Act, emphasized the role of state and local government in noise control. State and local ability was limited, however, in a court case between the Association of American Railroads vs. Costle. The Washington D.C. Court of Appeals held that section 17 of the Act preempted noise control by state and local government. Later, the Consolidated

FIGURE 14.5

Methods

Continued.

Rail Corporation vs. City of Dover case invalidated a local ordinance to control noise at a rail yard.

The combined effects of these cases chilled the exercise of local government controls. Consequently, in 1982, the revised federal budget cut the EPA's noise control program.

PART TWO

THE COUNTY FIGHTS TO CONTROL NOISE POLLUTION

County Ordinance PL G101.7 began as many other noise control ideas proposed by local governments around the country. Recently, many trend-setting

Figure 1.   Levels of Some familiar Sounds.

Source	Aural Effect	Sound Level in Decibels
Shotgun blast Jet plane at takeoff Firecrackers, exploding	Human Ear Pain Threshold	140
Rock music (amplified) Hockey game crowd Thunder, severe Pneumatic jackhammer	Uncomfortably Loud	120
Powered lawn mower Tractor, farm type Subway train, interior Motorcycle Snowmobile Cocktail party (100 guests)	Extremely Loud	100
Window air conditioner Crowded restaurant Diesel-powered truck/tractor	Moderately Loud	80
Singing birds Normal conversation	Quiet	60
Rustle of leaves Faucet dripping Light rainfall	Very Quiet	40
Whisper	Just Audible	10

2

**FIGURE
14.5**

Continued.

areas have experienced success with these measures. It appears that noise control regulations are gaining ground lost by the earlier setbacks.

The passage of PL G101.7 will be further aided by three factors:

☐ heightened local concern
☐ residential support
☐ feasibility of controls

In a survey by the Committee on the Problems of Noise, noise conditions were what people wanted to change about their neighborhoods. Figure 2 shows the percentages of people in noisy neighborhoods who cited various sources of noise annoyances. Given this information, there is little doubt regarding the future of the ordinance.

CHANGES BROUGHT ABOUT BY PL G101.7

As it stands to date, the proposed ordinance will mandate the following:

☐ insulation for noise control in new, residential buildings
☐ controlled noise level on the outside of industrial facilities
☐ buffer walls for freeways and other high traffic streets

Specific restrictions and control requirements will be explored in the later stages of the ordinance development.

Figure 2.   Percentage Contribution of Various Noise Sources Identified by Residents in Noisy Neighborhoods.

Source	Percentage
Motor Vehicles	55
Aircraft	15
Voices	12
Radio and T.V. Sets	2
Home Maintenance Equipment	2
Construction	1
Industrial	1
Other Noises	6
Not Ascertained	8

FIGURE 14.5   Continued.

Results

SOUNDPROOFING AND WHO WILL NEED IT

PL G101.7 mandates specific restrictions, controls, and recommendations in the following areas: residential, industrial, and transportation. From the nature of the ordinance and information gained in a recent survey, the following data, illustrated in Figure 3, are available regarding possible markets.

We turn, first, to the residential world.

SOUNDPROOFING ON THE HOME FRONT

Several factors have contributed to the growth of residential noise control problems. With more people moving to the cities, families are forced to live in apartment buildings, or in homes built very close to one another, allowing noise to travel easily from one dwelling to another. Sound levels of household appliances that frequently contribute to the noise problem are listed in Figure 4.

The ordinance stipulates that new residential buildings will require noise control technology. The most feasible idea seems to be an insulation that will function for noise control as it does for temperature control. A brief description of this insulation appears later in this report. The interior wall represents another possible mode of controlling noise at the residential level. Some detail of its construction will follow.

The ordinance will not require existing residential structures to retrofit for noise control, although this action will be recommended. However, an overwhelming majority of homeowners surveyed indicated an unwillingness to retrofit, due to the associated costs. Many, however, expressed an interest in the idea if some type of government incentive were given.

Figure 3.   Possible Soundproofing Markets.

Area	Scope	Soundproofing Devices
Residential	New facilities; Retrofitting recommended	Insulation Interior Wall
Industrial	New and existing facilities; Retrofitting required	Insulation Interior Wall
Transportation	Major highways and streets carrying high traffic volume.	Buffer Walls

FIGURE 14.5   Continued.

Figure 4.   Typical Sound Levels of Appliances.

Appliance	Sound Level
Refrigerator	xxxxxxxxxxxx
Clothes dryer	xxxxxxx
Air conditioner	xxxxxxxx
Electric shaver	xxxxxxxxxxxx
Hair dryer	xxxxx
Clothes washer	xxxxxxxxxxxxxx
Toilet	xxxxxxxxxxxxxxxxxx
Dishwasher	xxxxxxxxxxxx
Food mixer	xxxxxxxxxxxxxxxxxx
Vacuum cleaner	xxxxxxxxxxxxxxx
Food blender	xxxxxxxxxxxxxxxx
Garbage disposal	xxxxxxxxxxxxxx
Home shop tools	xxxxxxxxxxxxxxxxxxxxxxxx

```
                    40          60          80         100
              A-Weighted Sound Levels at 3 ft.
```

At present, the ordinance carries no such provisions, although it is quite possible that a provision of this type would be added in the fine-tuning stage that will follow passage of the existing proposal.

QUIETING DOWN INDUSTRY

Industry has long been a culprit in the noise pollution problem, and has thus been a target for noise control regulations.

PL G101.7 stipulates that industrial customers will need to take appropriate measures to control the noise level outside their facilities. Retrofitting is thus indicated. We can expect that a noise level standard will be applied to these facilities in proportion to their vicinity to the residential world. Once again, application of insulation and interior wall designs are indicated. In a few industrial facilities, the buffer walls might also be appropriate. New industrial customers will rely only on the interior wall and insulation.

SILENCING THE ROADWAYS

Traffic contributes a great deal to the noise pollution problem. Figure 5 illustrates the sound levels exceeded by vehicles. Figure 6 shows the levels of annoyance caused by heavy traffic noise during the hours of 5:00 and 10:00 PM.

5

FIGURE 14.5   Continued.

Figure 5.   Noise Survey Results of California Highway Patrol.

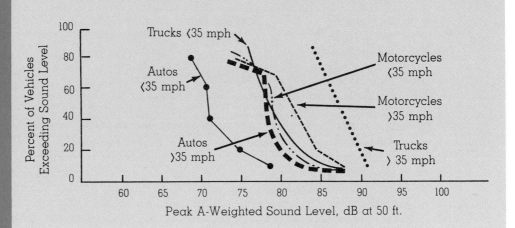

Figure 6.   Annoyance Due to Heavy Vehicle Traffic Between 5 and 10 PM.

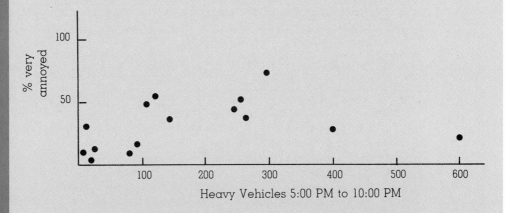

The ordinance specifies the use of buffer walls on major freeways and streets that carry high-volume traffic. An extensive market is created here. Although it may be some time before the implementation of this stage, local government will soon begin to seek competent contractors for these plans. Miska's experience in concrete will be valuable here. In the meantime, a solid foundation in other noise control applications will enhance your reputation.

6

**FIGURE
14.5**

Continued.

A summary of the market indications can be found in Figure 7.

LEGAL CONCERNS RAISED BY PL G101.7

The regulatory nature of noise pollution and control dictates several legal concerns. We can expect that guidelines will be set regarding the products used for control. Although the EPA has not yet entered this arena, we can expect such a move in the future. Meanwhile, local legislation, such as the proposed ordinance, will likely carry some production guidelines.

The buffer walls, for example, may be required to reduce noise level by a given amount. Perhaps they will be required to function for a specified period of time.

The interior walls may need to meet noise level requirements also. These guidelines will most likely accompany the latter stages of the ordinance. Future legislaton affecting noise control construction, as well as noise levels, may also come from the federal level.

The government may also offer loans to companies that wish to get started in the soundproofing industry. A needs assessment will need to be included on the loan application form. The government may also offer loans or incentives to homeowners who retrofit their homes with soundproofing devices.

The government documents section of the district or university library can be an essential resource. Consult the appendix for specific information regarding the use of this section.

Figure 7.   Expected Market Volume.

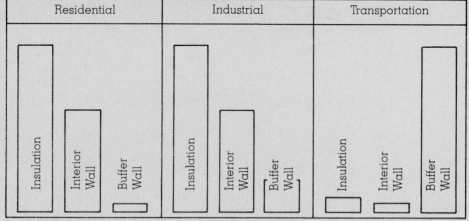

**FIGURE
14.5**

Continued.

PART THREE

THE THEORETICAL APPROACH TO NOISE CONTROL

A few terms will be helpful as we begin our discussion of design criteria. Sound is defined as a wave motion in the air produced by a vibrating body. Noise is undesirable sound. The sound intensity is usually expressed in units called decibels (dB). One dB represents the faintest sound the human ear can perceive. Frequency refers to the number of cycles or repetitions that a particular sound wave has per minute. Figure 8 demonstrates how sound travels from the source to the audience.

The purpose of noise control is to reduce the level or intensity of noise between the source and the protected audience. Three methods of accomplishing this goal are associated with PL G101.7. These are: insulation, the interior wall, and the sound buffer wall.

WHAT INSULATION CAN DO

In a typical context, insulation is used as an energy-efficient device to control and protect from heat and/or cold. Under the proper constraints, insulation can serve a dual purpose of noise management. Such factors as amount, density, and composition can influence an insulation's control capacity. This same principle applies to the wall depicted in Figure 9.

Figure 8.   Fundamental Elements of Noise Control.

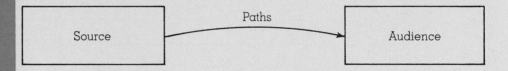

Figure 9.   Sound Waves Passing Through a Wall.

**FIGURE
14.5**

Continued.

With proper insulating devices, sound absorption can be greatly improved, as seen in Figure 10.

Some materials are classified as absorbants and effectively accomplish this goal. Some common absorbants are:

- ☐ mineral or glass wool blankets,
- ☐ molded or felted tiles, panels, and boards,
- ☐ plasters,
- ☐ sprayed on fibers,
- ☐ and foamed, open-cell plastics, and elastomers.

Technological research and advancement are growing in this area. This trend will continue as the issue spreads. Scientific periodicals should be consulted to keep abreast of any developments.

SOUNDPROOFING WITH THE INTERIOR WALL

Interior walls are seldom more than six inches thick. Dimensions will vary according to the function and environment.

The outer surface (neglecting paint, wallpaper, or other finishes) is generally a plaster or a wood or wood substitute. This layer serves aesthetic purposes, and, as we will see, can be important in noise control. An inner conglomerate of thin layers or laths serves as a base to the plaster or paneling. Artificial materials are often used here and wood or woven wire is also used. Upright inner supports called studs are mounted to the inner surface of the wall. Studs provide the framework for the outer layers and unify the two sides of the wall. The common 2 × 4 is a typical stud. Finally, the interior core may be filled with an insulating substance.

Design Procedure

In theory we could design a soundproof wall for every application. Unfortunately, situational constraints make this impossible from a practical stand-

Figure 10.   Examples of Noise Reduction by Two Walls With Different Transmission Loss.

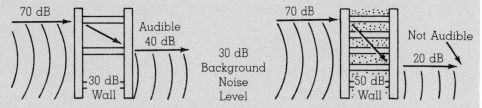

**FIGURE
14.5**

Continued.

point. Restrictions in size and cost will often affect the design. Assuming some degree of freedom, however, consider the following questions:

☐ What are the position requirements?
☐ What substances are to be used in the construction?
☐ What exterior design is best?
☐ What interior design is best?

Position
    The object here is not to determine the position of the wall unless this freedom has been given. The contractor must consider the position of the wall. In other words, how will noise control be affected by the position? What are the noise sources? Where are they located? The high frequency of a jet aircraft creates a need different from the need created by the low frequency of plumbing, although the decibel level may be the same in a given situation.

Composition
    As was demonstrated in the discussion on insulation, the material used in construction can be as important as how it is used. Certain materials are better for different frequencies of sound. The table in Figure 11 summarizes the data in its most general form.
    Standardized tables, like Figure 4, can indicate the relative frequencies for various noises. Such tables can also be found in the texts listed in the reference section of this report.

Exterior
    This discussion concerns the exterior surface of the interior wall. In other words, it is the surface that is seen from inside the building.
    A contractor is given little control over the exterior design of the wall. In a residential market it is generally not acceptable to vary from the ordinary flat surface. In a recording studio or a recital hall an irregular surface may be the norm. Is an irregular surface always best? No. Sound waves are not rigid. They are not "broken up" by irregularities in the surface. They may, however, change

Figure 11.   Construction Materials and Their Use.

Material	Works best at
Porous	high frequencies
Non-porous (such as paneling)	low frequencies

10

FIGURE 14.5

Continued.

direction. The exterior design, then, should attempt to change the direction of the sound waves. Figure 12 illustrates how an irregular surface might disperse sound waves.

Baseboards become important in dispersing sound waves also. The cracks and crevices of the floor can carry sound waves quite well. Always use some type of baseboard to seal the area between the wall and floor.

Absorbing materials, such as wallpaper and carpeting, also improve noise reduction. Layered materials, like paneling, serve the same purpose.

Interior

Often the interior of the wall is determined by the exterior characteristics. A solid concrete wall has the same internal and external construction. Changing the thickness would be the only way to enhance its noise control capabilities.

This layering effect, mentioned earlier, applies to the interior construction also. Layers successfully alter and absorb the sound waves much like the layers of winter clothing protect from the cold air.

Placement of the studs can be altered to facilitate noise control. In Figure 13a the studs are attached to both sides of the interior structure. Thus, the entire wall is acting as one layer. By staggering the studs as shown in Figure 13b, in effect, another layer is added to the wall, more effectively changing the vibrations which accompany all sound waves.

Figure 12. An Irregular Advantage.

Figure 13. Staggering the Studs.

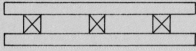   (top views)

a.   b.

FIGURE 14.5

Continued.

USING BUFFER WALLS TO KEEP SOUND WHERE IT BELONGS
    Buffer walls are generally best when used to control noise that originates close to ground level. Thus, they are ideal for the public transportation system. The extent of a buffer wall project dictates an unambitious criterion in composition and design. This is an ideal market for concrete construction. Height and thickness of these walls can be adapted to the project at hand. Generally, the walls will be from three to seven feet high, and approximately one foot thick.
    Varying the surface design is a feasible idea. First, the design will add to the appearance of the project.
    Second, the design can change the direction of sound waves, as Figure 12 illustrated. A thinner wall, using less material, can, therefore, accomplish the job of a much thicker wall.
    We have developed a design that can be utilized in the buffer walls (See Figure 14). Production of this design would not be difficult, as the mold can be easily constructed.

Recommen-
dations

PART FOUR

GETTING DOWN TO BASICS IN THE NOISE CONTROL ISSUE
    Proposed County Ordinance PL G101.7 follows a heritage of noise control attempts by local governments. Despite earlier setbacks, these measures are regaining prominence in the public eye. Given these circumstances, passage of PL G101.7 is almost assured. New markets in the residential, industrial, and

Figure 14.   Surface Design for the Sound Buffer Wall.

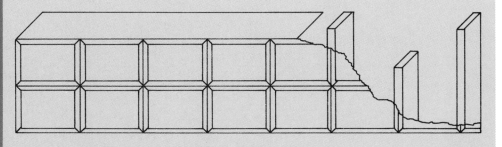

**FIGURE
14.5**

Continued.

transportation sectors will open for construction contractors. Miska should expect to find applications for sound insulation, interior walls, and exterior buffer walls.

Specific legal restrictions and guidelines will come from both the federal and local level. These regulations will have an important impact on the noise control contractor.

The following conclusions are relevant:

1. The proposed ordinance will pass, and new markets will be created.
2. The existing residential market appears unwilling to retrofit for noise control.
3. The new residential market is somewhat promising, particularly in insulation applications.
4. Industrial customers will be required to retrofit.
5. The public transportation system will be required to use buffer walls to control traffic noise.

WHAT YOU CAN DO IN THE NOISE CONTROL EFFORT

These recommendations are indicated:

1. Miska should use the submitted preliminary ideas for noise control to develop and produce control devices.
2. Miska should begin to advertise a strength in noise control contracting to gain public and industrial confidence.
3. Miska should begin to lobby for state and local transportation contracts, emphasizing an expertise in concrete construction.
4. Miska should look into government loans for noise control contractors.
5. Miska should investigate more fully new types of insulation and control methods through the use of scientific periodicals.
6. Miska should keep abreast of all regulation changes in noise control locally and through the use of the library's government document section.

Optional
Resource
List

PART FIVE

APPENDIX

Taking Advantage of Your Library's Government Documents Section

Follow these steps to keep up with changes in federal regulations regarding noise abatement. After locating the government documents section of the library,

13

**FIGURE
14.5**   Continued.

consult the Monthly Catalogue of U.S. Government Documents. Find the most recent issue or volume that contains the time period you are interested in. The
Catalogue is divided into categories such as author, title, subject, etc. Various
references will be given by numbers such as 85-21300. The first two digits represent the year. This reference number is used to locate the filing number of the
document. In the separate appendix volume, look up your number 85-21300. The
appendix will give you a filing number such as Y1.1/5:97-110. This is the number
that is actually listed on the shelves or on microfilm. Your librarian should be
able to assist you if you cannot locate the appropriate document.

REFERENCES

Bulletin of the Acoustical Material Association. 335 East 45th Street, New York,
    New York 10017.

Burns, William. Noise and Man. Philadelphia: J.B. Lippincott Co., 1973.

Close, P.D. Sound Control and Thermal Insulation. New York: Reinhold Publishing
    Co., 1966.

Government Document, Y1.1/5:97-110, 1985.

Harford, E.R. and Olishifski, J.B. Industrial Noise and Hearing Conservation,
    National Safety Council, Chicago, Ill. 60611, 1975.

Lipscomb, D.M. and Taylor, A.C. Noise Control: Handbook of Principles and
    Practices. New York: Van Nostrand Reinhold, 1978.

Magrab, E.B. Environmental Noise Control, John Wiley and Sons, New York,
    1975.

Maling Jr., G.C. Noise-Con 77, Noise Control Foundation, New York, 1977.

May, D.N. Handbook of Noise Assessment. New York: Van Nostrand Reinhold,
    1978.

"Noise Abatement Week Proclaimed," American City, October 1940, p. 60.

Yerges, L.F. Sound, Noise, and Vibration Control. New York: Van Nostrand
    Reinhold, 1969.

14

table of contents in upper- and lower-case letters instead of all capitals. A mixture of upper and lower case is easier to read.

Since this report reads well, Dorsey's final improvements may include his clients' suggestions. However, Dorsey still uses too much passive voice. The Miskas run a construction business and their work requires action more than contemplation. The active voice improves readability, increases comprehension, and heightens the report's appeal to them. It will also allow Dorsey to create a more complete level of detail, since the active voice enables him to use more descriptive verbs.

SPECIAL CONCERNS OF TECHNICAL REPORTS

You now understand that technical reports differ from term papers in many senses. We have already discussed the technical report's emphasis on an attractive appearance and on an easily readable format. Three other concerns of report writers are the table of contents, abstracts, and the citation or reference process.

Table of Contents

In Chapter 5, we discussed in detail how to build a table of contents. Refer to Chapter 5 for examples of tables of contents. As you do so, remember that although capital letters create emphasis for your headings, they are also very hard to read. Design a table of contents that uses both capital letters and lower-case letters in combination. To find the correct format for your table of contents and list of illustrations, consult the style handbook preferred by your publisher or firm, because the formats for these can vary considerably.

Abstracts

The abstract, in short, is a report summary. Why are abstracts important?

The abstract helps a librarian decide how to catalog your report, how to store it, and how to decide who should have access to it. Thus, in an age in which security systems, computer searches, and the nature of information itself all contribute to restrict a user's access, the abstract is critical. Your abstract must describe the report so accurately that no mistakes can occur in the cataloging, storage, or distribution of your effort. The accuracy of the abstract becomes even more important if the report is written for external publication.

There are two types of abstracts, descriptive or informative.

A descriptive abstract appears at the beginning of Bill Dorsey's report. This type of abstract is usually quite short (250 words or less), and describes, rather than summarizes, the report's content. The descriptive abstract, for example, may list the report's topics, its categories, its length, and other specific facts of the report's design or format, but it is no substitute for the report.

On the other hand, the informative abstract is truly a summary of the report's content. With a length equal to about 10 percent of the report, the informative abstract states the report's problem and scope, describes its method of investigation, enumerates results and conclusions, and lists recommendations. Figure 14.6 contains two examples of informative

**FIGURE
14.6**

Sample Abstracts.

8703788
Samaria, William J. (U Toronto, Toronto M5S 1A1), **Demythologizing Plains Indian Sign Language History,** UM *International Journal of American Linguistics,* 1987, 53, 1, Jan, 65-73.
¶ Questions are raised about the existence of Plains Indian Sign Language (SL) prior to Amerindian contact with Europeans. Sources for statements about SL are examined. It is suggested that gestural communication systems arise only in unusual circumstances, & that as the existence of trade networks before European contact did not give rise to gestural langs, the likelihood that SL was postcontact development is increased. The old histories of SL are found to be romanticized. 29 References. B Annesser Murray

Courtesy of The University of Chicago Press.

Impaired fat deposition in pregnancy: an indicator for nutritional intervention[1,2]

OAC Viegas, MD, TJ Cole, MA, PhD, and BA Wharton, MD

ABSTRACT   Unselective dietary supplementation of pregnant mothers can have adverse effects. It is essential to predict which mothers are nutritionally at risk and would benefit from supplementation. Preliminary studies indicated that mothers who increased their triceps skinfold by $\langle$ 20 μm during the second trimester were nutritionally at risk of having a small baby. In this study anthropometric indicators were explored in 81 Asian mothers living in Birmingham. Single (distance) measurements (eg, weight at 28 wk, etc.) were not effective predictors. Changes in measurements of the arm during the second trimester were predictive, and triceps skinfold remained the most effective predictor of poor fetal growth. This measurement probably reflects the balance of energy intake minus energy expenditure during the critical second trimester when the mother would normally be laying down extra fat in anticipation of later fetal demands.   *Am J Clin Nutr* 1987;45:23-8.

KEY WORDS   Triceps skinfold fat increments, intrauterine growth, pregnancy, fat deposition

From the *American Journal of Clinical Nutrition.* Courtesy American Society for Clinical Nutrition.

abstracts that researchers may read in the library to determine a given research paper's usefulness in their own search of the literature about a subject.

Here are some items your abstract should contain:

Title — the exact title of your report.
Identification — whatever identification your publisher may require, including authors' names, place of publication, date, identification or file number, number and type of illustrations, and number of pages.

Statement of the problem — the exact topic and scope of the paper or report.

Methods, results, and conclusions — a brief description of each, depending upon the nature and requirements of your report.

How to Write an Informative Abstract

You can follow some simple guidelines to draft a consistent, logical abstract.

1. Find out how long the abstract must be. If you cannot obtain this information, then you know the abstract must be 10 percent of the document's total length.
2. After reading the article once for overall comprehension, reread the article, and underline the author's thesis.
3. Read the article again and highlight every unique term, procedure, or method that should be noted.
4. Examine the article for sections or memorable words you think the abstract should quote verbatim.
5. Compose a short outline with the thesis at the top and the main ideas of the article ranked either by importance or by their order of appearance in the article.
6. Draft the abstract and proceed with the usual review process until the document is polished.

References

Each discipline has its own handbook for the process of "citation." Citation is the attribution of credit to the sources of your text's direct quotations, summaries, or paraphrases. For example, Bill Dorsey, Kim Ash, and Mark Graves use the humanities reference system in *A Manual of Style, 13th ed.*, published by the University of Chicago Press. A comprehensive manual of this type can answer many of your questions about criterion techniques and requirements. Before drafting your report, decide which reference system to use. Doing so will save you hours of time in the transposition of endnotes to footnotes within the text, or in the arrangement of your reference pages.

FIGURE 14.7   A Short Working Bibliography Arranged According to the Chicago *Manual of Style*, Form A for the Arts and Humanities.

A SAMPLE REFERENCE LIST

Reference Works about Desktop Publishing

Campbell, Alastair. *The Graphic Designer's Handbook*. Philadelphia: Running Press, 1983.

Lucas, Amy, ed. *Encyclopedia of Information Systems and Services*. Detroit: Gale Research Company, 1987.

Seybold Report on Office Systems. Seybold Publications, P.O. Box 644, Media, PA, 19063.

Sigel, Efrem (publisher). *Computer Publishers and Publications: An International Directory and Yearbook*. Larchmont: Communications Trends, 1985.

University of Chicago Press. *Chicago Guide to Preparing Electronic Manuscripts*. Chicago: University of Chicago Press, 1987.

SUMMARY

In this chapter, you learned to compose long reports such as closeout, investigative, and feasibility reports. You learned that long reports allow you opportunities to create a permanent record, to display your completed work, and to win supporters for your views. After reading this chapter, you should be able to:

- ■ Use a checksheet to examine a report you read, write, or edit to make certain the report contains all the usual elements.
- ■ Modify a checksheet to meet the requirements of your own reports.
- ■ Use a generic outline to compose a long report.
- ■ Modify a generic outline to meet the organizational requirements of the closeout, investigative, and feasibility reports you compose.
- ■ Describe for your classmates the ways in which technical reports differ from term papers.
- ■ Choose when to omit and when to modify the report elements listed in the generic outline.
- ■ Discuss the elements that create an appropriate approach and justify to others your decisions about the composition of your own report.

ASSIGNMENTS

Select a topic from the list we proposed in Chapter 3, or choose your own topic for a technical report. Follow the steps we list here.

1 Prepare for your instructor a memo which explains your plan of research. What questions will you study in the library? What questions will you answer through active research, including tests and measurements, on-site inspections, interviews, and other practical information-gathering techniques? How long will the investigation take?

2 Copy our generic outline and modify it to suit your report. Will you compose an investigative or a feasibility report? If employed, you may have the chance to write a closeout report. If your instructor wishes, submit your memo and outline for his or her review and comments.

3 Select a partner who will be named as the second author on your report's title page. The second author should read and mark all your drafts, and use our four-part review process to comment on how you can improve the drafts.

4 Prepare your report and work with your second author on its revision. At the same time, you may work with another student in the class as his or her second author. Thus, you will be commenting on a report for an associate as you compose your long report.

5 Prepare a cover letter that assumes a personal orientation toward your instructor. Why should your instructor be anxious to read your report? What new information have your partner and you unearthed? Persuade your instructor that your report is a valuable product. Submit your beautifully designed and formatted report, with its cover letter, to your instructor for evaluation.

CHAPTER
15

Describing Mechanisms and Drafting Instructions

This chapter introduces you to the description of mechanisms and processes. You will learn how to analyze a task, so you can write simple instructions for performing it. After reading this chapter, you will be able to:

- Choose a point of emphasis—physical, functional, theoretical, or a combined emphasis—for your planned description of a mechanism.
- Draft a description of a mechanism that reflects an appropriate emphasis for its audience.
- Design instructions with a specific audience in mind.
- Develop your instructions with a conventional plan of organization, including: introduction and theoretical description, physical description of the mechanism, list of parts, sequence of tasks, cautions and warnings, and emergency procedures.
- Edit your instructions to conform with appropriate style conventions.

CASE STUDY   A Description of the TREE Command

Ved Krishnan, an accountant, maintains the financial records for the Eastwood, Inc., grain elevator in which his brother, Lal, has a financial interest. Because of his involvement with Eastwood in the dual capacities of accountant and family member, Mr. Krishnan answers, with good humor, many questions relating to records-keeping, systems management, and computer programming.

Eastwood recently purchased a new computer system that allows the managers to quickly access their current records. Along with the hardware, the corporation purchased a reliable spreadsheet for the operation's various bookkeeping and inventory functions. Unfortunately, Lal and the other managers do not understand how the computer's disk operating system (DOS) classifies and stores information. When Ved finishes the books, the managers are afraid to make new entries. They fear losing the files which record a farmer's or a firm's transactions with Eastwood.

The computer's DOS stores information within a simple TREE system the managers seem to neither learn nor understand. Ved must compose a short set of instructions for Lal and the others, so they can monitor their daily records without damaging the files and without repeatedly interrupting Ved. Although quite a few books on DOS explain how to use the TREE command, Ved knows they are too difficult. His set of instructions must be short but theoretical enough that Lal and the other managers can regularly use the TREE command for their different customers and various bookkeeping requirements.

TYPES OF DESCRIPTION

Consider the various types of descriptions Ved might write about his brother's computer. We have listed some of the descriptions that are possible:

- ☐ Purpose—what the machine is for.
- ☐ Physical—how it looks.
- ☐ Electronic—how the cards, chips, and circuits work.
- ☐ Systems—how the disk drives, power supply, and other components work.
- ☐ Language—how hexadecimal and machine languages work.
- ☐ DOS—how the system allows the machine to communicate.
- ☐ Software—what programs the machine runs and why.

No doubt you can think of other aspects Ved might describe. In fact, the study of how people describe objects and processes consumes volumes of analysis. For this reason, read Chapter 6, Methods of Development, for an appropriate rhetorical background. In the field of technical communication, we like to concentrate on three broad categories of description for objects and processes—physical, functional, and theoretical.

Although this chapter shows examples of physical, functional, and theoretical descriptions, remember that the categories often overlap because writers describe mechanisms and processes for real audiences with specific requirements.

This chapter shows you two texts that discuss the TREE command. The first passage is a theoretical description written by one of Ved's business associates. The second passage, Ved's instructions for using the TREE command, explains the theory by leading Lal through the correct logical process with examples drawn from his accounts. Obviously, Ved's TREE can be explained not only as a theoretical analysis, but also as a set of instructions.

More simply, if Ved describes a new computer for the grain elevator, as he might in, for example, a request for a price quotation, he should also include a physical description of the product including its dimensions, the color, the specific design of the keyboard, and the other important features of the machine's appearance. At the same time, Ved may have to write a description that emphasizes the product's functional aspect, for example, the technical specifications of one- and two-speed computers. Thus, your emphasis is related to the purposes and the audiences for which you describe a mechanism or process.

Description of Appearance or Construction

Study the "living screens" description of shrubs and bushes for sale at a nursery. Although the information sheet about trees and shrubs candidly promotes Frank's Nursery, it also offers an effective example of the description of objects, in this case, trees or shrubs as screens, with a primary emphasis on their appearance. Consider the analysis the writer had to make:

Audience—Lay person, gardener-hobbyist, homeowner. Great interest in the topic but little time.

Education—Not important; all are lay audience.

Knowledge of Topic—minimal because all kinds of gardeners buy at this store. Wants to know *if*, not how, the plant will work as a screen.

Task—1) purchase the trees or shrubs; 2) plant them; 3) maintain them.

Information Needed—1) how the different plants look; 2) height and width of various plants; 3) decorative features; 4) how to plant; 5) how to fertilize; 6) watering schedule; 7) any other unique, critical maintenance features.

Tips for Describing a Mechanism by its Appearance or Construction

1. Describe all the physical characteristics, including the mechanism's length, width, depth, density, weight, material, color, sound in operation, or odor.
2. Describe all the characteristics of its construction, including the brand names of special parts, the unique parts that define or characterize the mechanism, the names of the unique processes that mechanism performs, and the list of its technical specifications. For example, here are the electrical specifications for a sump pump that fits the description shown in our example for this chapter (see Figure 15.2).

FIGURE 15.1

A Simple Chart Shows How to Classify a Group of Objects by Shape or Size.

Let Frank's help you choose materials for

LIVING SCREENS

The use of trees and shrubs for screening in residential areas has become popular for several reasons. Plant screens can block direct view of objectionable sights from outdoor living areas, give needed privacy from the bustle of streets and highways, and provide sun and wind control. The living beauty of plants also offers a softer, more colorful appearance than walls or fences. Described below are a few of the more popular plant varieties used for this purpose. Consult the description, then visit your nearby Frank's for further assistance and a wide selection of sturdy screening plant specimens.

PYRAMID ARBORVITAE - A compact evergreen species, it grows best in cool, moist locations. Its foliage is dense, waxy and fragrant. Consider it when providing shelter for more tender plants. Pyramid Arborvitae are also well displayed in borders and hedges.

BORDER FORSYTHIA - Here is a rapid-growing deciduous plant that assumes a dense, broad, upright stature, standing some 9 feet at maturity. Forsythia will be covered with yellow blooms before the leaves appear in early spring.

SPIREA - A member of the Rose family, the Spirea is a small to medium sized shrub with a rounded habit. Several excellent varieties exist, which bloom at different times. Most produce graceful, white blossoms. Spirea grows best in a rich, moist loam.

BLACK HILLS SPRUCE - A dense form of Spruce which often reaches heights of 20 to 40 ft. and requires less trimming than most other Spruce varieties. In good soil conditions it sometimes acquires a bluish color, but grows well even in poor soil.

LALAND'S FIRETHORN - This popular Pyracantha is a vigorous semi-evergreen spiny shrub with sprays of orange-red berries in the fall. Mature height and breadth are 6 to 10 feet. It tolerates poor soil nicely and withstands a variety of weather conditions.

ZABELI HONEYSUCKLE - A deciduous plant which is widely used for hedges and screens, the Zabeli Honeysuckle is noted for a profusion of fragrant reddish blooms in spring. This vigorous and hardy shrub also produces an abundance of red fruit in late summer.

SPREADING COTONEASTER - This is a deciduous plant with an attractive upright spreading habit. Graceful and open branched, with an abundance of bright red fruit in the fall, it is considered by most experts to be one of the finest Cotoneasters.

RED AND PINK WEIGELA - These are deciduous shrubs of the Honeysuckle family with beautiful spreading habit. Their pendulous branches bear beautiful clusters of bell-shaped flowers in the spring and early summer. Be sure to keep their soil moist!

SWEET MOCK ORANGE - An informal deciduous plant, Mock Orange is considered ideal for dry climates and soils. It grows upright to a height of 9 ft., has numerous twigs, and requires trimming only infrequently. Its flowers are white and very fragrant.

BURNING BUSH - A variety of Euonymus, the Burning Bush is a large native deciduous shrub that frequently attains a height of 10 ft. It thrives in sun or partial shade, and is not particular as to soil. Its leaves are flaming red in fall, and it has bright orange berries.

LILACS - A fragrant and colorful mass of spring blooms have made the Lilac an all-time garden favorite. These vigorous plants are available in several types including our own native Lilac and beautiful French hybrid varieties in white and shades of pink, blue or purple.

HICKS YEW - A tall columnar evergreen it is the narrowest of all Yews in general use today. Whether planted in a formal or an informal setting, this outstanding evergreen adapts well to trimming.

Frank's *Nursery & Crafts, Inc.*

007-0 99627 • Inst. 77 © 1971

Motors
    ⅓ HP—1 PH—60 Hz—115V—1725 RPM
    Split Phase, Overload Protected
    ½ HP—1 PH—60 Hz—115V—1725 RPM
    Split Phase, Overload Protected
Fuse Capacity
    15 amp. plug type or 8 amp. time delay type — ⅓ HP
    20 amp. plug type or 12 amp. time delay type — ½ HP

3. If you describe a new product or mechanism, edit the text for consistent terminology. For example, if you refer to a keyboard on a computer, decide which verbs—"type," "keyboard," "press," or "enter"—to use throughout the description.
4. Be sure to describe everything the mechanism is designed to do. For example, if the keyboard is for a computer, but you can also use it as a typewriter keyboard with a printer, outline this feature as well.

Description of Function or Use

Sometimes the user needs to understand what a particular mechanism can do for him or her, that is, how it operates. Thus, you may add to, or substitute for, the physical description a functional analysis that helps the user appreciate and use a mechanism properly. For example, Figure 15.2, A Homeowner's Guide to Sump Pumps, was written by students who work at a plumbing supply company. The students noticed that they were continually explaining to their customers how a sump pump operates because the pump's actions cannot be seen when it is operating. Because the customers sometimes do not understand the mechanism's functions, they often buy pumps that overheat and burn up or they install their backup pumps improperly.

    Note that to hold the description to a readable length, the students minimize the physical description and concentrate on how the sump pump performs its functions. Also, they explore the full capabilities of the product by explaining that some homeowners use two pumps, an original and a backup, in case the first mechanism fails during a rainstorm. Then they concentrate on how to install a backup, assuming that the customer already has one sump pump in the basement. Here is the students' analysis of the functional description project:

Audience—Lay person, homeowner, some mechanical ability.
Education—Not important; all are lay audience.
Knowledge of Topic—May think he or she understands how a sump pump works. Also, may have no knowledge. Wants to know which pump to buy as a backup.
Task—1) Select and purchase the right capacity pump; 2) buy the right equipment for installation; 3) install the pump as a backup.
Information Needed—1) How the pump works as an original or a backup; 2) what the parts are and what they do; 3) how to install a backup pump.

(Text continues on p. 401.)

FIGURE
15.2

Description of a Sump Pump With a Functional Emphasis.

A HOMEOWNER'S GUIDE TO SUMP PUMPS

This brochure explains how your sump pump works, so you can use and maintain this important appliance effectively. If you install your pump properly, it should give many years of reliable service.

How A Sump Pump Works

The domestic sump pump is especially designed for basement drainage. The different domestic sump pumps are classified according to the maximum gallons per minute they can drain.

The Impeller

Most sump pumps come designed with a non-clog impeller (the opening in the pump's base). Two principles apply here: the larger the impeller, the larger the solid that can be passed through the pump; the larger the impeller, the more expensive and elaborate the pump.

The Capacity

Two factors determine the gallons per minute the pump can dispel: the total head in feet and the pump's pumping capabilities. The total head in feet refers to the height to which fluid is to be raised and the amount of water the pump can deliver at a given height. A table illustrates the capacity in gallons per minute that the sump pump can dispose of at a given height.

total head in feet (height)	5	10	15	20	22	24
capacity in gallons per minute	46	42	36	30	27	22

The Features

Simple sump pumps offer adjustable on and off switches as standard equipment. More technically advanced pumps have optional floats or pressure switches that provide 6″ to 35″ of adjustment.

The most popular sump pump has an all-plastic construction, with a ⅓ horsepower motor. Such pumps also have cast-iron construction and an oil-filled motor for cooling. The thermal overload protection prevents burnout. Other more advanced and more expensive pumps come constructed of stainless steel, bronze, and corrosion-protected steel.

What to Buy

The construction of a pump should offer high strength, stiffness, impact strength, insulation, and heat resistance to 120 °F. These conditions should meet the needs of each drainage problem. Hydromatic Pumps publishes a bulletin, Pump Sizing Made Easy, that includes charts, graphs, and step-by-step instructions that allow the user to select the proper pump for a given application. This booklet is available from a plumber or a regional Hydromatic Pumps dealer.

**FIGURE
15.2**

Continued.

Why to Buy a Back-up Pump

To prevent water damage, many homeowners install a second pump as a backup. Installation of a second pump can create ease of mind for the homeowner, since it lessens the probability of the basement's flooding, causing water damage and expensive repairs. Back-up pumps seem well worth the investment in view of the dollars they can save during bad weather.

How the Parts of a Sump Pump Work

Base.   The base assembly rests at the bottom of the sump (Pump Housing) and supports the rest of the sump pump. The base contains the intake ports, where the water is drawn into the system; the pump mechanism, which forces the water through the system; and an output port, through which the water is pumped out of the system and out of the basement.

Check Valve.   The check valve connects into the line that carries the water out of the sump. It keeps the water in the pump drain line from returning to the sump after the pump turns itself off. **You must buy this part separately.** You can buy a check valve at any local plumbing supplier.

Electric Cord.   The electric cord connects the motor to the household electric current. You can connect the cord to any regular household electric socket (120 volts).

Electric Motor.   The pump shown here has a heavy-duty, ⅓ horsepower, single-phase, 60 Hz motor which runs off regular household current (120 volts). This motor can pump water at a capacity of 2800 GPH (gallons per hour) for 5 feet vertically, and the pump can also pump water up to 20 feet vertically, under light load conditions.

Electric Switch.   The electric switch turns the pump on and off.

Fixed Vane Rotary Pump.   The fixed vane rotary pump moves the water out of the sump and out of the basement. The pump mechanism connects to the electric motor through the pump drive shaft, which rotates the pump and causes the pumping action. The pump draws water into the system and, because of the rotary action and vane shape, pushes it through the output port and out of the basement.

Float.   The float rises as the water level in the sump rises. When the float rises to a certain height, it activates the switch, which turns on the motor and pumps the water out of the sump. As the pump forces the water out of the sump, and the water level lowers, the float will also lower. When the float lowers to a certain point, it will turn off the switch and stop the pump.

Float Rod.   The float rod connects the float to the switch on the motor.

Float Rod Guide.   This device keeps the float rod in a vertical position and guides it as the float moves up and down.

Pump Drive Shaft.   The pump drive shaft connects the electric motor to the pump mechanism. The shaft transmits the rotary action produced by the motor to the vane pump.

Shaft Housing.   The shaft housing protects the pump drive shaft from the water. It is basically the plastic tube that contains the shaft.

FIGURE 15.2   Continued.

Switch Actuators.   The switch actuators consist of two rubber stops attached to the float rod that turn the switch on and off. You can adjust the stops to allow the pump to activate and deactivate at different water levels.

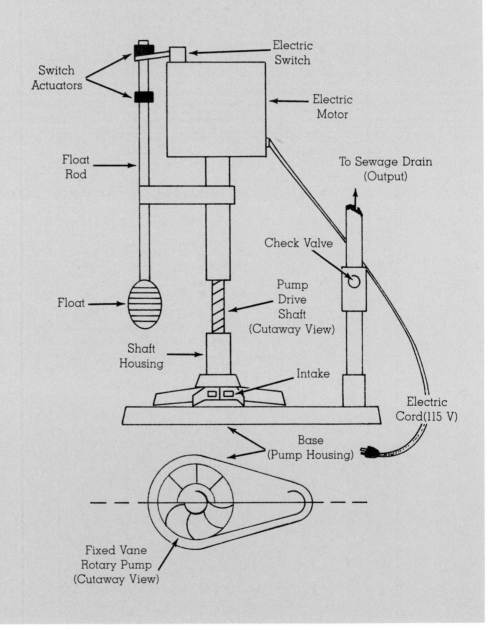

Switch Actuators

Electric Switch

Electric Motor

Float Rod

To Sewage Drain (Output)

Check Valve

Float

Pump Drive Shaft (Cutaway View)

Shaft Housing

Intake

Base (Pump Housing)

Electric Cord (115 V)

Fixed Vane Rotary Pump (Cutaway View)

| **FIGURE 15.2** | Continued. |

Suggestions for Use as a Back-up Pump

To set up an additional pump as a back-up system to another, follow these steps.

1. Compare the height of the two switch actuators on the primary pump to those on the secondary (back-up) pump.
2. Adjust the switch actuators on the secondary pump to the same height as those on the primary pump.
3. Move the lower switch actuator (the one below the switch) on the secondary pump about two inches down the float rod.
4. Test the two pumps by raising the floats to where they activate the switches. The float of the secondary pump should be about two inches higher than that of the primary pump.
5. Place both of the pumps into the sump, and make sure that the two pumps do not interfere with each other's operation.
6. Hook both pumps to the drain. You can hook up both pumps to the same drain line; however, make sure that each pump has a separate check valve system.
7. Observe the pumps in operation. When the primary pump activates, the secondary pump should not run. If the secondary pump does become activated, lower the bottom switch actuator another inch.

The Four Phases of Pumping

After the pump is placed in the well and plugged in, you are only four easy steps away from understanding the operation of a sump pump.

Phase 1.   The sump pump is turned on and off by a floating device. At the top of the float pole, there are two rubber stoppers. These stoppers turn the on/off switch to its appropriate setting, depending on the water level. The lower stopper turns it on, and the upper one turns it off.

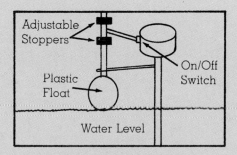

FIGURE 15.2   Continued.

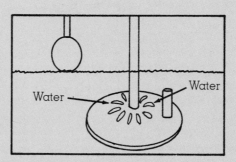

Phase 2.   The second phase begins when the on switch is flipped to the up position. This starts the ⅓ horsepower motor. The motor drives the suction at the base of the sump pump. The sump pump can drain 2800 gallons of water per hour.

Phase 3.   Next, the sump pump pumps the water out. The water passes through the tubing hooked to the outtake pipe at the base of the sump pump. Because you supply the tubing, make sure that the tubing you purchase has a ball check system. The ball check system prevents a backwash into the pump. Such a backwash can damage your motor.

Phase 4.   Once the water level drops enough, the top rubber stopper will turn off the sump pump. The process will repeat itself once the water level rises enough to turn the switch back on.

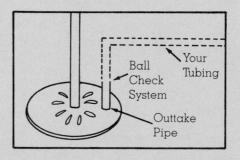

Tips for Describing a Mechanism by Its Function or Use

1. Describe all the operations the mechanism can accomplish, with an emphasis on the function the user needs to understand. For example, the students describe first how the sump pump works as a main pump, then how to install the pump as a backup.
2. Define the parts of the mechanism by explaining how they operate and how they help the entire device do its job. Note that the students define each pump part, not by its appearance but by its function.
3. In functional descriptions, emphasize how each part helps another part to perform. For example, the students explain that the rubber stopper begins and ends the pumping.
4. Remember that a functional emphasis is often necessary to explain moving mechanisms that users cannot see. For example, you might use a functional description to explain nuclear reaction, internal combustion, enclosed pumps or sprays, or certain sterile chemical or biological operations. Whenever possible, illustrate your description, preferably with line drawings that allow you to emphasize the operations you define. If you cannot illustrate the mechanism, create a verbal illustration for the user.

Description of Theoretical Aspect or Operating Principle

As you study Ved Krishnan's description of the TREE command, you can see that it is impossible to visualize a TREE command in normal operation, just as it is almost impossible to see a sump pump operating beneath a basement. The theoretical analysis shares this difficulty with the functional description. As you read this description, note that the author deals with the problem of abstraction in two ways. First, he defines the object, the TREE, generically, that is, in terms of any or all directories. He discusses the broadest possible application of the theory. Then, he develops the description with several examples, including paths, branches, directories, and other ideas related to the concept of a tree. He then sketches schematics or flowcharts to show how a TREE operates. In these senses, the description is a bit esoteric, but the author has to create a logical structure that allows his students to see the idea of the TREE.

Tips for Describing the Theoretical Aspects of a Mechanism

Sometimes, a topic is inherently theoretical; that is, you must explain a principle of operation in order to describe a mechanism. For example, the need for theoretical explanations arises with topics in the fields of computer science, electronics, the sciences, or mathematics. Always you face the twin problems of creating images and dealing with special jargon as you describe the topic.

1. Introduce special jargon in context. For example, the author of the TREE command description introduces the MKDIR command by explaining the expression for which it is a mnemonic.
2. Illustrate copiously. Note that the author illustrates each new command with a TREE. Do not expect the user to generalize from one illustration to another.

**FIGURE
15.3**   Description of the TREE Command with an Emphasis on Theory.

TREES AND ROOT DIRECTORIES

If you have a hard disk, Bernoulli box, or other large capacity, fast access memory storage such as virtual disks, and the need to store and access a complex file system the MKDIR, CHOIR, RMDIR, PATH, and external TREE commands are the answer. These commands exist in all DOS systems. Different DOS systems incorporate them in slightly different manners and use some additional commands. However, all are essentially the same.

The system of directory construction is a TREE form as follows.

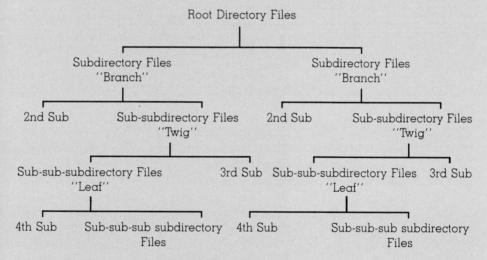

The process of constructing a TREE is simple and minimal in its concept. The MKDIR command, a mnemonic for "make a directory" or its abbreviation, MD, for "Mk Dir" is all that is necessary. The user may create a subdirectory at any level by entering the directory the user wishes to divide into subdirectories. The MKDIR command creates a directory in the location where it is executed. For example, if the user is in a fourth-level directory and has a C⟩ on the screen, the MKDIR command begins to partition the directory at that point.

The MKDIR command partitions memory registers at whatever address in memory the command lies. All register locations after or below that address will exist under that address in that register.

The disk space is partitioned in the same way. A 20-megabyte disk with one root directory has 20 megabytes of space in that ROOT file. Add one subdirectory, and the subdirectory now "takes away" 20 megabytes. Add a second subdirectory, and the two subdirectories will "store" the 20 megabytes. The sharing is open. That is, all 20 mega-bytes could be used by one subdirectory if no entry of data is made in the second subdirectory. The actual file size is limited by the DOS version being used, as is the size of memory and storage device memory size.

| **FIGURE 15.4** | Analysis Form for Technical Descriptions. |

ANALYSIS SHEET—TYPES OF DESCRIPTION

Background
    Audience Analysis Summary
      Education

      Skills

      Use of Product

      Presumed Skills, Ability, Knowledge, and Use—Summary

Fill in each of the columns below about the mechanism.

| Physical | Functional | Theoretical |
| How Does It Look? | What Does It Do? | How Does It Operate? |

Which column will you emphasize? Circle one item. If you must circle two items, creating two points of emphasis, explain briefly.
    My emphasis will be: physical   functional   theoretical
    My explanation for two areas of emphasis is:

The goal of my description is (use one sentence):

The audience will read my description to find out:
1.
2.
3.

3. Use a consistent format. For example, the author highlights the difficult concepts by placing them alone on lines and by using quotation marks and capital letters.
4. Use redundancy to create a uniform explanation. Introduce the new terms or mechanisms, and then define, illustrate, and summarize them. Even a skilled reader needs repetitions to understand and remember a new idea more easily.
5. Balance your text with your illustrations and surround both with ample margins. The description gains in visual appeal and is actually easier to read and understand with more white space.

CONTENT CONVENTIONS FOR DESCRIPTIONS OF MECHANISMS

Mechanism descriptions tend to conform to one pattern whether you write a one-page description for a plumbing supply company or a two-paragraph description as the first part of a set of instructions. Users expect certain types of information in the introduction, body, and conclusion of your description.

Introduction

It is sometimes useful to offer background or consumer education about a new product in the introduction's first paragraph. For example, the TREE command explanation offers such a background statement in the first paragraph. If you describe a product, you may wish to use the first paragraph to reassure or congratulate the consumer for his or her choice.

If you write a brief background statement, the first sentence of the second paragraph should introduce the product with its specific brand name. The second sentence should describe the class or category of objects to which the mechanism belongs. For example, the TREE definition fits this pattern because Ved Krishnan defines the TREE as part of the machine's DOS, a system the user can program to create various kinds of directories.

Here is a less formal version of an introduction for a typical wall telephone.

This telephone is designed to offer you the best quality and convenience for home and business use. This telephone includes these following features:

☐ Tone or pulse dialing (switchable)
☐ 32-digit last number redial
☐ Pause function
☐ Mute function
☐ Electronic tone ringer with volume control
☐ Desk to wall convertible
☐ Fully modular connectors

This introduction says little about the classification of the telephone, but instead gives the specifications a buyer is eager to identify. You have some flexibility in your introduction, depending upon the description's purpose and audience.

Body

The body of your description should contain several types of explanations, including:

Physical, functional, or theoretical principles of the mechanism
Identification and explanation of each part
Explanation of relationships among the parts
Unique features or attachments
Limitations of use

For example, the sump pump description contains a brief functional narrative, moves to a list of parts, and then explains the four phases of pumping. The use of the backup is explained as a unique feature. Note, too, that the limitations are outlined; you must use a valve on the hose or a backwash could destroy the pump.

Conclusion

Study the TREE description for an example of an effective conclusion. Quickly, the writer reviews the use of the TREE, and offers suggestions for security in the use of this command. This example shows that a conclusion need not be long. However, it should review the principles of the description, reinforce the point of emphasis, and reiterate any last-minute cautions or limitations of the mechanism.

CREATING EFFECTIVE INSTRUCTIONS

We use the term "instructions" to refer to processes that users must perform on or with mechanisms. Some instructions take the form of procedures, step-by-step explanations of what occurs and why during a given process. Other instructions, directions, tell the user how to complete a certain process with a sequence of actions. Refer to Chapter 5 for a thorough background in the rhetoric of process explanations. Because instructions are so important in technical communication, we devote the rest of this chapter to them, and we do not discuss process explanations. The following informal table (Figure 15.5) outlines the most obvious differences between procedures and directions.

As with descriptions, certain conventions govern your instructions' content and style. So that you can see how the conventions apply, and when it is appropriate to omit certain sections, we refer you to Figures 15.6 through 15.10.

FIGURE 15.5	How Procedures and Directions Compare as Instructions.

INSTRUCTIONS

	Procedures	**Directions**
Purpose	explain the steps in the process	tell or list steps in the process
Topic	show that or how human beings are parts of the process	show what human beings must do in the process
Point of View	list the steps in the process	list the steps in the process
Audience	tell user 1) how the process works, and 2) how to complete the steps	tell user how to complete the steps
Level of Detail	explain steps for understanding; list steps for performance	list steps for performance
Example	Trimmer may be used for two purposes: 1) shrubs and hedges 2) grass Trimmer should be kept free of debris.	To prune hedges, set trimmer blade on 3. To trim grass, set trimmer blade on 4. Clean blade and shield after each use.

Content Requirements

All instructions should contain five parts:

Introduction and theoretical description
Description of the mechanism
List of parts
Sequence of tasks
Cautions and warnings
Emergency procedures

Sometimes one of these sections is deleted. For example, Figure 15.9 contains neither a theoretical nor a verbal description of the ski rack as a mechanism. However, since the instructions are written to fit on a one-page bulletin, the authors have chosen to create this

| FIGURE 15.6 | Ved Krishnan's Instructions for Using the TREE Command. |

UNDERSTANDING AND USING YOUR COMPUTER TREE

Large capacity disks, such as fixed multi-megabyte Winchesters, can store a very large file or a very large number of smaller files. With one or two large files, access is not a problem. One or two first-level subdirectories with no additional subsections are easy to remember. However, a dozen or more subdirectories with additional "sub" subdirectories in each is a problem in filing logistics.

The "TREE" command is the DOS command designed specifically to address the problem of multiple subdirectories and sub-subdirectories. The term "TREE" is taken from the logical (and visual) structure of the command's form shown here.

Basic TREE Structure.

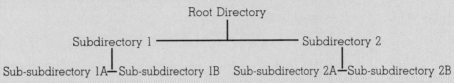

1. Root Directories

You can make a root directory on a disk with the command "MKDIR," the mnemonic for "make a directory." You type MKDIR or the abbreviation MD followed by one space and the name of your directory and [return] as follows:

```
See          C>
Type         MKDIR Root Directory Name 1 [return]
See          C>
```

This command will create a root directory beneath which you may add subdirectories. When you see the C> after the [return], you will be in your new root directory. This structure will be logically as follows:

The Logical Structure of the Installed Root Directory.

Root Directory Name 1

2. Subdirectories

You may now add subdirectories under root directory Name 1. with the same command, that is MKDIR or MD Subdirectory Name 1 and [return] as follows:

```
See          C>
Type         MKDIR Subdirectory Name 1 [return]
See          C>
```

**FIGURE
15.6**

Continued.

This will create a subdirectory beneath which you may add a sub-subdirectory. When you see the C⟩ after the [return], you will be in your new subdirectory. This will be logically as follows:

The Logical Structure of a Root Directory with One Subdirectory.

Root Directory Name 1

Subdirectory Name 1

3. Sub-subdirectories

You may now add sub-subdirectories under Subdirectory Name 1 with the same command, MKDIR or MD Sub-subdirectory Name and [return] as follows:

See	C⟩
Type	MKDIR Sub-subdirectory Name A [return]
See	C⟩

The file structure is called a TREE, and the TREE command can immediately access any branch. The file structure is the same as the "Generic File," only you have installed your file names. For example:

See	C⟩
Type	MKDIR Wheat Accounts [return]
See	C⟩
Type	MKDIR Received [return]
See	C⟩
Type	MKDIR Mys Ltd. [return]

You have constructed this TREE:

Root Directory Name: Wheat Accounts

Subdirectory Name: Received

Sub-subdirectory Name: Mys Ltd.

FIGURE 15.6   Continued.

4. Added Categories

To add a "Sold" category for seed wheat, from computer start up execute as follows:

See	C⟩
Type	CD/Wheat Accounts [return]
See	C⟩
Type	MKDIR Sold Seed [return]
See	C⟩
Type	Mys Ltd [return]

You now have the following TREE:

Root Directory Name: Wheat Accounts

Subdirectory Names: Received ———————— Sold Seed

Sub-subdirectory Names: Mys Ltd ⌐   └— Mys Ltd

To add additional "accounts" at the sub-subdirectory:

See	C⟩
Type	CD/Wheat Accounts/Sold Seed
See	C⟩
Type	MKDIR Ziecek Int. [return]
See	C⟩

You now have the following TREE:

Root Directory Name: Wheat Accounts

Subdirectory Names: Received ———————— Sold Seed

Sub-subdirectory Names: Mys Ltd ——⌐   Ziecek Ltd ——┴— Mys Ltd

FIGURE 15.6   Continued.

5. Accounts

To access an account, use the "CALL DIRECTORY," (CD), command as shown here.

```
See      C>
Type     CD/Wheat accounts/received [return]
See      C>
Type     MKDIR Ziecek Int [return]
See      C>
```

You now have the following TREE:

Root Directory Name: Wheat Accounts

Subdirectory Names: Received ———————————————— Sold Seed

Sub-subdirectory Names: Ziecek Int—— Mys Ltd   Ziecek Int—— Mys Ltd

To access an account:

```
See      C>
Type     CD/Wheat accounts/Received/Ziecek Int [return]
See      C>
```

You are now in Ziecek Int, Received, Wheat Accounts

You can expand the tree to essentially any size with any combination of branches and any number of root, sub, sub-sub, and lower level directories. Limitations are a function of the storage space and the data field size.

CAUTION: For confidentiality or security, use coded numbers, or letters, or a combination of both. Typically you can use eight characters which you may follow by a period followed by three more characters as a maximum name length.

description with a detailed line drawing of the rack. As with any narrative, the emphasis chosen for a specific audience determines how you organize or classify the material.

Introduction and Theoretical Description. The introduction helps familiarize the user with a product, as well as outlines the instructions' scope and plan of organization.

You need to "sell" a new product to the customer while preparing instructions for its assembly or installation. Since your instructions are part of the product, your introduction

(Text continues on p. 421.)

**FIGURE
15.7**   Do-It-Yourself Instructions for Installing a Garbage Disposer.

INSTALLATION INSTRUCTION FOR GARBAGE DISPOSER

YOUR DISPOSER

Congratulations on your purchase of a durable, 1/3 horsepower garbage disposer. Your disposer will keep your kitchen clean by dissolving excess waste foods such as table scraps, fruit and vegetable peelings, eggshells, coffee grounds and other ''light'' food substances. Your disposer will not, however, grind large objects such as fruit rinds, ham, or steak bones.

Your disposer does not use knives or blades to chop garbage. Instead, it uses a spinning wheel with two blunt metal projections to grind the garbage much as a pencil sharpener grinds the wood off a pencil.

PREPARING FOR INSTALLATION

To install your garbage disposer, you will need these tools:

Flathead screwdriver (medium size)
Phillips screwdriver (medium size)
Pipe wrench
1/4 inch wrench
Pipe sealant compound
Pipe clamp
1 1/2 inch hose clamp
Wire stripper
Wire caps
Flexible conduit
111 V 14 gage grounded wire

If you are installing the disposer in an older house, first you must disconnect the old drain and drain pipe from the sink. (Place a bowl under the disconnected pipe to catch any excess water that may leak.) Be sure that the drain line from the sink to the sewer is not clogged with debris.

Once you have removed your old drain and disconnected the drain pipe, you are ready to install the garbage disposer. Your garbage disposer can be installed in any stainless steel sink that has a 3 1/2 " to 4 " drain opening without any alteration to the hole.

INSTALLATION OF YOUR DISPOSER

Refer to the diagram throughout the installation.

1. Insert a screwdriver under the latch (1) and pop the latch open. Open hooks (2,3) and lift off the entire top of disposer (4-10).
2. Remove disposer cover and set aside (10).
3. Separate pieces 4 through 9.
4. Place seal (8) around drain opening. Insert disposer drain (9) into drain opening.

**FIGURE
15.7**   Continued.

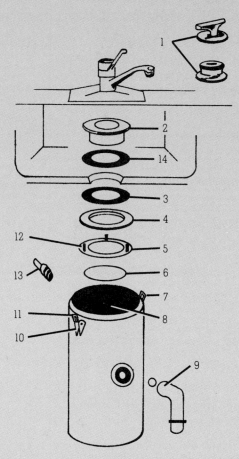

5. Place second seal (7) and metal ring (6) around bottom of disposer drain. Screw support plate (5) onto disposer drain until the three screws touch the bottom of the metal ring. Turn the screws until they are tight against the sink.

6. Lift disposer and hook (3) onto support plate. Turn disposer to align drain pipe (11) with sink drain pipe.

7. Snap hook (1) to secure disposer.

8. Remove screws and plate from drain pipe opening (12). Insert disposer's drain pipe (11). Screw plate back on disposer.

9. Connect sink drain pipe to disposer's drain pipe (11). (You may need a 1½ inch pipe clamp to make the connection. Apply sealant compound to both pipes before tightening connection.)

10. Connect dishwasher hose to disposer (13) if needed. (A hose clamp will be needed for this process.)

11. Run water through the newly installed garbage disposer to check for leakage.

12. Your installed disposer should resemble one of the following two drawings:

**FIGURE
15.7**

Continued.

TYPICAL INSTALLATIONS Height of drain must not exceed height of disposer discharge opening.

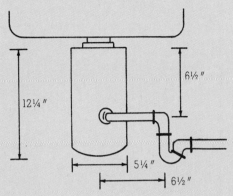

Typical "P" Trap Connection

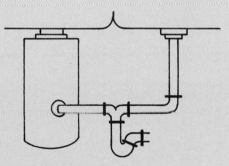

Connection With Straight-Away
Drain Spout

CONNECTING ELECTRICAL WIRES

Refer to the diagram of the bottom of your disposer.

Warning: Shut off electrical power supply before connecting disposer to wires.

1. Cut flexible conduit to length for amount needed to run wire from electrical box to switch and from switch to garbage disposer.
2. Run 110 V 14 gage grounded wire through conduit from electrical box to switch. Connect at both ends. (Connect hot to hot, cold to cold, ground to ground.)
3. Run 110 V 14 gage grounded wire through conduit from switch to disposer. Connect wires in switch together. (Connect hot to hot, cold to cold, ground to ground.)

FIGURE 15.7

Continued.

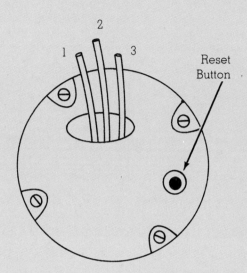

4. Connect hot wire from 110 V line to hot wire on disposer (1). Connect cold wire from 110 V line to cold wire on disposer (2). Connect ground wire from 110 V line to ground wire on disposer (3).
   CAUTION: Secure all wire ends with wire caps and black electrical tape.
5. Ground all wires.
6. Secure all loose ends of flexible conduit.

IMPORTANT OPERATING INSTRUCTIONS

1. To be sure of the proper installation of your new garbage disposer, please follow the instructions given in this pamphlet.
2. Be sure to have a continuous flow of cold water at all times while the disposer is in use.
3. Do not attempt to grind such objects as glass, metal, or plastic in your garbage disposer.
4. **Warning:** Do not put fingers into your disposer at any time.
5. If the disposer becomes jammed, turn machine off and attempt to unclog machine with the use of a wooden object. Remove any objects from disposer with tongs or pliers—**not with your fingers.**
6. When disposer is in use, leave drain cover on to prevent objects from shooting out or falling in.

Courtesy of Brian Chambers.

FIGURE 15.8   Service Manual for the Lawn Spreader Model FH-22.

Lawn Spreader

Service Manual & Parts List

Model FH-22N

Maintenance instructions before use:

- Lubricate with a good grade of oil or a WD-40 type spray.
  - Axle bushings
  - Both ends of control rod
  - Flow bar, if tight
- Check to see that all nuts are kept tight.

Maintenance instructions after each use:

- Hose the complete unit very thoroughly.
- Allow to air dry for all parts that cannot be wiped with rag, etc.
- Lubricate thoroughly by following the same steps as in the "BEFORE USE" instructions.

Red Devil

PRECISION PRODUCTS INC.
2415 S. Grand E., P.O. Box 2546, Springfield, IL 62708

Printed in U.S.A.

FIGURE 15.8

Continued.

ASSEMBLY INSTRUCTIONS FOR MODEL FH-22N

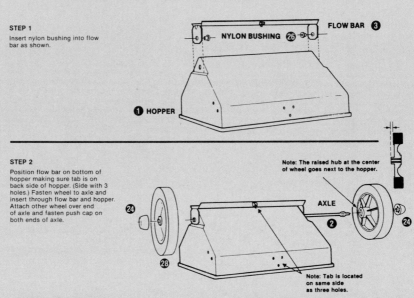

STEP 1

Insert nylon bushing into flow bar as shown.

NYLON BUSHING ㉖   **FLOW BAR** ➌

➊ **HOPPER**

STEP 2

Position flow bar on bottom of hopper making sure tab is on back side of hopper. (Side with 3 holes.) Fasten wheel to axle and insert through flow bar and hopper. Attach other wheel over end of axle and fasten push cap on both ends of axle.

Note: The raised hub at the center of wheel goes next to the hopper.

AXLE

㉔   ⑫   ㉔

㉘

Note: Tab is located on same side as three holes.

STEP 3

Bolt the two pieces of agitator bar to either side of axle as shown.

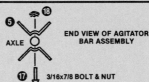

➎   ⑱

AXLE

END VIEW OF AGITATOR BAR ASSEMBLY

⑰   **3/16x7/8 BOLT & NUT**

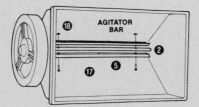

⑱   **AGITATOR BAR**

②

⑰   ➎

SETTING YOUR LAWN SPREADER

*   Select desired "NUMBER SETTING" for Product to be used, by one of the following methods.

1. Check reverse side of bag of material to be used for Precision Setting.

2. Consult attached Application Chart.

3. If product that you have purchased has setting for Scott Spreader, it may be used on the approximate same setting with your Precision Spreader.

*   Turn "NUMBER SETTING" as selected toward bottom of spreader, to point where dial and control rod receiver make contact.

FIGURE
15.8

Continued.

REF. NO.	PART NO.	NAME OF PART ( ) REQUIRED PER ASSEMBLY
1	1100	Hopper (1)
2	1119	Axle (1)
3	1123	Flow Bar (1)
4	1138	Leg Stand (2)
5	1113	Agitator Blade (2)
6	1131	Dial (1)
7	1134	Control Rod Receiver (1)
8	1137N	Control Rod (1)
9	1030	Control Lever (1)
10	1031	Control Lever Cover (1) Vinyl
11	1141	Lower Handle (1)
12	1145N	Upper Handle (2)
13	1028	Handle Grip (2)
14*	1038	Wing Nut (1)
15*	1051	¼" Nut (12)
16*	1043	¼" x ½" Carriage Bolt (1)
17*	1177	3/16" x 7/8" Bolt (2)
18*	1170	3/16" Nut (2)
19*	1179	3/16" x ½" Bolt (1)
20*	1180	3/16" Lock Nut (1)
21*	1176	¼" x 1" Bolt (8)
22	1133	Dial Bracket (1)
23*	1049	¼" x 2½" Bolt (1)
24	1040	Hub Cap (2)
25	1036	Control Lever Housing
26	1842	Bushing (2)
27	1176	¼" x ½" Bolt (3)
*28	see note	Wheel (2)

*Common Hardware Can Be Purchased Locally If Replacement Needed

*NOTE: Please State Diameter And Composition Of Wheel

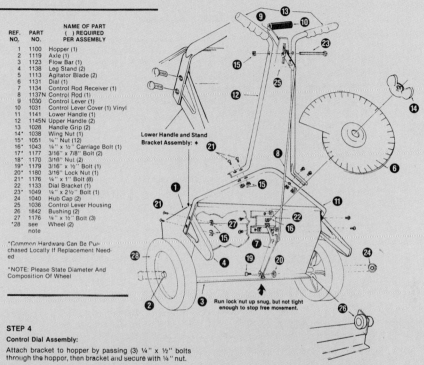

Lower Handle and Stand
Bracket Assembly: *

Run lock nut up snug, but not tight
enough to stop free movement.

STEP 4
Control Dial Assembly:

Attach bracket to hopper by passing (3) ¼" x ½" bolts through the hopper, then bracket and secure with ¼" nut.

Insert narrow, formed end of Control Rod Receiver into Bracket Slot and attach other end to Flow Bar on Bottom of Hopper with 3/16" Bolt and Lock Nut.

Insert ¼" Carriage bolt through center (square) hole in Bracket, place Dial on Bolt and secure with Wing Nut.

STEP 5
*Lower Handle and Stand Bracket Assembly:

Place Flanged edges of Leg Stands against hopper; Lower Handle goes on outside of Leg Stands.

Pass ¼" x 1" Bolts through Lower Handle; Leg Stands, sides of Hopper and secure with ¼" nuts provided.

STEP 6
Upper Handle and Control Rod Assembly:

Fasten Upper Handle Sections to Lower Handle with ¼" x 1" bolts and ¼" nuts provided.

Place Control Lever through slot in Control Lever Housing. Insert one end of Control Rod into hole in Control Rod Receiver (see diagram). Place other end into lower hole of Control Lever.

Fasten Upper handles together by passing ¼" x 2½" bolt (in succession) through side hole in left handle; one side of Control lever housing; center hole in control lever; other hole in housing and out through right handle; secure with ¼" nut.

Control Rod must be on right side of Control Lever (check from rear).

Slip plastic sleeve over Control Lever and Handle Grips over Handles.

Operation:

To Open—Mover control lever firmly forward, toward hopper.

To Close—Pull control lever backward toward operator.

The no. 1 enemy of your lawn spreader is the material put through it. A complete and thorough cleaning after each use is necessary. Wash, dry and place a light oil o all moving parts. The agitator blades in this unit are removable. We suggest that they be taken out for easy access to bottom of hopper when cleaning.

**FIGURE
15.8**

Continued.

APPLICATION CHART
Model FH-22N Lawnspreader

Product	Setting	Product	Setting	Product	Setting
O.M. Scott*		Nutro Golf Green Turf Food	8	K-Mart 20-10-5	6
Turf Builder Normal	9	Nutro Golf Weed & Feed	8	K-Mart 10-6-4	10
Light	6	Nutro Plant Food Pellets	20	K-Mart 27-3-3	6
Heavy	12	**W. Atlee Burpee Co.**		**Heritage House**	
Turf Builder Plus 1 Normal	6	Burpee Weed Killer	10	Crabgrass Preventer Plus 3	9
Heavy	9	Crab Grass Killer	10	Lawn Food 20-10-10	5
Turf Builder Plus 2	6	Insecticide	6	Spring Crabgrass Prevent	12
Windsor Grass Seed	6	Triple Threat	10	Weed & Feed Plus 11-4-7	9
50% Windsor	9	Lawn Food 20-10-15	10	**International Mineral and Chemical**	
Family Brand	9	Marion Ky. Blue Grass	6	Thrive 12-6-6	10
Amchem		Supreme Bluegrass	10	Thrive with Weed Killer 18-3-5 Plus 2, 4, D	8
Weedone Lawn Weed Killer, 2, 4, 5, TP	8	Shady Nook	14	**Milorganite**	11
Weedone Crab Grass Killer	12	**California Chemical Co.**		**Northrup King Company**	
Armour		Ortho Lawn & Garden	5	Lawn Food 22-5-9	11
Vertagreen 18-6-9	6	Ortho 20-10-10	10	Lawn Food 24-6-3	11
Vertagreen Turf & Tree	8	**Wards**		Triple Tonic F-3 18-3-5	13
Lawn & Garden-Pebble	10	Turf Formula	6½	Triple Tonic 23-3-5	11
Lawn & Garden-Pulverized	9	Turf Formula II	5½	Crab Grass Killer with Fertilizer 13-3-5	13
Asgro		**Germains, Inc.**		Winter Life 15-5-15	11
Easy Grow Lawn Food 20-10-8	7	Lawn Food 22-8-6	8	Winter Life 10-15-5	11
Weed-Out Lawn Weed Control	4	Lawn Food with Weed control 6-0-0	9	**Seaboard Seed Co.**	
Gro-Sod 22-8-4	7	Lawn Weeder 20-10-5	9	Weed & Feed 10-6-4	10
Turf Fungicide Control	10	**Greenfield Products**		Lawn Builder	8
Bandini		Triplex Crabgrass Killer, Fet., Insecticide	9	Triple Action Crabgrass	15
Plush	9	Crabgrass & Broadleaf Weed Killer	8	**Stadler Fertilizer**	
Blade	9	Green Power	6	10-6-4	10
Gro Pellets	12	Two Way Green Power	6	Heavyweight	20
Weedilizer	9	**Grass Seed**		20-10-5 Lightweight	6½
Blackstone Guano Company		Bent Grass	3	10-6-4 with 2, 4, D	7
Plantation 10-6-4	12	Marion Bluegrass	3½	**Vaughn's**	
Borden (Smith Douglas)		Kentucky Bluegrass	3½	Feeder/Weeder 20-10-5	10
Green N Gro		Bermuda Grass	6	Turf Food 20-10-5	8
Green N Greener Super Turf Food 32-4-4	8	Fine Seed Mixture	7	Prekill	12
Green N Gone Super Weed and Feed 30-4-4	9	Coarse Seed Mixture	7	Four Way Beauty	7
Green N Gro Turf Food 27-3-3	9	Rye Grass	9	One Shot Spring	7
Green N Gro Turf Food 15-5-10	10	Fescue	9	Fall	6
Green N Gone Turf Food and					
Weed Control 25-3-3	12				

The following settings can be used to apply products not in the application chart.

FERTILIZER		GRASS SEED		HYPONEX		SCOTT 4 STEP		GLORION	
Pounds Per 1000 Sq. Ft.	**Setting**		Setting			Step #1	9		Settings
2.5	8½	Blue Grass/	10	Weed & Feed 2837	9	#2	8	10-6-4 Weed & Feed (15# Bag/5000 Sq. Ft.)	9
3.	9	Bent Grass				#3	9		
3.6	9½	Fine Mixture	12	Lawn Food 2948	9	#4	9	10-6-4 50% Organic (40# Bag/5000 Sq. Ft.)	10 x 2*
5.	10½								
If a higher setting is required we recommend a double pass in a criss cross pattern.		Coarse Seed (Rye, Fescue, Bermuda)	14					32-3-3 (15# Bag/5000 Sq. Ft.)	9
								*Two passes	

**IMPORTANT NOTICE: Being only human, it is possible a part was missed when packing your merchandise. If so, we are extremely sorry for the inconvenience. However, if you will jot down the Model, Reference, and Part Number, and/or the description below, it will be forwarded prepaid to your choice of address the same day.

Please give us the inspector's name located on the front of these instructions.

Send to:　Missing Part Div.
　　　　　P. O. Box 2546
　　　　　Springfield, IL 62708
　　　　　(217) 528-1311

Courtesy of Precision Products, Inc.

**FIGURE
15.9**

Installation Instructions for Barrelcrafters SR-71 Roof Mounted Ski Rack.

Barrecrafters

SR-71
ROOF MOUNTED SKI RACK

PARTS LIST

2-3975 Inner Slides
4-6101 Final Assemblies
4-7876 Raingutter Clamps

1-7888 Wallet Card
1-7889 Envelope & Keys
4-8004 Lock Washers
4-8057 1" Carriage Bolts

4-8074 Wing Nuts
4-8121 Flat Washers
4-8173 Weld Nuts (flat)
4-8174 ½" Hex Head Screws

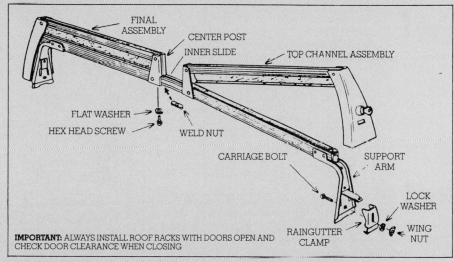

FINAL ASSEMBLY
CENTER POST
INNER SLIDE
TOP CHANNEL ASSEMBLY
FLAT WASHER
HEX HEAD SCREW
WELD NUT
CARRIAGE BOLT
SUPPORT ARM
LOCK WASHER
RAINGUTTER CLAMP
WING NUT

IMPORTANT: ALWAYS INSTALL ROOF RACKS WITH DOORS OPEN AND CHECK DOOR CLEARANCE WHEN CLOSING

READ ENTIRE INSTRUCTIONS BEFORE ASSEMBLING RACK

INSTRUCTIONS

1. Slide a weld nut inside a center post on a final assembly so that the hole in the center post lines up with the hole in the weld nut. Put a flat washer on a hex head screw and thread into weld nut loosely through hole in center post. Repeat on remaining three final assemblies.
2. Turn final assembly upside down and slide the inner slide (also inverted) into it, making sure that the weld nut enters the slide. Repeat on remaining final assemblies.
3. Open top channel assembly and loosen the wing nut that holds the raingutter clamp. Do not remove the nut.
4. To attach to car, place assembled racks on roof from 24 to 30 inches apart so that the support arms rest in the rain-gutters. Put raingutter clamps around raingutter and tighten wing nuts securely. Center inner slides and tighten hex head screws securely with ³/₈" hex head wrench.

ALWAYS CARRY SKIS WITH TIPS TO THE REAR OF THE CAR.

WARNING: This rack is to be used only on cars with substantial raingutters. Do not use on cars with only thin rainshields.

Courtesy of Shelburne Corporation.

NOTES:

1. Periodically check screws and wing nuts for tightness.
2. Move racks closer for more holding pressure on skis, apart for less.
3. On some cars it may be necessary to remove the vinyl on the raingutter clamps and replace it with a piece of electrical tape for door clearance.
4. Barrecrafters recommends that you keep your rack closed at all times when traveling.

ORDER PARTS FROM YOUR DEALER. PART NUMBERS AND RACK MODEL MUST BE INCLUDED.

FIGURE 15.10   Section of Instructions on How to Install Fiberglas® Building Insulation Blankets.

Installing Fiberglas insulation

On basement walls

Tools needed

- Heavy duty stapler
- Sharp knife
- Straight edge
- Gloves
- Hammer, nails, saw
- Masonry nails, small sledge hammer
- Furring strips or studs
- Safety Glasses

Recommended insulation

Use kraft- or foil-faced Fiberglas insulation with attached vapor barrier (**or** unfaced insulation with a separate polyethylene film).

NOTE: If basement walls are damp or wet spots are visible after rains, it is recommended that inside walls be sealed with a paint made for this purpose.

Installation instructions

1. Before insulating, check to make sure that your basement walls have been properly sealed and that no moisture is getting through from the outside. Install framework of furring strips or studs to masonry walls. Nail bottom plate directly to floor and top plate to joists above. Nail furring or studs to framework, either 16″ or 24″ on center.

2. Install kraft- or foil-faced Fiberglas insulation between furring strips or studs and staple facing flange to strips or studs. (Unfaced insulation requires a separate vapor barrier such as polyethylene film.)

NOTE: Polyethylene should not be left exposed. It should be covered with an approved interior finish as required by local codes as soon as the insulation has been installed.

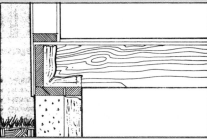

3. Cut pieces of unfaced insulation to fit in band joist between top plate and sub floor.

4. Once insulation and vapor barrier are installed, wall finish should be applied directly to furring strips or studs. **DO NOT LEAVE FACED INSULATION EXPOSED.** The facings on standard kraft- and foil-faced insulations will burn and should be covered with an approved interior finish as required by local codes as soon as the insulation has been installed.

Courtesy of Owens-Corning Fiberglas Corporation.

should convince the customer that he or she has made a wise choice. For example, note that Brian Chambers, author of the garbage disposer instructions, reassures the buyer that he or she has made a wise purchase. He uses the first sentence to describe the product, the second to outline its strong points, and the third to note the mechanism's limitations.

Brian presents a brief theoretical description of the garbage disposer in the second paragraph. Note that he emphasizes only one aspect of the mechanism's design—its construction with grinders instead of blades—to explain the theory of operations. He then uses an analogy—the comparison with a pencil sharpener—to clarify this explanation. By focusing on one major detail with a figure of speech, Brian Chambers practices excellent rhetoric in drafting a description of a mechanism with a theoretical emphasis.

Neither the orientation section nor the theoretical description should slow the user with too much reading. Note that some of our other examples substitute several illustrations for the introductory text. For example, the instructions for assembling the lawn spreader lack both of the sections we have just analyzed in the garbage disposer directions. Sometimes, to save space or to minimize the amount of reading, simply show an accurate line drawing of the mechanism and start with the other parts of the instructions on the first page. The audience analysis, the task analysis, and the policy requirements of your employer should all help you to determine the product's proper introduction and description.

Description of the Mechanism.   Typically, this section contains a physical or functional description of the mechanism. (For more information, consult the first part of this chapter.) This section should be well illustrated to help the user focus on the task at hand.

To minimize the user's reading, be sure the physical or functional description is well-focused. Note that Brian Chambers does not describe the whole disposer. Instead, he tells the user to examine and prepare the drain pipe assembly. In the "Preparing for Installation" section, he differentiates between the drain pipes of old and new houses, and he describes only the features relevant to installing a disposer.

Note that several of the examples show a less traditional, but quite readable approach—illustrations instead of verbal, physical, or functional descriptions. Thus, the lawn spreader instructions display an exploded view on the upper right portion of page three, the visually strongest and most appealing section of an open single-fold sheet of descriptions. The Fiberglas Insulation Blanket instructions also move directly to pictures of the work surface in place of physical descriptions in steps one and two. Sometimes the publication's size can limit your opportunities. Then you must choose either a verbal description or an illustration. In that case, draw on your task and audience analyses for the best choice. If these two phases of the preparation suggest you are writing for an audience of "doers" rather than "readers," use illustrations, and keep the words to a minimum.

List of Parts.   You need to include two lists of parts in your instructions. First, list all the parts the users will need to provide. Second, list all the parts the manufacturer includes with the product.

No doubt you have purchased, brought home, and prepared to assemble a product—after the stores have closed. How annoying it is to find that you lack an essential washer for a drain assembly on your dishwasher or the right size nut to finish installing a ceiling fan. Offer the customer clear lists of the package's contents. List separately, and near the front of the instructions, the items and the specifications or sizes of the tools the customer must supply.

Note that Brian Chambers' instructions for the "do-it-yourself" disposer installation list each item needed to successfully complete the job. The spreader assembly instructions display a parts list with the illustration. Thus the consumer knows he or she can assemble the mechanism using the most ordinary household screwdrivers and wrenches. The list of tools needed appears right on page two of the Fiberglas Insulation Blanket instructions, so the user can gather these tools before unrolling the insulation and exposing the product to possible damage.

Sequences of Tasks.   This section tells the user how to complete the job you have defined. Therefore this section, the very core of your instructions, must be specific. As mentioned elsewhere, you cannot tell the user everything that can be known about a product or process. To properly focus your performance instructions, draft them after completing these preparation phases.

- ☐ Complete research and personal understanding of the mechanism.
- ☐ Obtain clear directives about the instructions' language style.
- ☐ Obtain clear directives about the instructions' format.
- ☐ Conduct audience analysis.
- ☐ Conduct task analysis.
- ☐ Complete a checksheet that outlines the task analysis.
- ☐ Obtain clear directives about the illustrations for the instructions.

We refer you to the sections of this chapter that discuss task analysis and style and format considerations. Also consult the chapter on illustrations for appropriate guidelines.

The content of your performance instructions should be divided, numbered, and arranged in obvious numeric sequence. Restrict the items in the sequence to fewer than 10 for the general, nontechnical user. In sequences of more than 10 items, divide the sequence into several phases, and number each step within its phase.

For example, the Fiberglas Insulation instructions use a number for each "chunk" of information. The lawn spreader has a three-step procedure; all the other "steps" appear under individual headings on page three. Ved Krishnan's instructions also rely on an enumerative organization; they use repeated illustrations and short paragraphs or "chunks" of text to create easy reading.

Cautions and Warnings.   These two messages contain different types of information. Warnings explain to users the injuries possible if they fail to follow the instructions properly. Cautions tell users that the machinery may be damaged if the users fail to follow the instructions. Set the terms "Warning," and "Caution" in boldface type. Always use these two terms, not synonyms. For examples of warnings, study the disposer installation directions which warn against electric shock and injured fingers. The ski rack instructions also contain warnings. Here is an example of a caution from the Operating Instructions for Model No. 257.857952 Craftsman Electric Lawn Edger-Trimmer:

CAUTION: Disconnect cord before adjusting or servicing your Craftsman edger. Form No. 88-304-00, page 4.

Cautions and warnings protect the user and the product, but they also protect a manufacturer from liability when users are injured by products. The cautions and warnings you compose may actually help to protect a manufacturer from charges of negligence by providing adequate consumer information.

For this reason, make the user aware of safety throughout your discussion. Do so by using boldface type and a clear series of user instructions. Rather than overload the user with cautions and warnings, the Craftsman edger manual contains a special safety section, which we include as an example of well-prepared safety instructions.

 FOR SAFETY AND EFFICIENCY:
DON'T leave edger unattended while cord is plugged in.
DON'T adjust edger positions while motor is running or blade is still turning.
DON'T permit observers to stand in front of rotating blade.
DON'T permit blade area to pack with mud and grass.
DON'T permit clutch and blade to slip excessively by pushing edger too fast.
DO use in rain. Store indoors after use.
DO unplug cord when changing blades or doing maintenance in the blade rotating area.
DO clean out grass and debris that collects in fan area (see Maintenance).
DO disconnect cord before performing any maintenance.
DO operate edger from position where eyes cannot see cutter blade.
DO wear safety glasses for added protection.

Operating Instructions for Model No. 257.857952 Craftsman Electric Lawn Edger-Trimmer, Form No. 88-304-00, page 5. Courtesy Sears, Roebuck & Co.

Emergency Procedures. Also called "troubleshooting," this section tells the user what to do when the mechanism malfunctions or breaks. Most computer users are familiar with troubleshooting procedures because they consult them when "error" messages appear on the screen. Here is a simple troubleshooting message for a computer user.

 PROBLEM: There's a write-protect tab on the disk containing your file.
SOLUTION: Remove the write-protect tab from the disk.

From *Wordstar 2000 Reference Guide Release 2.0*, p. 210. Courtesy of Micropro International Corporation, San Rafael, CA.

This simple message has two user parts. The "problem" section duplicates the printed message the user finds on a monitor when a problem occurs. The "solution" section tells the user how to correct the problem.

Some troubleshooting or emergency procedures divide the presentation of the problem into columns for an easily read presentation. The manual may have a section that reads:

PROBLEM	CAUSE	SOLUTION
write-protect error	write-protect tab on disk	remove tab from disk

Some troubleshooting sections use more informal diction as this example does:

WHAT YOU SEE	WHAT WENT WRONG	WHAT TO DO
write-protect error	write-protect tab on disk	remove tab from disk

Your particular audience's and company's requirements may determine your choice of approach. Remember, the emergency procedures section should define every problem a user may encounter, so that users can repair their own problems without calling either the supplier or the manufacturer. If company policy allows, include a telephone number and the times when technical assistance from the manufacturer is available.

Style Requirements

Each company tends to develop its own guidelines for the directions and procedures published with their products. Learn these style requirements before producing instructions for a product. This chapter does not address the style conventions of manuals, or lengthy instructions. Our guidelines are generic conventions typical of the types of short instructions written in industry today.

Format. A company's publications policies may determine much of the format, including the length of instructions. Nevertheless, certain customs prevail. Use short paragraphs of one to three sentences to describe tasks, cautions, and warnings. This technique, called "chunking," makes your instructions easy to read because the short length of each segment allows the user to look up any needed information quickly.

Use as many illustrations as your budget permits. For instructions, concentrate on exploded and close-up views that show the particular section or parts the user must work with. Use human figures in instructions for the general public, to make the materials personal and interesting.

Use design features such as boxes, side bars, shading, boldface type, italic type, various type sizes, indentions, and lists to emphasize the most important materials and to maintain the users' interest.

Language. Although some procedures are written in the third person, indicative mood, most procedures and directions today are written in the second person, imperative mood. Use active voice verbs, short sentences, uniform language, and grammatical parallelism for lists and short sequences of tasks. For example, if you write instructions for formatting a computer disk, decide on the correct verb to use. For example, Ved Krishnan used the verb "type" instead of mixing in "press" or "hit."

Redundancy. Repetition is well-known as an effective teaching technique. Use repetition in your instructions, so the user can see the same concept presented several different ways. For example, if you describe how to prepare a Chinese recipe, tell the user in the introduction that you will explain how to prepare potatoes with ginger. Then show a picture of a wok and the correct tools, a picture of the correct thickness of potato slice, and a picture of the amount of peanut oil needed. Then explain the steps in the process. Some overlapping of information occurs when you write such instructions; repetition allows the new user to think about and to organize the concepts and tasks you describe.

How to Plan Instructions

Examine the process Ved uses to prepare his short instructions. First, he researches his topic thoroughly. He reads about disk operating systems, and, as an experienced user of DOS, he is prepared to write. If less experienced in a field, prepare your topic using these methods:

☐ Read about the topic in your company's product literature.
☐ Obtain a model of the mechanism to examine, sketch, and try out.
☐ Interview the designers or users of the mechanism.
☐ Obtain or make the decisions about the format, illustrations, and style.
☐ Conduct an audience analysis.
☐ Complete a task analysis and checksheet.

Here is Ved's audience and analysis.

Audience—Educated lay persons, some computer experience.
Education—All BS degrees; some graduate business courses.
Knowledge of Topic—Limited computer knowledge in word processing and bookkeeping; good knowledge in accounting; strong ability in mathematics; weak in language.
Task—1) Understand relationship between TREE and directories; 2) create directories, sub-directories, and sub-subdirectories.
Information Needed—1) Logical operations of directories; 2) logical structure of commands for directories; 3) how to modify the TREE command to make sub- and sub-subdirectories.

Just as the audience must be defined, so also must be the goal of your instructions. Ved's goal is to show Lal how to construct a TREE and access it with the TREE command. This activity has a further goal, to facilitate Eastwood's bookkeeping procedures. But Ved's goal is not to teach bookkeeping or business practices. Therefore, he states his goal in broad terms, but in terms specific to the task at hand.

Goals of TREE Instructions

Action	**Knowledge**
Access a TREE	TREE access commands
Build a TREE	TREE construction commands

After defining the actions and knowledge in general terms, Ved begins a task analysis of the materials he must present. First, he reviews his assumptions:

☐ The audience knows computer basics.
☐ The audience knows some bookkeeping.
☐ The users understand a little about how commands for software are written.
☐ The readers are used to jargon and symbolic language.

It is important for Ved to state these assumptions because he must define exactly what actions he will describe. His assumptions allow Ved to concentrate on how to build a TREE, not on

**FIGURE
15.11**
Analysis Form for Instructions.

ANALYSIS SHEET—INSTRUCTIONS

Background
    Audience Analysis Summary
       Education

    Skills

    Use of Product

    Presumed Skills, Ability, Knowledge, and Use—Summary

First Task:

Second Task:

Third Task:

The goal of my instructions is (use one sentence):

The audience will use my instructions to find out how to:
1.
2.
3.

any extra skills. If a user needs to know how to "boot" the computer's system, for example, he will not find this information in Ved's directions.

    After clarifying his assumptions, Ved begins with the goal, the building of a TREE. He states this action first; then he lists in reverse order each task that leads to this goal. Then he lists in reverse order the types of knowledge that must lead to the successful completion of the tasks. Here are Ved's lists.

FIGURE 15.12   Ved's Completed Analysis Form for Instructions.

VED'S ANALYSIS SHEET—INSTRUCTIONS

Background
   Audience Analysis Summary
      Education—Lal—college plus; others—some college or business courses

      Skills—all good in bookkeeping or accounting some computer knowledge

      Use of Product
         use the computer; monitor accounts
      Presumed Skills, Ability, Knowledge, and Use—Summary

First Task: Form a root directory
[type] C)MKDIR "root directory name" [return]
[see] C)

Second Task: Form a "branch" directory
While in root directory, that is, the C) in first task above:
[type] C)MKDIR "branch directory name" [return]
[see] C)
Third Task: Form a "twig" directory
While in branch directory, that is, the C) in second task above:
[type] C)MKDIR "twig directory name" [return]
[see] C)
The goal of my instructions is (use one sentence): I will tell how to make a TREE (for filing bookkeeping information).

The audience will use my instructions to find out how to:
1. Create a root directory
2. Create branch directories
3. Access directories—roots and branches

Task Analysis of Building a Computer TREE

Action	Knowledge
Build a TREE with three levels	Tree construction commands
Build a sub-subdirectory	MKDIR sub-subdirectory name
Build a subdirectory	MKDIR subdirectory name
Build a root directory	MKDIR root directory name

Ved writes the actions and knowledge lists in reverse order to avoid overlooking steps in the task analysis. After reviewing the list for omissions, he writes the entire list — including each step — in its natural order.

Task Analysis of Building a TREE

Goal	Action	Knowledge
Make root directory	[Type] MKDIR root name [return] [See] C⟩	First entry of MKDIR will be root directory when subdirectories are added.
Make first level: subdirectory	[Type] MKDIR subdirectory name [return] [See] C⟩	Entry by MKDIR within a previous MKDIR entry will be a subdirectory.
Make second level: sub-subdirectory	[Type] MKDIR sub-subdirectory name [return] [See] C⟩	Entry by MKDIR within a previous MKDIR subdirectory will be a sub-subdirectory.

With the example above explained in detail, any number of root, subdirectories, and sub-subdirectories can be created on an infinite number of levels. To access the TREES constructed in this manner, the full TREE command is used.

Access 3 levels of a directory	[Type] TREE \Wheat \Received \ Mys Ltd	Levels may be accessed by naming a PATH after the TREE. '' \'' is begin first name directory; '' \'' to separate from second word subdirectory; '' \'' to separate from third word sub-subdirectory; '' \'' to separate for all sub-subdirectories.

With his task analysis now complete, Ved can plan and complete an accurate set of instructions. In summary, here are the steps to complete a task analysis.

1. Research. Study the task you must describe. You can attend the training classes given by your employer, interview the engineers and product specialists in that area, read the existing literature about the product, experiment with the mechanism yourself if possible, and study its technical specifications. Research the audience as well, using the techniques discussed elsewhere in this text.
2. Eliminate. List all the related tasks the user will need to know to use your product, and eliminate those you will not discuss. List every task you can imagine.

3. Define. State the broad goals of your instructions. State the goals as actions and as types of knowledge the users will need.
4. List. State all the actions and types of knowledge the users will need to perform the task. You may start with a freeform jot list if this helps you to think, but eventually your list must become a reverse chronology. First, list all the tasks the user must know. Then list all the types of knowledge the user must have to do the job.
5. Organize. List each action and its component knowledge from the start of the task to the finish or to completion of the goal.

Now you can draft and edit the actual instructions as you would any technical text. The task analysis, audience analysis, and understanding of the product are the real work of composing directions and procedures. If you use the techniques for planning this chapter demonstrates, you will significantly reduce your preparation and writing time.

CREATING ADMINISTRATIVE DOCUMENTS

Administrative documents such as employee handbooks, management policies, supervisory guidelines, and procedures for job performance compose the fastest growing segment of technical communication. Known sarcastically as "administrivia," "paperwork," "bureaucracy," and other less savory names, these documents govern vast spheres—one employee's performance evaluation, a corporate division's profit sharing, a multinational company's international image, for example. Thus, you should approach administrative technical writing with concern and accuracy. But first, since administrative technical communication has a wider scope than descriptions or instructions, here are some examples of this rhetorical subspecies.

Performance evaluation—a supervisor's annual, semi-annual, or quarterly written commentary about an employee's completion of assigned tasks, attitude, accomplishments, and areas needing improvement. In some companies, this report determines the employee's entire pay increment. The document may be completed on a form or in a letter.

Job description—the narrative text that defines a job. The job description should outline the position's scope, list all the tasks the employee must perform in order of importance, state who will supervise the employee, and specify the pay range. The job description serves as the institution's contract with the employee; he or she can be terminated for failing to do the tasks listed in a job description.

Operations policies—usually lengthy, detailed texts that explain in the third person how a firm or institution conducts its business. For example, a firm may have a policy to explain how its hourly employees should schedule their flex time. Another firm may have a dress code. A university may have a policy explaining how professors should schedule and conduct their office hours. A hospital policy may require nurses to schedule their shifts in three-day blocks of time. An airline policy may prohibit its pilots from consuming alcohol for 24 hours before they fly.

Operations instructions—usually more specific than operations policies, these documents should use a second-person narrative to tell individual employees how to perform specific tasks within a given corporate or institutional culture. There are many kinds of operations instructions, including documents that explain how to: complete a payroll firm, submit a travel reimbursement form, prepare a purchase requisition, determine a budget, sign out a company car, and so forth.

GUIDELINES FOR ADMINISTRATIVE WRITING

Often the corporate environment affects the style of administrative writing. For example, since a firm's attorneys must review a personnel or insurance policy, the law restricts the document's style. Sometimes, after repeated legal, personnel, and managerial reviews, such documents become difficult to read and, unfortunately, must remain so. But you can also compose easy-to-read administrative documents.

Study Figure 15.13 for an example of a readable operations policy. The policy shows several techniques you may find useful when composing longer documents.

1. Focus. Figure 15.13 does not describe all ten jobs in the medical office. It concentrates on two, receptionists and bookkeepers. Thus, the author can say something specific about each group.
2. Action. Throughout Figure 15.13, the writer uses descriptive verbs in the active voice to clearly delineate the employees' responsibilities. The writer begins each bulleted item with a grammatically parallel action verb to reinforce the thrusts of the action statements. The active voice also improves the document's readability.
3. Design. The writer uses "chunking" techniques to lead the eye through the document. The two job categories, reception staff and billing staff, stand alone and are easy to see on their lines. Almost in outline form, the task descriptions line up indented beneath them. There is an adequate level of detail that does not overwhelm the reader. Note, too, that instead of numbering the tasks, the writer used bullets to attract attention to these items.

Now examine Figure 15.14, the operations instructions for the medical office receptionists, billing staff, and insurance personnel. Note that the focus is even more specific than that of the operations policy. Because this document is essentially a job description, the writer first performed a task analysis to determine the duties of the general medical office personnel. A task analysis of the type shown earlier in this chapter allows the writer to determine the importance, frequency, and difficulty of the responsibilities assigned to an employee. The writer then used the second-person, imperative mood to address the staff directly about the important issue of confidentiality. As you compose more lengthy operations instructions, remember these suggestions:

1. Focus. Limit the topic to one idea, procedure, or policy to which the employees must adjust. Thus, this entire document is about how to maintain patient confidentiality, not patient confidentiality and billing. Remember, you compose operations instructions to tell a specific audience how, not necessarily why, to do something.
2. Action. As in the operations policy, use active voice, descriptive verbs. Use the second-person, imperative verb whenever possible to speak directly to the reader.

FIGURE
15.13
A Sample Set of Procedures a Medical Office Manager Can Use to Evaluate the Employees.

OPERATIONAL PROCEDURES

Duties of Reception and Billing Staff

Unlike the supervisor, whose responsibilities frequently change, the billing and reception staff have specific tasks they must complete daily.

The reception staff should:

☐ Concentrate on producing error-free work.
☐ Handle patients' questions courteously and efficiently.
☐ Answer phone calls and take messages.
☐ Schedule appointments correctly.
☐ Refer routine patient education questions to the nurses.

The billing staff should:

☐ Handle patients' questions about their accounts courteously and efficiently.
☐ Access all account balances quickly for review.
☐ Understand all third-party billing procedures.
☐ Understand all state and federal reimbursement procedures.
☐ Complete all insurance and government reimbursement forms.
☐ Keep all the accounts up to date.
☐ Maintain the billing, financing, and collection cycle.

3. Level of Detail. Note that this document goes into greater detail than the operations policy does. Figure 15.14 refers to specific job-related problems such as setting up a patient's history or coping with a distressed patient. Again, since the operations instructions are more specific than the operations policy, the greater level of detail is useful.

4. Design. Use "chunking" techniques, so readers can see the tasks, or groups, that make up their responsibilities. Enumerate if your list is ten items or less. Describe the tasks as specifically as possible. For example, don't say, "Avoid letting patients hear you" when you can write, "Do not take patient calls at the reception area desk. Transfer these calls to the inside office."

OTHER TYPES OF ADMINISTRATIVE WRITING

In Figure 15.15, you see a checksheet, yet another of the many forms of administrative writing you may use in your work. The checksheet offers many advantages:

**FIGURE
15.14**

Instructions for Confidential Medical Information Show the Importance of Administrative Technical Communication.

OPERATIONAL INSTRUCTIONS

Keep All Patient Information Confidential

Because medical information is strictly private, follow these simple rules to guarantee confidentiality for our patients.

1. Discuss confidential information out of earshot of the patients. For example, when you phone the pharmacy, use the phone in the enclosed medical records office, not the phone at the receptionist area desk.
2. Do not take patient calls at the reception area desk. Transfer these calls to the inside office.
3. Do not joke or complain about patients within sight or earshot of patients. Use special care in the halls leading to examining rooms since patients can hear you in these rooms and as they come and go in the hall.
4. Discuss a patient's history only if it is necessary. For example, you may need the history to set up a record; don't discuss the history simply because you find it of interest.
5. Use a low tone when you speak with patients on the phone and in the halls.
6. Protect the patient's file. Store records for updating in the daily file; after records are updated, return them to the records office. Do not leave records sitting on any of the desks.
7. Do not hand a patient's file to anyone but the doctors, nurses, and other office management personnel. Do not discuss a patient's history on the telephone.
8. Do not hand a patient's file to him or her to read. Place the file on the outside of the door of the examining room for the doctors to read. The doctors will carry the files out of the room when the examination or consultation is over.
9. If you expect a patient to be upset or troublesome, conduct him or her to one of the conference rooms and close the door. Have the patient wait there for the doctor.
10. Do not permit patients to discuss their concerns about fees at the receptionist's desk. If they need to discuss fees, conduct them to the accounting office.

Remember—Patients experience many stresses; they may feel great fear and anger about medical procedures and personnel. Your sensitivity and discretion can reassure them and facilitate their cooperation with our office's procedures to maintain their health.

☐ Ease of use. The user doesn't have to do any extensive writing, and yet he or she has a record of an employee's accomplishments during each shift. Also, this document is easy to duplicate and modify for other jobs.

☐ Ease of reading. Sometimes, employees are poor readers. Even fluent readers dislike extensive reading for supervision. A weak reader or harried manager can study this checksheet easily and quickly understand its meaning.

FIGURE
15.15

A Checksheet for the Manager to Use in Supervising Restaurant Personnel.

ADMINISTRATIVE CHECKSHEET

Before you go home, what do you need to complete?

_____

At the end of your shift, you will have to complete your daily clean-up. It is important to the employees of the next shift that you do a good job and complete your clean-up properly. The manager on duty will use this checksheet to make sure your clean-up is complete before you leave.

_____

Cleanup Checksheet

Under the Counters
_____**1.** Check supply of children's bibs
_____**2.** Dust register area
_____**3.** Clean underneath cash register
_____**4.** Stock ice cream line

In the Lobby
_____**1.** Clean cigarette machine
_____**2.** Clean menus
_____**3.** Wipe down all windows
_____**4.** Empty ashtrays

Near the Register
_____**1.** Sweep floor mat
_____**2.** Sweep register area
_____**3.** Sweep foyer
_____**4.** Sweep restrooms

In the Meeting Room
_____**1.** Restock jelly beans, candy, toothpicks
_____**2.** Wipe down chairs
_____**3.** Clean chalkboard base
_____**4.** Clean podium

_____   _____
Employee   Manager   Date

☐ Level of detail. You have to explain exactly to an employee how to complete a job. You can't say simply, "We want you to clean the restaurant." What does "clean the restaurant" mean? Does it mean waxing? Or steam cleaning? Restocking? Putting the chairs back on the floor or leaving them up on the table? A checksheet allows you to specify exactly what a task means. You can also add and delete items as needed. You then have a basis for discussing an employee's work record.

The checksheet shown here is simply one of the vast assortment of administrative documents used in corporate or institutional culture. As technical communications, these texts can affect our lives dramatically even as you may find their composition troublesome. Remember, as a technical communicator, you have an obligation to the employees you supervise to write operations procedures and operations instructions to suit their—not necessarily your—reading level.

SUMMARY

In this chapter you learned how to plan and write a description of a mechanism with an emphasis on its physical, functional, or theoretical operation. You learned, as well, to plan and write short instructions with an emphasis on either procedures or directions. After studying this chapter, you should be able to:

■ Research a description or instruction assignment by reading the product literature, interviewing the designers or users, and observing the object yourself.
■ Complete an analysis sheet that will help you decide your point of emphasis for a description—physical appearance, function, or theory.
■ Write a description of a mechanism, select the appropriate line drawings for illustrations, and suggest the format of the finished description.
■ Identify directions and procedures.
■ Identify the six sections of a set of instructions.
■ Conduct an audience analysis for an instruction assignment.
■ Complete a task analysis for an instruction assignment.
■ Compose each of six sections of an instruction assignment you might be asked to complete for your class.
■ Draft instructions, select illustrations, and make written suggestions about the format to be used when the instructions are set in type.
■ Recognize the verbal and graphic style of instructions, so you can use the appropriate styles when you refine and edit your draft of instructions.

ASSIGNMENTS

1 Prepare a description of an object that you use regularly at work or in the field you are studying. Try to choose a mechanism unfamiliar to the other students. Fill out an analysis sheet, write, and illustrate your description.

2   Incorporating the steps of the first assignment, prepare a set of instructions for the class that describe how to use the mechanism you have chosen. Select a focus for your instructions; for example, concentrate on assembly instructions or maintenance instructions. Try not to explain everything about the mechanism.

3   Have you ever tried to obtain accurate directions to a point of interest in a strange city or town? This experience can be most frustrating. Prepare instructions that will enable a visitor to find a major tourist attraction in your city. Include maps, route information, and anything else you consider important to the user.

4   Bring to class a set of instructions that you used recently. Prepare to evaluate these instructions in a group discussion session. What are their strong points? Where did you have difficulty? Be prepared to critique these instructions, according to the guidelines of this chapter, for a five-minute oral presentation.

5   Draft a letter to the manufacturer whose instructions you have brought to class. Explain what went well and what went wrong. Make specific suggestions for revisions.

6   Hand in your instructions and your letter. Your instructor may now redistribute the instructions with their attached letters to students who have not yet analyzed or worked with them. You should revise or rewrite the instructions your instructor has given you. Then write a letter to your instructor explaining why you selected your revisions. Submit your instructions and letter to the instructor.

7   Have a five- to ten-minute conference with the student who brought the instructions to class. Have you corrected the problems that student identified? Obtain some feedback, and try to convince the student that your revisions are in his or her best interests.

8   Complete the planning, research, audience analysis, and task analysis, for a set of instructions for an audience you do not know well. Consider some of the following groups: men, women, older people, children, nonreading adults, or students from overseas. As you prepare this assignment, keep a journal. You might consider some of these questions in your notes.

☐ What types of questions did you find yourself asking?
☐ What assumptions were you able to make?
☐ What facts did your research unearth about your group?
☐ Were there any problems unique to your group?
☐ What were the special needs of your group?

Compose a one-sheet visual aid about your audience's needs as users of your instructions; make a transparency of your aid to use on an overhead projector, and be prepared to explain your project to your class for ten minutes.

9   Prepare the operational procedures for an activity at home or work; for example, the policy for use of the family telephone or the policy to govern the use of office supplies.

10 Now draft operational instructions for one job your operational procedures mentioned.

CHAPTER
16

Oral Communications and Presentations

This chapter introduces you to the preparation, organization, and delivery of the oral presentations which are so necessary in the technical communication profession. After reading this chapter, you will be able to:

- Select an appropriate type of oral communication for your purpose.
- Analyze your audience.
- Select the correct information and approach for your group.
- Organize your presentation.
- Plan and write the support materials you will distribute to your audience.
- Design your visual supports.
- Improve your planned presentation through practice sessions.
- Deliver your presentation, and enjoy doing so.
- Conduct and enjoy a discussion session after your presentation.

CASE STUDY   **An Unavoidable Presentation**

Ted Wilson, a graduate of a premier technical communication program, is a successful agricultural writer at the local university's campus center for the state's agricultural agents. In fact, at "Ag. and Mining," the shortened name for his campus, he receives healthy annual merit raises for the many quality publications he produces: news stories on farming trends, agricultural economics, and local farming developments; a series of newsletters that the state's agricultural agents distribute to the farmers in their districts; and all the 4-H materials for agriculture that originates in his state. Because of his success in his field, Ted has built a following among the readers, who trust his news stories, and the farmers and gardeners, who trust him to translate technical terms into concepts a lay audience can understand.

Ted's followers elected him President of the university's chapter of the Society for Technical Communication. Honored as he is, Ted is now out of his secure world of the office, word processor, and library. He has a distinct fear of speaking publicly; in fact, in the past, he has almost become ill before speaking in public. Now he must travel to a rival university, State College, to represent his chapter of the Society for Technical Communication. The people at State College are considering starting an STC chapter, and they have invited Ted Wilson, one of the state's outstanding technical writers, to speak about the opportunities in the profession.

PURPOSES OF ORAL COMMUNICATIONS

Ted Wilson, who can no longer avoid oral, public communication, calls to mind the concerns many Americans share about speaking before an audience. In fact, in a survey reported by Toastmasters International, most respondents stated they feared public speaking more than they feared illnesses, financial losses, or pain. For most people, an unavoidable public presentation is the worst thing that can happen.

Yet most people who seek success in their careers will find that public speaking is essential. If you perform your professional tasks with excellence, as Ted Wilson does, you may move toward a leadership position in your chosen field. You will be asked to explain your professional interests and concerns. For example, you could enter a managerial position and become a spokesperson for your company because of your technical expertise. Within your community, you may become active in a group of citizens concerned about technical issues, such as the quality of the local water or the control of the zoning ordinances as your suburb expands. Even as a parent or "quiet" citizen, you may find yourself drawn into public debates to speak at a town or school board meeting about your views of obscenity, creationism and science, or the various technical aspects of floating a bond to expand your community's school.

These are only a few examples of technical oral communications. Whether you appear at a rival school, at your local church or synagogue, or on the *McNeil/Lehrer News Hour*, you will speak to explain a product or process, to persuade about a topic or concern, or to solve a problem. In any case, you will want to represent your concern or to describe your topic in the best light possible. You will want to appear authoritative but relaxed, confident but empathetic, as well as poised and prepared. You can accomplish these goals by learning not only what kind of speech—description, persuasion, or argumentation—to prepare, but by learning, as well, how to select and organize your information. We also know from our experience and available research that if you know your topic and rehearse your presentation, you can learn to relax and, eventually, to enjoy speaking with people.

You can expect to be asked to prepare any of three types of oral technical communication: formal address, small group discussion, or informal presentation. Because speech/communication is a sophisticated discipline with a sizable body of literature and research about all three types of addresses, we concentrate on the informal or somewhat formal presentation, the form of address that technical and business communicators seem required to deliver on most occasions. Because technical communicators spend much, by some estimates 30 percent, of their working time talking with others to obtain information and to present results and new products, we discuss informal presentations at some length.

PREPARATION OF YOUR PRESENTATION—EIGHT ESSENTIAL STEPS

Most communication experts agree that preparation is the answer to a novice speaker's key problems: sounding knowledgeable and appearing relaxed. Not only does a thorough preparation enable you to think through most of your topic's ramifications, but, more importantly, realizing that you have taken enough time to prepare properly will build your confidence. We have coached many individuals like Ted Wilson through their fear of public speaking. The Ted Wilsons of this world often become convincing speakers in their specialized areas when they concentrate on the eight essential steps any effective speech coach would require.

Step 1—Analyze Your Audience

Use the audience analysis sheet shown in this chapter whenever you prepare a topic. You may see similarities between this form and the form you completed in Chapter 1 because both sets of questions ask you to focus on your primary group's education level, socioeconomic composition, and interests. This knowledge determines the information you will present and the appeals you will use to convince your listeners.

Ted Wilson knew a great deal about his audience through the "grapevine" of fellow university and agricultural employees. But you can research your audience very effectively with any of three approaches.

Visit the Site. A technical writer visited a rural community to represent her firm's interest in constructing a toxic waste facility there. Although it took an entire day, she traveled to the community. There, she discovered much greater hostility to the project than even her employers had anticipated. The company took extra measures to prepare for the presentation, including factual visual aids, a favorable market study, and an encouraging geologist's report that the speaker distributed with her proposal at the presentation.

Find a Liaison. If you cannot visit the presentation site to develop an audience profile, locate a local contact, and obtain that person's assistance by telephone. Often the person who invites

FIGURE 16.1

Ten Questions You Should Ask to Prepare an Audience Analysis for an Oral Technical Presentation.

You should analyze your audience as the first step in preparing an oral technical presentation. Here are ten questions you should answer in order to understand the needs of your listeners.

1. How old are your average listeners?

2. Is there an age range that represents the group?

3. What percentage are men? women?

4. What ethnic/racial mix is represented?

5. What are the salary ranges of most of the listeners?

6. What occupations are represented?

7. Do these people have a dominant religious/political view? If so, describe the view briefly.

8. How literate are your listeners about technology?

9. How literate are your listeners about science?

10. Determine your audience's attitude toward your presentation topic by circling one item on each line below.

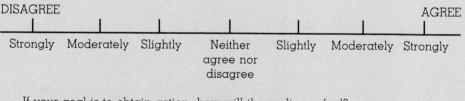

DISAGREE   AGREE

  Strongly   Moderately   Slightly   Neither   Slightly   Moderately   Strongly
                                     agree nor
                                     disagree

        If your goal is to obtain action, how will the audience feel?

    Opposed to acting   Inactive   Ready to act   Take action

the speaker will automatically assume the liaison function. Ask the questions from your audience analysis sheet, and ask how the listeners feel about your topic.

Use the Library. Often novice speakers forget that both public and college libraries offer the latest information on corporations, their earnings, successes, difficulties, and sensitive issues. The library is an excellent source of factual information if you sense you will address a hostile audience. Use the indexes to current periodicals and professional journals to find recent news about your listeners. Some libraries will even assist you by telephone.

After completing the audience analysis sheet—but before gathering your presentation's facts—classify your audience as favorable, neutral, or hostile. Your awareness of your listeners' attitudes should help you to select the facts, arguments, and examples you will offer them. In this case, consider Ted Wilson, whose school has a friendly rivalry with State College. Early in his presentation, Ted intends to use this rivalry as an "icebreaker." If he does not, his neutral to slightly favorable audiece may turn hostile because they think he is too authoritative about both his school's and his own accomplishments in the fields of football and technical communication.

Step 2—Define What Your Audience Should Know

Only after analyzing and classifying your audience should you begin gathering the information you will present. As you research your topic, your information and your audience analysis will help you determine an appropriate point of view for the presentation.

Make a Jot List. Ted "brainstormed" his subject to compose a jot list; we recommend that you "brainstorm" your topic as well. By "brainstorming" we mean that for a specific time, for example, fifteen to twenty minutes, you write down every fact, idea, feeling, concern, or random thought about your topic.

If brainstorming your topic in a group, remember several tips that will simplify your efforts. Appoint a member of your group as the recorder and leader, who calls on members as they signal to speak. The leader should use a blackboard or flip chart to write down each idea, verbatim. The leader should encourage the group to volunteer ideas, and none should be criticized or challenged. Remember, your group brainstorms to dig up every possible idea about the topic. You can delete useless ideas later, but you must have a list from which to extract the useful information for your presentation.

If you brainstorm alone, as Ted does, use either of two methods. For the first, select a large, unlined sheet of paper, for example, a sheet of newsprint, and write down all of your ideas as they come to mind. Work steadily, and don't stop until your time is up. Cover the paper with your ideas and attitudes about the topic. For the second, keep on hand a pile of standard-sized scrap sheets quartered to about the size of $4'' \times 6''$ index cards. Writing one—and only one—idea or fact per sheet, fill a whole pile of sheets. In either case, avoid writing your ideas on standard-sized, lined paper. The brainstorming process is for idea collection—not organization.

When your time is up, flip through your sheets or study your newsprint to see if your ideas fall into an apparent order. For example, Ted, who used the small sheets method, found that he had three types of categories about the technical communication profession.

FIGURE 16.2

Appeals for Use With Favorable, Neutral, and Hostile Audiences for Oral Technical Presentations.

Public relations experts refer to three kinds of audiences—favorable, neutral, and hostile—for any topic or presentation. Now that you have completed your audience analysis, can you classify your audience in one of the three groups? After you do so, organize your presentation by using the appeals listed here.

Favorable Audience

1. Use emotional appeals to intensify your listeners' support.
2. Ask or motivate your audience to make a public commitment.
3. Outline several specific alternatives for action.
4. Prepare your audience to carry your message to others.

Neutral Audience

1. Stress attention factors with an uninterested audience.
2. With an uninformed neutral audience, emphasize material that clarifies and illuminates your views.
3. For an undecided or unresolved neutral audience, build your credibility by presenting new arguments that blend logical and emotional appeals.

Hostile Audience

1. Set realistic goals for a single speech.
2. Stress common ground.
3. Use clear logic and extensive evidence to present your view.
4. Work to establish a believable image.

☐ the technical communicator's education
☐ the industries that employ technical communicators
☐ the technical communicator's wages

After rereading his sheets, Ted still had extra sheets that seemed important but that somehow did not fit in his three categories. They were:

you have to be patient and like to listen to people
tech. coms. in the Sunbelt of the West earn more
the aviation industry
wars—good for the tech. com. biz.
MA or BA—does it matter?

who is better: journalist or English major?
what about computer science?
don't forget to get list of schools w/programs from STC
you'll work overtime
60% of work time on manuals
30% work spent talking to others
learn how to punctuate and spell
Chinese diagrams at the museum, beautiful
a lot of tech. com. in great lit.
images of technology in society

After rereading and reshuffling his papers, Ted decided to speak about three other categories, the skills and skill levels of technical communicators, the history of the profession, and the personal qualities of successful technical communicators. He checked his six idea-groups against his audience analysis sheet. Since his audience was neutral, primarily college-aged and secondarily middle-aged, middle-class individuals who would probably know little of the profession's history, he decided to supply that background information. He also added the two other qualities because he assumed his audience would be career-oriented and would like to hear what personality traits can contribute to a technical communicator's success. Ted's final list contained six items:

the history of the profession
the specialties in which you can work
the education you will need
the wages
the skill types and levels
the personal qualities of successful technical communicators

After composing a list of information for the presentation, Ted was ready to consider his point of view.

Select an Approach. Ted had three choices—to inform, to persuade, or to debate—that would define his attitude toward the subject and create his rapport with the listeners. Because he had defined his audience as neutral to slightly favorable, he decided to persuade and to inform. He would persuade the audience at State College that the technical communication industry offers outstanding career opportunities, and show them how the profession has been a source of growth for others.

Note that Ted decided to discard some interesting ideas from his jot list. For example, his "images of technology in society" idea sounds intriguing, but it simply did not fit this collection of categories, his approach, or his point of view.

After defining his approach and eliminating inappropriate ideas, Ted wrote a thesis statement, the topic sentence on Card One: "A career in technical communication offers an excellent future for a writer, but to see if you are qualified, you should examine six ideas." Ted could now group his ideas by their inherent order.

In preparing an oral presentation, think of the organization process in two phases. For the first phase, select the rhetorical pattern or plan of organization for your information. For the second phase, prepare an organized collection of $5'' \times 8''$ cards with your speech plan, any cues or reminders about audio-visual aids, and any extra material you want to remember to introduce. Use Ted Wilson's note cards to help you design an effective plan of organization.

Organizational Structure. You can organize your presentation with any of six methods: chronology, problem/solution, order of importance, comparison/contrast, deduction or induction, or spatial relationship. For more details about these methods, check Chapter 5 on organizational patterns.

Your chosen pattern should reflect the inherent characteristics of the information you present. For example, a comparison/contrast plan lends itself well to a demonstration of how your product compares with a competitor's. A chronological plan is logical if you want to explain how a particular technology, for example, a programming language, came into existence. You may, of course, use a combination of methods, for example, a problem/solution approach to explain why the community should endorse water fluoridation, with a chronological pattern to explain the phases that will complete the installation of a fluoridation system.

However, we caution you in combining these approaches. Because the listeners cannot see your plan of organization and must rely on you repeatedly to mention it, keep your plan simple. Use only one method of organization. When grouping your information, use only first-level headings and divisions of information. For example, Ted Wilson's first card contains his plan of organization. He states a thesis in a sentence and only six first-level headings, so he can skim through his plan quickly as he refers his listeners to his central thesis.

Note Cards. To assist you in this phase of organizing your presentation, we include Ted Wilson's note cards as models. Please examine them closely. We have noticed that novice speakers often do not know how much information to include on note cards. We will "walk you through" Ted's cards to show you how powerful a tool a well-designed set of cards can be.

Really useful note cards:

1. Point and are keyed to the thesis.
2. Combine the key ideas and selected supporting facts.
3. Show by their arrangement the relationships among the ideas.
4. Show annotations that prompt the speaker with ideas, facts, visual aids, icebreakers, or reminders.
5. Use readable, even oversized, type so the speaker can see the text without squinting or looking strained.

Ted's Cards. Card One contains Ted's thesis and his six first-level headings for a plan of organization. (Note that he has a card following on each of the six items.) This card also contains two other reminders: Ted's "icebreaker" (the line about the journalists) and his audience reminder (the boldface section about the approach of the football game). Card One shows that a good organizer consciously formats a card to emphasize certain items. The football game reminder, for example, is in boldface, and the topics of the plan of organization are in-

FIGURE 16.3   Card 1 of 8 Showing the Introduction of a Presentation.

SAMPLE CARD 1

THESIS: A career in technical communication offers an excellent future for a writer, but to see if you are qualified, you should understand
- ☐ the history of the profession,
- ☐ the specialties in which you can work,
- ☐ the education you will need,
- ☐ the wages,
- ☐ the skill types and levels, and
- ☐ the personal qualities of successful technical writers

JOKE: Tech. coms. work for real wages/journalists sometimes don't—remember to kid the journ. majors in a friendly way.

Ted: Don't forget—State College is "enemy turf;" you are speaking three days before the State College/Ag. Mining football game—State College is favored. Let them know you understand the friendly rivalry.

dented for optimal visibility. Use such cues as boldface type, underlining, color coding, and indentions for levels of meaning, to prompt your memory and your delivery of the material.

A further word about Card One may demonstrate how truly useful cards are. Ted's host never really specified a length for the presentation. In one conversation, she suggested fifteen minutes, but later she suggested 30 minutes; thus, Ted developed an expandable or condensable plan of organization. He listed six topics on his first card, but if he receives more specific instructions when he arrives at State College, he can merely cross out several items in his list, slip those cards to the back of the stack and proceed with the somewhat abbreviated but still organized address. A plan of organization typed on a sheet of paper is not manipulated as easily as cards.

Cards Two through Seven contain the ideas, facts, and other prompts Ted needs to sound organized. As you scan these cards, note their readability, even with the abbreviations, sentences, and sentence fragments used as lists, and several other inconsistencies of usage that would be substandard in expository prose. Ted Wilson, not an instructor, is their audience, and the cards communicate directly with Ted using terms and reminders he understands best. These cards are not research notes you would submit to your instructor.

Nonetheless, an effective collection of cards has an obviously logical format. For example, Cards Two and Six both indent to show a classification system and the inherent relationship between an idea and the facts which support it. Thus, Card Two has its topic, "Where Tech Com Started," printed by itself on line one. Under that you see, "Issues relating to development" on its own line, meaning that Ted intends to discuss the history of technical communication as a set of issues. Then, indented for subordination, Ted lists his four supporting reasons.

| **FIGURE 16.4** | Card 2 of 8 Showing Part One of a Discussion of Technical Writing, the History of the Industry. |

SAMPLE CARD 2

HISTORY—Where Tech. Com. Started
Issues relating to development
    Aviation industry WW II and after
    Cost-effectiveness of communication
    Writing as a product—manuals for sale as numbered parts
    Writing as a product—within firms, structure of system

Ted: If time allows mention Chaucer's *Treatise on the Astrolabe* as an example of tech. wrt. down through the ages. Also, mention blueprints and building instructions of ancient China at the museum.

| **FIGURE 16.5** | Card 3 of 8 Showing Part Two of a Discussion of Technical Writing, Where Technical Writers Work. |

SAMPLE CARD 3

SPECIALTIES—Where tech. com's. work—where to look

agriculture, automotive, aviation industries (Sundstrand)
business communication (Mid-Am Bank)
basic industry (Dana—hoses)
computers—hardware, software (GE Info. Sys., Intel, Hewlett-Packard)
insurance (Erie Insurance)
education (teaching, training)
medical (journals, patient education)
proposals (waste disposal)
publishing science (Argonne, Bell Labs)
training (R.R. Donnelly)
U.S. gov't.

| FIGURE 16.6 | Card 4 of 8 Showing Part Three of a Discussion of Technical Writing, Where to Find Information on Education. |

SAMPLE CARD 4

EDUCATION—mention 2 sources

Ted—define tech. communicator, not tech. writer

1. Levering, *The 100 Best Companies to Work for in America, 1984.*
2. *Technical Communication,* STC, Wash., D.C., Vol. 31, No. 4, 4th qt. 1984, Spec. Issue on Education in Technical Communication.

Typical MA or MS—35 hours   list of
   2 to 5 electives   schools avail.
   internships   at my office

Typical BA or BS—21 to 30 hours
   2 to 5 electives in technology
   internships

| FIGURE 16.7 | Card 5 of 8 Showing Part Four of a Discussion of Technical Writing, Some Average Wages of Technical Communicators. |

SAMPLE CARD 5

WAGES—**Learn to negotiate** (Suggest tech com. more money than journalism)

	Entry	Advanced
Midwest	22,000—30,000	30,000–35,000—33,000–39,000
NY and East	25,000—32,000	30,000–40,000—35,000–45,000 +
S'west and Mtn.	22,000—30,000	30,000–34,000—33,000 on up
Pacific	28,000—35,000	35,000–40,000—41,000 +

**FIGURE
16.8**

Card 6 of 8 Showing Part Five of a Discussion of Technical Writing, Necessary Technical Skills for a Writer.

SAMPLE CARD 6

SKILLS NEEDED

Entry—Writing—sp., punc., usage
    Speaking, interviewing, conversing

                    plus

    Computers—programming
         equipment

    Technology—blueprints, schematics

    Design, layout, graphics

Managerial—budgets, supervision, scheduling, leadership

**FIGURE
16.9**

Card 7 of 8 Showing the Transition to the Conclusion of the Presentation Topic, the Technical Communication Profession—a Career of Possibilities.

SAMPLE CARD 7

PERSONAL QUALITIES—**Ted, ask them to self-evaluate.**

Curious—Are you willing to learn about new products: hoses, or pumps, or computers?

Adaptable—Can you work with others who might not want to talk with you: engineers, program and systems analysts, floor workers, etc?

Resourceful—Are you a problem solver?

Socialized—Are you business oriented (willing to wear a suit, work reg. hours)?

Motivated—Are you willing to work overtime? You will be well paid, but they need you now. Can you take the pressure of meeting deadlines?

On Card Six, Ted again uses indention to create "blocks" of text he knows are more readable than a manuscript format at a speaker's podium. Thus, the entry-level communication skills appear grouped in one block with the word "plus" freestanding to lead him to the computer, technology, and design skills. Ted obviously uses indentions to create his classification system. Can you think of a color-coding system that would be effective for the same purpose? Some of our students like to use contrasting highlighter pens to create the classification pattern. Again, because your cards communicate with you, feel free to design your own prompts.

Note as well that Ted uses his cards to create drama and emphasis in his presentation. Card Seven has a very rigid parallel grammatical format because Ted wants to reinforce for his audience the idea that all of the personal qualities he discusses are necessary and inseparable in a successful technical communicator. To create this effect of useful repetition, he lists all of his terms as one-word nouns, and he attaches to each word a very personal sort of question. Thus, each noun and its question indirectly point to one of Ted's overall themes, "You should think about whether you are right for the technical communication profession."

Ted also uses his cards to record specific facts that he fears he may forget or overlook at the podium. Thus, he closes by offering to provide the key name and address for the Society for Technical Communication. Note, too, that on Card Three he records the names of some companies where friends in the profession work. He decides, wisely, not to memorize these

---

FIGURE 16.10   Card 8 of 8 Showing Information to Be Presented in the Conclusion to Encourage the Audience to Join the Profession.

SAMPLE CARD 8

KEY NAMES

National

William Stolgitis, Executive Director
Society for Technical Communication
815 Fifteenth St., NW
Washington, D.C. 20005

Local

Mention name of chapter representative
Also mention: American Business Communication Association
                    Women in Communication
                    IEEE

names, but he wants the specific facts to be at his fingertips, so that his presentation will offer an adequate level of detail. Again, the cards allow you to record rather than memorize the facts. You can color code such facts so they really stand out when you need to refer to them.

Studying Ted Wilson's note cards, you may ask, "How do I use the cards?" "Should I hide them?" "Should I try not to look as though I'm looking at them?" We have some very specific suggestions, but we refer you to the section on practicing your presentation for that information. Since you should create your support materials and visual aids as soon as you have prepared your note cards, we treat those sections next.

Step 4—Plan and Write Your Support Materials

We all like to receive "premiums," the free gifts that banks, manufacturers, credit card companies, and others offer to persuade us to purchase their products or services. Remember this psychology as you prepare to speak with your audience. Bring samples of your products, product literature, or other kinds of handouts with you, to convince your audience that you stand behind your presentation. Here are some premiums or support materials that our clients and students have used.

Single-Sheet Handouts. Ted Wilson brought materials obtained from the Society for Technical Communication and copied at the office: membership application blanks, brochures about charting a career in technical communication, an article about finding technical communication jobs and an article showing the organizational structure of a major computer manufacturer's communication group, and, of course, copies of his own agricultural writing.

Booklet or Brochure. A student of ours recently applied for a position with a major manufacturer of accounting software. For part of his interview, he had to give a presentation about his internship. During the presentation, he distributed copies of a brochure he had prepared for the computer services department of a university. By examining the booklet, his listeners learned in a glance that he can complete a professional project.

Complete Report. You may work with a group or team to complete an investigation or a special project. Even though the details of that effort may have great meaning for you, they may not for your listeners. Use a presentation to tell your managers about the beneficial results of your efforts; then to explain the scope and efforts of the entire project, present a limited-distribution, formal report to your managers at the start of your presentation. This document is sometimes referred to as a summary report. For more details on its contents, consult our chapter on reports.

Slides. People enjoy learning through pictures. If you have a unique product or process to explain, you may be unable to leave a sample or a set of specifications. But you can have attractive slides with close-up views of the results your process can achieve. For example, if you need to explain what your graphics software package can do, slides can show the results your product delivers—without distributing proprietary information.

Software. If you must give a presentation to a small group, such as six employees of your firm's client, a software package may be a perfect "leave behind." For example, a student of ours brought to an interview with a major oil company the intramural scheduling software she

| **FIGURE 16.11** | The Benefits of Visual Aids to an Oral Presentation. |

Professional communicators have a special vocabulary for the visual aids they associate with oral presentations. They point out that different types of visual aids work well together. Try to mix your aids if possible. Also, keep the message that you illustrate, the concept, simple and direct. This list outlines the benefits of certan aids.

VIDEO

You can tailor video to fit the needs of your audience. Video summarizes the key points while being direct and concise. When used effectively, it will hold your audience's attention in an immediate and personal way. Video works well with other support visuals. Video can show your audience a variety of things by "bringing it to them." Videos are expensive, and they require production by a staff of professional film-makers.

SLIDESHOWS

Think of video without motion. Slideshows can show an event or process in detail. Although slideshows lack the advantage of motion, they work very well with supporting materials. Slideshows work most effectively when they are well-timed to keep the attention of the audience.

HANDOUTS

These "low-technology" aids offer the advantage of simple, low-cost preparation. Handouts can be prepared quickly and distributed in large quantities. They keep you within your budget, and they do not require special audio-visual equipment. Another major benefit of handouts is their "leave behind" capability; listeners like to see things in front of them and to take something with them from your presentation.

TRANSPARENCIES

Clear vinyl sheets with information and drawings on them, transparencies are also inexpensive and effective. You can use them with a large group, and you can produce them on the spot with freehand sketches and lettering if need be, although a projector and, usually, a screen are required. Remember to keep your transparencies simple to be effective.

FLIP CHARTS

These "low-technology" relatives of handouts and transparencies are large, visible, and easy to use. You need remember only a newsprint pad and an easel to be ready with a flip chart. You may use detailed, prepared flip charts, like the ones used in military field training, or you can prepare them on the spot. Remember: to use flip charts best, keep the message simple, and spell correctly. Use wide-point markers with a good supply of ink.

BROCHURES/MANUALS

These print materials go well with video or a slideshow. They can present details about your subject, allowing you to use the video or slideshow for the ideal purpose of stating your message briefly.

wrote for her college. She left copies for their programmers and analysts to examine. Of course, they hired her. We have seen software used to explain the uses of new products and new services, for example, in banking and in education.

Video. Communicators use video regularly to explain the benefits of complicated hardware and manufacturing systems for the same reasons that we recommend software or slides. This "leave behind" tells a great deal about your product or service if the nature of your business or profession does not lend itself to the preparation of sample products. Although video is costly, it makes a lasting impression on the client. If you have an ample budget, consider video.

Remember, the support materials you design speak for your presentation after you have left. They attest to your firm's or your own credibility. Some reminders can help you keep your efforts coordinated:

1. Key your support materials to the sections of your presentation. Refer to your materials by name during your presentation, and ask your listeners to examine the materials as they listen to you.
2. Coordinate your support materials to give a unified and successful impression of your firm. For example, if you distribute several types of handouts, design them with similar formats for continuity. If your firm has a logo, use it often in a uniform design.
3. Coordinate your support materials with your visual aids. You may want the support materials to duplicate the visual aids, or to elaborate the key points you make with the aids. The support materials should flow from and refer to the visual aids.

Step 5—Design Your Visual Aids

We recently watched a presenter at a workshop on direct marketing. An expert in the area, he had an eager audience anticipating his insights about reducing the costs of direct marketing. Introduced to enthusiastic applause, he informed us that he would now show a video his associates had designed to answer our questions. As test patterns appeared on the hotel's television, he discovered that he had the wrong connecting wire for his television and video cassette player. At this point, he also discovered a restless audience who expected high-technology entertainment.

Never let this failure happen to you. Always, for safety's sake, assume your high-technology support and visual aids, such as slides, software, and videos, will fail, and have prepared a hierarchy of visual aids for your audience. By a hierarchy, we mean a collection listing the aids in a range from simple to complex. For example, if Ted Wilson were to have an appropriate budget, his hierarchy from simple to complex might include handouts about the Society for Technical Communication, brochures about the technical communication profession, transparencies that repeat the information on the handouts, a script for a video he wrote about new fertilizers, and the actual video he could show to the group.

Ted's hierarchy is a form of insurance. Assuming that the video equipment could fail (see our section on practicing your presentation), Ted still has a viable support system for his presentation. Although his listeners would be disappointed if they were unable to view his video work, he can still show them examples of his writing and his script. Furthermore, the listeners will admire him for his preparedness and his fortitude. If he retains his poise, his neutral-to-slightly friendly listeners may become a truly favorable audience.

To assist in your preparation, we have included a set of definitions of visual supports and a description of their uses. Refer to that list and to our chapter on graphics for details about developing visual aids. We have some tips on the uses of visual aids:

1. Keep each aid simple, with one idea per aid. Novices tend to crowd a transparency or slide with too much information. For example, if you want to show a transparency in a presentation that defines eight new terms, prepare eight transparencies, one for each definition.
2. Use visual aids to reinforce your plan of organization. For example, good presenters show their audience an outline of the material they will cover. They periodically refer to this outline, to mark their progress. Other effective speakers show an outline of the whole presentation and then show outlines of each section as they move through the discussion.
3. Use aids for emphasis. For example, if Ted Wilson's listeners were journalists, he could show a slide with the wages a successful technical communicator can earn. Since the average successful technical communicator earns more than most journalists, these figures might be impressive. Better yet, Ted might show a line graph comparing the average earnings of journalists and technical communicators.
4. Use visual aids especially to communicate anything mathematical or scientific to your audience. Americans, especially, have trouble reading texts that discuss or analyze numbers. Because there may be many mathematics illiterates even among educated groups, design visual aids that explain how to interpret the figures you present. Do not assume that your listeners can analyze or speculate about data, equations, or financial figures.
5. Use visual aids to create transitional sequences between the parts of your presentation. For example, you can show illustrations of the object you intend to discuss next as you conclude a discussion of one object.
6. Use audio-visual aids to inject humor into your presentation. For example, to win a slightly hostile or almost neutral audience to your view, show a cartoon that describes a difficulty you and the listeners share. You may also decide to use a cartoon to show how confusing a technical problem is. Laughter will relax both you and your listeners.

At all points, remember that your visual aids and presentation must complement each other. You can have the finest video produced, but you must incorporate that element into your discussion with the audience. You — not your visual aids — are the center of the discussion. Relate your aids to your total message. Our next area of concern, practicing your presentation, can help you master the technique of combining your discussion and your visual aids into one smooth performance.

Step 6—Practice and Improve Your Presentation

If effective planning reduces your fear of public speaking, a successful presentation must include several thorough practice sessions. Toastmasters recommends that you practice your presentation six times, and that you rehearse your speech for the following objectives:

First Practice   Organization
Second Practice   Organization
Third Practice   Timing
Fourth Practice   Voice Control

FIGURE 16.12   Body Language That Betrays the Speaker—Seven Telltale Gestures of Nervousness or Fear.

Expert coaches of successful public speakers refer to seven deadly body language traits that betray your discomfort to your audience. Practice your presentation in front of a mirror and in front of caring friends so that you can overcome these signs of nervousness.

FIG LEAF—Arms crossed, hands clasped in front of the body below the waist—a gesture of weakness.

STERN FATHER—Arms crossed above the waist—creating a psychological barrier between your audience and you.

SWAY—Body moves back and forth while the feet stay still—reinforcing an image of monotony your audience will perceive about your presentation.

JANGLER—Coins in the speaker's pockets jingle around as he or she plays with them during the presentation—and the audience concludes that you are hyperactive and ineffectual.

ARM LOCK—Either arm sits rigid, permanently crooked and locked in front or in back of the body—telling the audience that you are rigid and uncomfortable with your presentation and, perhaps, with them as listeners.

PARADE REST—An old military term for a speaker who stands with both arms behind his or her back, the hands clasped, "at ease"—actually anything but at ease with your audience because with this position you betray your lack of authority.

PODIUM CLUTCH—White-knuckled, the speaker clings to the podium because there is nowhere else to go—and when you do this your audience knows you fear your material, them, or both.

Fifth Practice	Style
Sixth Practice	Style

Organization. It is one thing to feel organized; it is another to appear and sound organized. When our students evaluate each other's oral technical presentations, during both the practice and the final sessions, they most often recommend improving organization. The same students encouraged to improve their organization usually hear from their classmates that they should improve their knowledge of the subject as well. Actually, these students are often quite knowledgeable; their appearance of disorganization simply convinces the listeners otherwise. Organization, then, is the key to high scores on the evaluation form included in this chapter—and on the unconscious "evaluation forms" any audience uses.

How can you appear organized? Show your outline to the audience on a transparency or slide at the start of your presentation. Use a pointer or pen to indicate on the slide or transparency how you will move through the outline. As you finish each first-level heading, show your audience a small outline of that section's subjects before moving the pointer to the next section. This way you create a visual picture to accompany the aural impression of your presentation.

Know—but do not memorize—your outline. Practice the words you will use to introduce your presentation and to move from section to section. Then practice using a pointer on a screen to move from section to section of the visual aid showing your plan. Become comfortable with this procedure, so that you can point to your visual aids and talk simultaneously.

During the first two practice sessions, practice acting organized as you use your cards. Mark them with very clear numbers to indicate their sequence. During the first practice session, run through the cards as they are, and test for the places where you have trouble finding information or seeing transitional material. Correct your marks or add new information, read through the cards again, and practice the presentation the second time to become familiar with your "edited" cards.

Timing. Your practice for timing should really consist of two sections. First, deliver your presentation to a friend who can record how long each section takes. Do some sections seem to drag? Do others require expansion? Do you need to delete one or two parts to deliver your presentation during the allotted time? Remember, audiences like a presentation to average no more than twelve to fifteen minutes. Can you arrange your materials and deliver the message in that timespan? After you and your listener analyze the timing, deliver the presentation again to increase your confidence and to learn your changes.

Voice. Just as the appearance of disorganization betrays an amateur, so does a weak or unsteady voice. Incidentally, since voice control is a technique that professional actors work to develop throughout their careers, this book does not cover this highly complex art and science in great detail. But you can avoid the breathy, mouth-breathing delivery that dismays your audience, sabotages your authority, and actually creates further nervousness. Here is a simple exercise that slows your heart rate, lowers your pulse, and relaxes your autonomic nervous system. Closing your mouth, breathe slowly and deeply through your nose to the count of five; then, exhale slowly through your mouth, not your nose, to the count of five. Do this at least three times shortly before you begin to speak.

Many speech coaches use this deceptively simple relaxation exercise. You can learn to like this exercise because you can do it undetected while awaiting your turn to speak. This exercise will help you to speak in rounded, relaxed tones. During your presentation, remember to inhale through your nose, not your mouth. Inhaling through your nose slows your delivery just enough to keep your tone relaxed; you then exhale through your mouth as you speak. If during your presentation, you start to feel nervous, inhale through your nose. Finally, use pauses in your delivery to relax. Speakers, especially nervous ones, often deliver their presentations so rapidly that listeners simply cannot follow. Indicate the ends of your sentences with pauses. Use these breaks to glance at your cards for the next sequence as you inhale through your nose. All of this is quite basic, but we have seen presenters win awards for public speaking because they have mastered their breathing, and thus their voice control.

Style. During these two practice sessions, work on all of the items on the evaluation form for an oral technical presentation. For your delivery, use an elevated conversational style that requires you to speak in complete sentences but of the simple, subject-verb-object variety. Run off and distribute copies of the form to the friends, relatives, or roommates who will evaluate your rehearsal. At the end of the first rehearsal for polish, collect and review the forms. Then correct your style, and rehearse your presentation for the last time. We will present the list quickly:

1. **Eye Contact.** Avoid the "eye dart," or glancing everywhere at once; the "triangle," or looking at the same three places or persons throughout the presentation; and the "copyreader," or looking only at your cards and avoiding any eye contact with the audience. Look carefully around the room at all the listeners, and maintain each eye contact for about four to five seconds. Don't look above their heads; look them in the eyes.

2. **Poise.** Since your entrance is crucial, establish your poise and authority with the way you enter the room and stand to face your listeners. Again, this may sound simplistic, but we have seen first-rate engineers sabotage their own effectiveness by slouching to the front of the room. Your audience will decide in the first ten seconds whether they are with or against you. After that, you either confirm or try to reverse their decisions. Scan our list of body language gestures that betray the speaker, and then try to avoid these behaviors. Stand erect, or if at a conference table, sit erect. If you use hand gestures, gesture above the waist since gestures below the waist connote weakness.

3. **Enthusiasm.** Greet your listeners, and, if it is appropriate, smile. Do the voice control exercise so you can project your voice well enough to sound enthusiastic. Work to overcome either a monotonous delivery or a nasal tone.

4. **Audience Analysis.** We have said a great deal about this already. Doublecheck during this rehearsal for any content or any meaning that your audience may find offensive.

5. **Knowledge.** Although this achievement will flow naturally from a strong impression of organization, know your topic so well that, if necessary, you can deliver it without note cards.

6. **Organization.** Do you show your listeners a plan of organization? Do they know when you have completed each section of your presentation? Can they anticipate the end of your presentation as the logical outcome of the information you discuss?

7. **Level of Detail.** Do you include enough facts, examples, illustrations, reasons, and visual aids to make your topic vivid and memorable to your audience? The author Alex Haley once delivered a lecture on the family, a well-worn topic. But his level of detail as he described his parents and grandparents conveyed, in a memorable and vivid way, the message that he learned his racial history from his family. If discussing a densely technical topic such as mutual funds, use tables, equations, verbal prompts, and easily readable graphs to communicate your message.

8. **Use of Visuals.** Have you selected the right visual aids for the job? Do you present a hierarchy of visual aids? Can the people in the front and in the back see your visual aids?

9. **Effective Conclusion.** Do you tell the listeners that you have progressed to the end of the outline? Do you work in a summary statement? Do you deliver a closing statement that relates to your introduction and to the audience as interested listeners?

FIGURE 16.13	A Sample Evaluation Form for an Oral Technical Presentation.

Name _____ Topic_____

Factors	Rating					
	Excellent	Very Good	Satisfactory	Below Average	Poor	Comments
1. Eye Contact						
2. Poise						
3. Enthusiasm						
4. Audience Analysis						
5. Knowledge						
6. Organization						
7. Level of Detail						
8. Use of Visuals						
9. Effective Conclusion						
10. Effective Question/Answer						

10. **Effective Question/Answer.** Do you conduct this session with empathy for your audience and with poise if you are challenged?

Step 7—Deliver Your Presentation

You may be nervous. Even experienced speakers feel some excitement before facing an audience. For them, however, the excitement is all positive because they have learned to use the tension a public appearance creates. You can deliver a presentation successfully by doing some last-minute preparation.

First, practice a relaxation technique known as visualization. Imagine yourself in the room where you are to deliver your presentation. The audience turns to greet you, and they smile warmly. You, looking your best, stride confidently to the front of the room. At the podium or conference table, you take command of the audience, who look interested in your presentation. You imagine your attractive visual aids, your simple, clear plan of organization, and the audience nodding in agreement as you present the point of view they will appreciate. You imagine warm applause or eager questions as you finish your presentation, and you imagine, as well, how much you will enjoy the question-and-answer session, how much you will enjoy talking about your topic in a less formal way.

You can fill in the details of this visualization exercise to suit your own presentation. In short, visualization is a process that encourages you to imagine a successful outcome. Experiment with this process of building your confidence by imagining as many details as possible of your success while you prepare to deliver your actual presentation. The theory is that if you concentrate on a successful presentation, you will deliver a successful report.

To ensure your presentation's success, you can also do some more practical things. For example, always try to visit, ahead of schedule, the room where you will deliver your presentation. Check the room's size. How much will you need to project your voice in a roomful of listeners? Will your visual aids be visible in this environment? These are some of the problems an advance visit can reveal and solve.

Also arrive early for your presentation. An early arrival enables you to check and, if necessary, replace your audio-visual equipment. At this time, check whether you need a microphone. If so, find out how to use it and how to adjust its height so you can speak through it without straining to see your note cards and the keys to your visual aids.

Finally, arrive early so the moderator or host knows that you are present and willing to deliver your presentation. Remember, the moderator, too, may be anxious for the presentation's success, and your presence can be reassuring. Remember also to thank the moderator after you have delivered your presentation because the moderator has assumed some responsibility for managing the audience, presenting the speaker(s), and assuring that the presentations move smoothly. Because we gave suggestions for a successful presentation during our discussion of your preparation, we assume you now know what to do, and we want you to understand how a well-managed question-and-answer session can cap a strong presentation.

Step 8—Control the Question-and-Answer Session

Customarily, speakers at informal presentations conclude their discussions by asking the audience for question and discussion points. These question-and-answer sessions can be enjoyable if you remember these pointers:

Listen. Overcome your nervousness or excitement long enough to pay close attention to the question asked. Don't try to think up an answer; listen, first. As you listen, look at the questioner, so he or she believes that you listen with interest.

Repeat. Repeat the question. Research on teaching effectiveness has shown that students do not listen to other students' questions—even the most useful ones—unless the teacher repeats them and, therefore, makes them "official." Use this technique with your audience. You can say, for example, "I believe you are asking (insert rephrased question) . . . and if that is what you ask, I would suggest the following (answer or reply). . . ." By repeating the question, you grant the listener authority, you direct the audience to your interaction with them and your desire to answer them, and you have a chance to rephrase the question in a manner appropriate for the instruction of the group.

Empathize.   Create what is called a shared purpose with your audience. Use the terms "we" and "us," not "you" and "I" in developing a shared purpose with your listeners. Explain to your audience that you have some common concerns. Let them know that you identify with their preferences or concerns. Use this bridge to build answers to your questions. For example, in a question and answer session with your supervisor after a presentation, allude to your mutual interest in cutting costs and increasing profits as you provide further information about your project and its expenses.

Answer.   Answer — don't avoid — the question. If you cannot answer the question, if, for example, someone asks for figures you do not know or cannot remember, tell the truth. Offer to obtain the information, or tell the questioner where to get the answer easily. Although your listeners expect you to be knowledgeable, they don't expect you to know everything. If you retain your poise as you offer to find the information, you can increase your credibility with the group.

Ask.   After you answer or discuss a fairly lengthy or complex question, turn to the questioner, and ask, "Did I answer your question?" Pause until that questioner nods or asks another question. Again, your concern for accuracy will help you to convince the listeners of your sincerity about the topic.

If you practice these techniques, you need not fear questions. One caution: it seems any group has a disruptive questioner. This person may try to involve you in a long dialogue, or to take over the floor by giving a lengthy, personal view in response to your presentation. You can watch the audience's reaction to this person. It is best to acknowledge and invite the questioner to speak with you in detail about your mutual interest after the questions and answers are complete. By extending such an invitation, you acknowledge the person's question, which your audience will want you to do, but you also maintain the pace of your discussion, which will retain the enthusiasm of your audience. Without the whole audience to "play to," a formerly disruptive questioner may be a joy to meet.

Finally, the question-and-answer session should be more relaxed in tone and demeanor than the actual presentation. If it is appropriate, use humor to deal with difficult questions. If you use humor effectively, the audience will relax. Humor can also signal that the presentation is drawing to a close now that the ten minutes allotted for questions have indeed passed.

SUMMARY

In this chapter, you learned that technical communicators are often requested to make informal presentations to audiences about their special areas of training. You learned, also, that a nontechnical communicator may find it necessary to give a presentation about some aspect of science or technology After reading this chapter, you should be able to:

■ Conduct an audience analysis using our prepared form so you can classify your audience as favorable, neutral, or hostile to your topic.
■ Select the appropriate information for your listeners.
■ Define the most appropriate approach to the topic for your listeners.
■ Develop the plan of organization for your presentation.
■ Prepare 5″ × 8″ note cards that contain the plan and all the necessary supporting information for your presentation.
■ Design an assortment of support materials to distribute to your audience.
■ Prepare a hierarchy of visual aids to complement your presentation.
■ Rehearse your presentation six times to create a professional style of delivery.
■ Deliver your presentation within the time allotted.
■ Maintain control of your audience during a discussion session.

ASSIGNMENTS

1 Compose a list of five individuals in public life who you think speak well. For example, you might get some ideas for this assignment by watching the "Focus" segments on the *McNeil/Lehrer News Hour* on your local public television station. Complete the oral presentation evaluation form for each speaker. Bring the forms to class prepared to explain the strengths and weaknesses of your favorites.

2 At the end of Chapter 12, you selected a topic about which to develop a proposal. Can you explain to a group, perhaps the class or a group of residents in your community, why this topic should receive funding? Complete the audience analysis and the selection of appropriate information and an approach. Bring this information to class.

3 Ask several classmates to review the information you have collected. Perhaps you and three others can work as a group to complete the reviews of each other's work. Can your classmates suggest some appropriate support materials and visual aids? By the end of your collaboration session, you should have listed a hierarchy of materials for each person in your group.

4 Select a partner, in a sense a co-author of your presentation, who can work behind the scenes to help you rehearse. You in turn can rehearse your partner. Complete the six rehearsal steps suggested in this chapter. Perhaps your partner can choose some understanding classmates or associates who can sit in on a few rehearsals and fill out the evaluation forms for your analysis. Work together to perfect your presentations.

5 You and your partner should now deliver your presentations to the class. Your partner may act as your assistant for distributing support materials, setting up visual aids, or dealing with any last-minute emergencies. Convince the class in no more than ten minutes that they should fund your proposal.

6 After completing your presentation, distribute copies of the oral presentation evaluation form to your classmates. Ask them to complete and return the form.

7   You and your partner should now study these forms to see how you can improve your presentations in the future. How did the listeners perceive you? Did they understand your topic? Did they endorse your proposal?

8   Remember your group? How did they fare with their oral presentations? How many of them convinced the class to fund their proposals? Perhaps now that you have all assisted each other, the other group members will share their feedback about what they learned.

PART THREE

Management

CHAPTER
17

Understanding Corporate Production of Documentation

This chapter introduces you to the way a publications group functions within a corporation. After reading this chapter, you will be able to:

- Identify the division of a corporation to which a publications group might report.
- Identify the vertical structure of a publications group.
- Name the tasks that the members of a publications group perform.
- Describe the steps that move documents from the research to inventory stage.
- Discuss the advantages of both desktop publishing and external, offset printing in terms a corporate middle manager can understand.

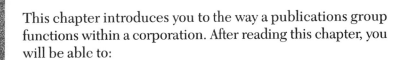

CASE STUDY   An Intern Prepares for Corporate Writing

As a first-semester senior at a leading New England university, Maryann Hasama has received an internship offer in technical writing from a major accounting software firm. Her excellent background in technical writing, editing, and computer science helped her win the position.

Although the interview went well, Maryann returned home somewhat uneasy, because the personnel manager spoke a great deal about corporate life and the company's policies. The recruiter questioned her extensively about her ability to work with others, her college experiences of co-authorship, and her understanding of corporate structure. For three years Maryann has developed an excellent portfolio through enrollment in writing—not business management or production—courses.

As many technical communication students do, Maryann assumed that in her chosen career she would work alone, writing and editing technical documents. She feels unprepared for the corporate life, and knows very little about the communication channels in a modern business. She understands, as well, that her successful internship can lead to permanent employment with the firm. Now she must learn quickly who will supervise her work, how her job will be defined, and what method to use to complete her assignments. In fact, she does not even know what to ask in order to understand the differences between college and corporate technical writing assignments.

THE WRITER'S ENVIRONMENT IN A CORPORATION

Despite warnings from their placement offices, many college students fail to learn about the corporate environment—in particular, how that setting will determine their jobs and their pursuit of their professions. Entry-level technical communicators often report their surprise at learning how much time they spend on corporate communications, simple interviews, and ceremonial conversations in order to maintain their project schedules. The Society for Technical Communication reports that a technical communicator may spend 25 to 45 percent of his or her time conversing with associates in order to develop and maintain a project schedule. Repeatedly, our graduates report that they are well prepared in their fields, but that they would also like to have learned more about how the world of business affects the technical writer.

This chapter, then, is an orientation to the methods of publications groups; their supervision; their writing, editing, and production schedules; and some of the technical communicator's professional concerns. This chapter also explains what you can expect from an individual writer or a publications group when you contact them for help.

WRITING AND EDITING IN A CORPORATION—
THE FUNCTIONS OF A PUBLICATIONS GROUP

This organizational diagram is a composite of several firms that include publications groups. This chapter discusses three aspects of the publications group:

- ☐ Placement within the organization.
- ☐ The group's hierarchy, or vertical organization.
- ☐ Responsibilities—that is, the types of tasks the members perform.

FIGURE 17.1   A Flowchart Showing a Typical Plan of Organization for a Technical Publications Group.

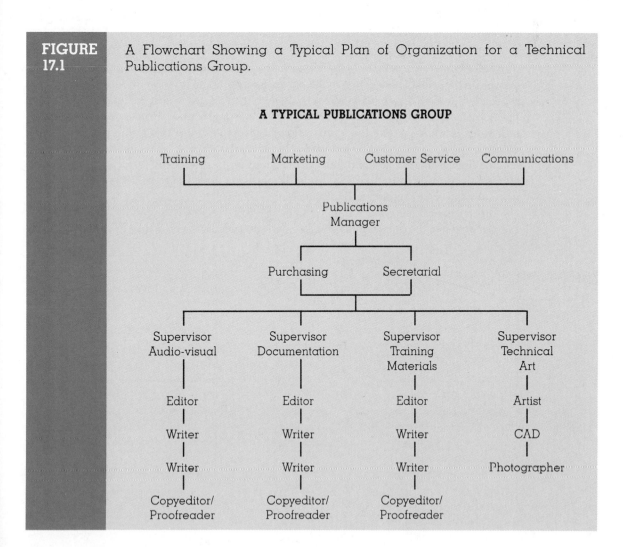

A TYPICAL PUBLICATIONS GROUP

First, a word about terms is necessary. We use "publications group" to refer to a working unit of technical communicators in an industry. Groups of technical writers in industry are also called "Technical Communications," "Documentation," "Documentation Specialists," "Customer Communications," and "Product Information Services." The terms for the technical communication groups vary with the names and organizational charts of the industries. We use "publications group" because it is a convenient, well-known term. We prefer "technical communicator" for a member of a publications group because the term suggests the full range of publication services, including training and educational materials, videos, and software tutorials, that the modern technical writer offers in industry.

Placement

Our generic flowchart shows that a publications group may work within any of four corporate divisions. Often, the location of the publications group stems from the firm's franchise or its business purpose.

For example, a firm that invests heavily in the training and development of its labor force may place the publications group in the personnel or training division. In some newer firms, especially among software manufacturers, the publications group is in the marketing division because manuals and users' guides critically influence product sales. In the more "traditional" industries, such as automotive corporations, the communications group may be in the customer service area, since these manufacturers market their services and their training for mechanics. Finally, in a large firm, the publications group may either be located in an entire communications division that handles legal matters, public relations, technical and other communications, or stand alone if it is still more sizable.

Because the technical communication profession is relatively new (about forty years old), the placement of the publications group sometimes reflects the growth of an individual firm,

FIGURE 17.2   A Flowchart Showing How a Technical Publications Group May Interact With the Other Divisions in an Industry.

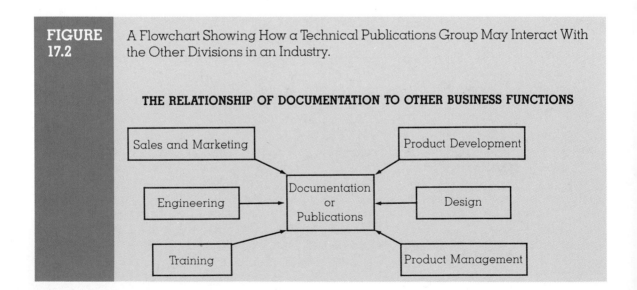

THE RELATIONSHIP OF DOCUMENTATION TO OTHER BUSINESS FUNCTIONS

rather than the quality or "status" of the publications group itself. For example, in certain defense industries, the publications personnel actually work in several of the engineering departments. In some software companies, too, technical communicators receive the title of "software engineers," because they work closely with the design and applications personnel in developing new products. In short, publications divisions sometimes report to their managers through seemingly inappropriate channels. When the firm was small, technical communications was placed within a particular division that has not reorganized as the firm has expanded.

Hierarchy

Our organizational chart shows a five-level hierarchy through which the publications group may communicate.

Division.   We show four divisions, training, marketing, customer service, and communications, as examples of a corporation's major working organizations. Usually run by a corporate vice-president, the division represents the highest level of authority to which a publications group may report.

Manager.   As a member of the division's middle management structure, the publications manager may report to a division vice-president. The manager assumes total responsibility for the publications group's budget, hiring, and projects. Typically, the manager only reviews and authorizes projects that others complete.

Supervisor.   Our chart shows four supervisors reporting to the publications manager. Typically, supervisors — that is, employees with some seniority in the company — evaluate and keep records of the projects under development by the unit's editors, writers, and artists. In some companies, the supervisors continue as working technical communicators who guide others while they themselves develop special or costly projects such as videos or training films.

Editor.   In some companies the term "editor" describes a senior writer whose experience with the firm and its products has prepared him or her for a start in supervision and management. In such cases, the editors use the company's systems to prepare documentation for the expensive and exacting processes of production and distribution (for more details on editing systems, see Chapter 18). The editors ensure the standardization and quality of a document's appearance, information, and style. Often the editors act as project managers for the document writers. The editors communicate with supervisors, production personnel, and writers and coordinate their activities.

We deliberately place the editors at the top of the fourth hierarchy level to suggest two important aspects of the editorial job. In some companies, an employee can become an editor only after working as a junior, then senior technical writer. Thus, the editor's job at these companies is a first-level promotion involving increased responsibility for the completion of projects within their allotted budgets. But not all companies define the editor's position as a promotion above writers. Many firms hire writers as editors or writers. The titles have equal rank but different job descriptions. Thus, each company specifies its editorial jobs in its hierarchy.

An editor may manage the production of one major project on an intensive basis or of as many as 25 to 30 projects simultaneously.

Writers. Like editors, writers appear in the fourth level of the hierarchy since they, too, may be responsible for one major project, several large projects, or many smaller documents simultaneously. Company policy determines how many projects a writer develops. The writer may research a subject by interviewing the product developers, by learning to use the product, and by reading about its operations. In companies whose writers specialize in certain products, a writer may also conduct the task and audience analyses, arrange for and conduct the testing of the manuals, conduct and incorporate the reviews, and prepare the document for production. Thus, a writer spends many hours in oral communication with engineers, product users, designers, supervisors, and, perhaps, production personnel as well.

Most companies differentiate between junior and senior technical communicators. A junior communicator usually has a bachelor's degree in technical communication, low seniority, and minimal knowledge of the company's products and policies. On the other hand, a senior technical communicator has substantial experience (probably three years or more) with the company and its products. Because the senior communicator has probably created an entire product line's documentation, the writer has expertise with one or more products and a working knowledge of all the company's products and documentation. Most senior communicators have master degrees in technical communication, business, or technology.

Artists and Photographers. Artists, computer-aided design (CAD) personnel, and photographers also appear on the fourth level of the hierarchy because, like writers and editors, the illustrators are typically experts in their fields. Although the development of computers and other equipment has rapidly changed the field of technical illustration, the typical "art department" still provides camera-ready line art for in-house publications. Art departments also create and maintain "clip art" files of the standard views and illustrations that the firm's manuals use repeatedly. In many firms, the artists design and complete the layouts for manuals and other important publications products. When special services, such as airbrushing, are necessary, the art department supervisor hires, supervises, and budgets for this service. The artists work with individual writers to create illustrations for documentation, and they communicate through their supervisor with the supervisors of the various groups of writers in the unit.

Copyeditors and Proofreaders. Often, new employees in a publications group start as copyeditors and proofreaders. Because copyeditors mark manuscripts with instructions and codes for the printers to set in type and because proofreaders examine all typeset copy for various types of errors, these jobs enable new employees to become familiar with a company's policies, products, and practices. Some firms hire freelance agencies or temporary employees to supply copyediting and proofreading.

Responsibilities

Our organizational chart shows that publications groups have an inherent hierarchy: a leader—whether called a "director," "manager," "supervisor," or "lead technical writer"—and reporting employees. However, all levels of the hierarchy must operate as a team—the corporate world's basic metaphor. No matter what its internal organization, the group must accomplish four responsibilities in order to maintain current technical publications:

1. Planning
2. Writing
3. Editing
4. Production

We define each term to show you how a typical firm divides these responsibilities into the actual tasks its technical communicators must accomplish.

Planning. In many companies, the unit supervisor plans new documents to be produced. In other companies, each writer completes the planning phase for his or her documents. No matter where in the hierarchy this task occurs, planning—so critical to a term's success—is always less expensive than rewriting, editing, or making changes after production has begun. A wise publications team, then, devotes extensive effort and testing to the planning phase of producing documentation.

Planning includes about ten different activities. To accompany a new product, for example, the planner must first write an overview. This document covers three areas. A product overview defines and describes the new product, its specifications, uses, physical description, and the support or customer services that accompany it. An inventory overview delineates all the types of documentation that must be drafted to accompany a new product to the market. For example, for a new ink-jet printer, the company may plan a user's manual, a fact sheet, and a command card. The content, physical appearance, and individual production and print characteristics of all three products must be described, so that cost estimates can be made. Finally, the overview must include an audience analysis that defines all the publications' possible readers, their demographics, their reading level, their familiarity with the product, interests, their reasons for using the product, and any other information the market researchers consider relevant.

Next, the planner must prepare a working schedule for the documentation. This schedule should be a milestone chart or timeline that shows the start and finish of each document's planning, writing, editing, and production phases.

With this schedule, the planner should then prepare a personnel estimate, a form that specifies the writers assigned to the document, the expertise needed, and how the publications group will meet the personnel demand. For example, should the group hire freelance writers or illustrators to meet certain deadlines, release writers from other projects, or use still another method to staff the project?

Finally, all of the forms related to the ink-jet printer project are compiled into a documentation plan, a comprehensive form that outlines all of the decisions regarding the documentation. Everyone working on the ink-jet printer project should review, comment on, and suggest revisions to the documentation plan.

Writing. The writer's responsibilities are many:

1. Create a schedule or timeline that the supervisor and the other writers can see and use.
2. Communicate with all the personnel on the product's documentation team: other writers, illustrators, engineers, designers, marketing personnel, and so on.
3. If the project requires many writers, work with the supervisor to divide and clarify the individual writing assignments and assist the other writers as assigned.
4. Obtain supervisory and editorial decisions about the structure, style, standard symbols, and terminology of the specific ink-jet printer documents.
5. Research the product.
6. Learn to use the product if possible, and interview other users.
7. Write a working draft of the document on schedule.
8. Arrange for appropriate reviews of the document by the other writers, the supervisor, and others in the corporate communication channel.
9. Arrange for users' tests of the document, or work closely to assist those assigned to conduct the users' tests.
10. Obtain the required written approvals during each planning, writing, editing, and production phase.
11. Inform all reviewers of the termination dates after which no reviews or changes can be made in the document.
12. Inform the supervisor of the progress on the documents and of any current or expected problems.
13. Deliver appropriate rough sketches or information for visuals to the illustrators within the deadlines.
14. Maintain an active file of information about the document's history and development, including written reviews, records of changes, results of users' tests, etc.
15. Approve all camera-ready copy and illustrations.
16. Deliver the entire manuscript for the ink-jet printer documents to the production personnel.
17. Read and sign off on all proofs.

Editing. In most publications groups, editors review documents for quality and accuracy.

The term "quality" suggests the human dimension of the editor's job. The editor works with the writer to obtain the standard of quality a firm requires. This effort includes improvements in language, style, grammar, and conformity to the corporation's publications standards. In managing a document's "quality control," the editor should also examine the document's approach, that is, its quality in terms of the more traditional aspects of writing: logic, organization, clarity, and tone in addressing its audience. Editors should work with writers to create a standard vocabulary for the documents. For example, in a computer manual, they must determine whether to use the verb "press" for keystrokes.

"Accuracy" refers to both the document's technical correctness and language usage. The editor should examine the document for missing or inadequate information and check that the writer adds the material to the manuscript before the document goes to production. The editor should also help the writer obtain any needed technical information or assistance once the writer has begun drafting the manuscript. Finally, the editor should remember the audience analysis during the technical accuracy review. Does the writer offer enough information for this group of users? Does the manual explain a new product clearly enough for the intended market?

Accuracy of language includes the standard checks and markings for language, style, and usage. It should include, as well, careful review of the document's illustrations for agreement with the text. Are the drawings labeled correctly and in sequence? Are the labels grammatically parallel and correctly spelled? Has the illustrator used the correct technical terms? Again, correcting these errors at the editorial stage can cut the document's production and distribution costs.

Production.   Production is the phase during which the document becomes a true product, a typeset, bound, or packaged publication. In some companies, the production responsibilities fall to the editorial department. In larger firms, entire production departments supervise the marking, printing, binding, and distribution of documents. We know of firms where the writers manage the production. Thus, the entire publications group must understand the production responsibilities since production is the most expensive phase of the documentation process. Here are the production activities that should be assigned to specific members of the publications effort:

1. Discuss and anticipate any typesetting problems a particular document may have. For example, if special typefaces, printing fonts, or type designs must be ordered, some individual should be responsible for this activity.
2. Formally accept the document for production from the writers or editors. That is, the production person ascertains that an appropriate member of the editorial staff signs and dates the documentation. That signature means that a complete document, including illustrations and annexes, has been turned over to the production staff. At this stage, the editorial staff understands that only the most critical and necessary changes can be made in the text.
3. Ascertain that all the drawings are coordinated with the text, that the drawings are complete and coded to match the text, and that the instructions for producing both are clear.
4. Work with the designers, illustrators, and editors to ascertain that the proofs conform to the firm's design and esthetic specifications.
5. Return proofs of the camera-ready copy to the writer for a review for their accuracy and any problems the writer may detect.
6. Maintain records of all changes and their costs. Keep the project within its budget and maintain written records of the manager's approvals of cost overruns.
7. Submit and maintain all the records of the copyright process.
8. Assure and obtain a prompt delivery of the completed documents and arrange the appropriate channels of distribution.

EDITING FOR PRODUCTION—THE TEN-STEP APPROACH

This section outlines the corporate documentation cycle—that is, the process that results in a finished technical document. Because corporations invest heavily in their publications, many individuals share the responsibility for their completion. To insure that a document receives all the necessary reviews, the publications manager must see that it passes through several clearly defined development phases. We define nine phases here, although a particular

| FIGURE 17.3 | A Simple Diagram Shows That Systematic Reviews Shape the Quality and the Content of the Document Produced Within a Corporation. |

HOW REVIEWS HELP A DOCUMENT TO EVOLVE

Although college classes call for individually authored papers, corporations rely on processes—in this case, the review processes—to shape and design their documents to meet the needs of the firm. Here is a generic diagram showing how many times a document may undergo reviews in a firm. Individual companies change this plan to meet their special needs.

Step	Description	Progress of Draft
1	Complete Research and Task Analysis	Begin
2	Draft Working Outline	Complete First Draft
3	Write Discussion Draft	
4	Review Research and Outline	
5	Insert Additions and Changes	Revised, Complete First Draft
6	Complete Final Organization	Working or Second Draft
7	Review All Facts	
8	Study Internal and External Reviews	
9	Add Final Changes to Draft	Revised, Complete Second Draft
10	Design Text with Art	Final Draft of Production Quality

firm may organize the specific development process somewhat differently. Nonetheless, in any firm's publication cycle, you should be able to identify most of the phases described here.

Research

During this phase the writer learns about the product. He/she may attend the special training classes for the service personnel who work on the product. He/she may study the notes and initial documentation prepared by the computer programmers as they write a new program. The writer should identify and interview the key product developers, and keep extensive notes.

Although not a marketing expert, the writer should keep track of similar products sold by the company's competitors. The writer's research also necessitates awareness of competing products. Here are some questions the writer should consider. To whom do the competitors sell the products? How does their market differ from ours? What design features do they incorporate in their documentation? How many types of documentation do they offer to support each product? In short, the writer should know the documentation of the competing firms quite well, so that he/she can advise the team about possible improvements in the user manuals for the product.

Planning

Planning includes several activities. (For an extended discussion of planning, see the first part of this chapter.) By the end of the planning phase, the writer should have completed the product overview and the document's inventory overviews. The writer should compose as well a documentation plan for the product's writing, editing, and design specifications.

Write D1 (First Draft)

First, the writer composes a detailed, finished outline for the supervisor's written approval. (For more details on outlines, review Chapter 5.) This outline should include all the headings that will appear in the document, and clearly show where all the details belong. The outline should be so detailed that a second writer, using it and the original writer's notes, could finish the documentation project. We emphasize the outline here because it is the last part of the developmental phase of the documentation cycle. Corrections in the outline are quick and inexpensive while corrections in the first and succeeding drafts are more costly and difficult to accomplish. Thus, all those with responsibility (or credit) for the project's success must examine, correct, and approve the outline.

At this stage, the writer also collects the artwork and photography. The text, then, can include notations about these items as it is written. First, the writer and artists must make selections from their existing art on file. Second, they must schedule and design any new artwork the documentation may require. Third, they must obtain supervisory approvals of their choices. Finally, they must check regularly that the new artwork remains within the production schedule and budget.

After composing the outline and designing the artwork, the writer prepares D1, the document's working draft, for group review. The easiest of the three tasks this step describes, this draft may actually be prepared by several individuals from one master outline and sets of notes. During this composition phase, the writer should concentrate on recording the overall picture of the product and how to use it. At this stage, the writer should not worry too much about language usage or the fine points of style. Later, editors or other writers who may be more objective about the document can correct these details.

Produce D1

During this phase, the writer or word processor prepares D1 for writer, editor, and technical expert reviews. At this stage, the entire document should be entered into a computer system that will permit multiple authors or editors. In addition, the document should receive a number or label for easy access. A backup copy should be prepared and stored. Since the writers and editors work as a team to prepare a new document, the document must have a standard format and method of access so that everyone involved can respond quickly to changes.

Production of D1 should include, as much as possible, the preparation of visuals so the reviewers can understand how the text and the illustrations interact. Writers should remember that D1 requires an entire review. Illustrations, like text, must be reviewed for their accuracy and appropriateness. The sooner the illustrations appear in the working draft, the less possibility of error in later drafts and in the final version.

After including the drawings, the writer prepares D1 for review. This process includes coding: the text for type sizes, for artwork to be included in this draft, for artwork to be included later; and the headings and captions for typefaces. Then, the writer prints and distributes D1 to all the scheduled reviewers. Although the writer must always contain costs, we recommend using "hard," or paper, copies, for reviews because the paper copy of the product is easier to read than the on-line version.

Internal Reviews

Writers should consider developing a form to send to reviewers at this stage of the documentation cycle. The forms allow for immediate record-keeping, and they save the reviewers time in reading and commenting on the document. A typical review sheet might include: a checklist for grammar, punctuation, or spelling, and questions about the reader's response to the document's organization, tone, and approach. It is more difficult to design a review sheet for technical accuracy. Since each manual may involve its own distinct set of technical problems, the reviews for technical accuracy should be solicited by a note asking the engineer or technician to read for information in his or her particular area of technical expertise. All review requests should contain a written statement reminding the reader to return the document by a deadline. Experienced writers usually call the readers several days before the deadlines to remind them of the deadlines and to ask if any problems can be resolved over the telephone. After receiving and compiling the reviews, the writer must make the changes in the computer file.

External Reviews

External reviews occur differently in different companies. In some cases, focus groups, carefully selected groups of typical users, test and comment on a manual during an in-depth interview with the writer and a marketing expert. In beta testing, another type of external review, users at a controlled site actually test the manuals. Several companies now use videotaping to analyze the review process. As the test subject uses the manual, cameras record the

FIGURE
17.4

A List of the Simple Design Elements You Can Use to Improve the Appearance of Materials Produced Using Desktop Publishing and In-House Printing Equipment.

DESIGN FEATURES FOR DESKTOP PUBLISHERS

You can "design" an attractive page or an entire project by using the inherent features of the text itself to improve the appearance of your document. Here are some features that writers often take for granted. Learn about these features, and arrange them carefully to improve the quality and readability of your copy.

Text
Justified or Ragged Right
Work Spacing
Line Spacing
Line Length
Levels of Headings
Margins
Page Design
    Headers
    Footers
    Placement of Page Number
    Column Width
Type
Choice of Typeface
Type Size
Type Variety for Headings
Paper
Text Stock
Cover Stock
Package
Binding
Color
Visual Cues
Size of Page or "Sheet" Size
Section Dividers
White Space

subject's reactions and methods. Such videos help the technical documentation team to focus on the users' needs as they revise.

At the end of the internal and external reviews, the documentation team may arrange a meeting to review all the changes recommended by the internal and external reviewers. After the meeting, the writer should have a clear idea of the changes the manuscript requires. The writer should record the recommendations made at the documentation team's meeting, and make those changes in the master copy.

Write D2 (Final Draft)

Draft two (D2) emerges after the writer makes the changes that result from the review cycle. The writer should also make all the necessary changes in the artwork, photography, and captions. In short, D2 is the improved version of D1, with all the details added to make a complete draft. At this point, the writer should edit and refine the prose for clarity, consistency of usage and tone, and appropriate style. The resulting draft should be the complete manual that can go into production.

The second draft may need no user testing if the documentation team has decided that the first reviews were adequate. However, the team's senior editor and the supervisor should perform a total review before the writer delivers the manuscript for production. Both the supervisor and the editor should study the document for correctness throughout, paying special attention to the codes used to instruct the printer. After reviewing the document, both readers should sign and date the manual to indicate their approval and its release for production.

Typeset D2

Releasing for typesetting used to mean that the writer would sign over the manuscript to the production personnel or a printing firm. The printer would then deliver "galleys," copies of the typeset manuscript for proofreading, correction, and final production. Thus, although involved in certain production aspects, the writer did not control the entire effort.

Desktop publishing has changed that cycle. Often now, the writer, together with an editor or a word processor, may work on the final copy of the document and see it through to completion, including even the binding and inventory control. In many firms, to avoid the expense of setting type, the writer uses a laser printer to produce "camera-ready" copy, that is, a document that appears to have been set in offset type. The printer then needs only to photograph, print, bind, and return the copy for distribution. Thus, as new printing equipment enters the market, the writer becomes more involved with the production—as well as the writing—of the finished manual.

Print D2

The printing of D2 is a "point of no return" for the writer, because printing marks a firm commitment to the manual's production. Any alteration at this stage is costly and slows the entire documentation cycle. Thus, a writer should remain quite involved with the documents he or she sends to production—even in a firm whose policy dictates that a production group oversee the manual's printing, binding, and distribution.

The writer should sign the document over to the production group. The writer's production manager should review all preparations, artwork, coding, and special requests before returning the galleys to the printer. With the returned galleys, the printer may prepare "blue lines" or "silvers," the first prints from the actual film or plates used to produce the manual. The writer and the publications manager should proofread and, in writing, approve the blue lines before the project proceeds.

Inventory

Using the blue lines as the guide, the printer prepares the final manual for the documentation group. The printing house may bind the manual or the documentation group may contract independently with a bindery for the bookmaking.

Even when the manual is produced, the writer has more work. In preparing the manuscript, the writer should be certain the document receives an inventory number. If the team does this systematically, the number should be placed on the book's back or inside cover so customers can reorder conveniently. The book must also receive a notation in a central catalog of the company's publications, and several hard copies should be filed for future use or for copying once the manual has gone out of print. Finally, the electronic copies of the manual must be carefully filed and stored. The group's attention to this effort can result in many recovered costs for the company because carefully stored electronic copies can be revised regularly, inexpensively, and with a minimum of the writer's effort.

The documentation cycle is an essential tool in any publications group because the production of technical documentation in a large business firm requires strategic planning and cooperative teamwork. This cycle ensures an organizational framework that produces finished documents. It also defines the essential communication channels, so that diverse individuals can contribute their expertise to a firm's written products. The key to a publication group's success is its effective use of the documentation cycle.

SUMMARY

In order to explain how corporations produce documentation, this chapter concentrated on the publications group. We discussed the need for student technical communicators to understand the corporate life and corporate structure. After reading this chapter, you should be able to:

- List the vertical structure of a typical publications group.
- Name the tasks that individual members of a publications group perform.
- Explain the differences between collegiate and corporate technical writing assignments.
- Define the terms "quality" and "accuracy" as they relate to corporate documentation.
- Describe four responsibilities of a publications group.
- Describe the ten-step approach to production.
- Describe the process a typical firm may use to review its documents.

ASSIGNMENTS

1. Draw an organizational chart of a firm where you hope to work when you finish college. If employed, draw a chart of the firm where you now work. Find out where the communicators belong on the organizational chart. Are they an entire division or part of a division? Sketch in an appropriate box to show their place in the corporate channel of communication and supervision. Prepare to write a one-page memo to your instructor explaining what products and services the communicators in "your" corporation offer.

2   How should the communicators of "your" business expand and upgrade their services? Prepare an oral recommendation.

3   Arrange to visit a local corporation with, if possible, a publications group. Interview a member of the group, questioning him or her about corporate life and the process that the corporation uses to produce documents. If the corporation does not have a publications group, select a manager from one of the following divisions: training, marketing, customer service, or communications. Summarize your interview in a report of about 500 words.

4   Create an organizational chart of your ideal publications group or division in which to work. Keep in mind lines of authority, the size of the firm, placement of your work group within the corporation, and your place in the scheme.

5   Reflect upon your experience in organizations—work, school, recreational groups, and your family. Are you a leader or a follower? A doer or a thinker? An energizer or a peace-maker? Identify your strengths and weaknesses in working with groups. Prepare for your resume a short section about your ability to work with other people.

6   Using Figures 17.1 and 17.3, contrast your experience of writing college papers with the steps in producing corporate writing projects.

7   Arrange to interview a local printer, so you can write a one- to two-page report about the experience for the class. What services does the print shop offer? What do they charge for revisions in the copy? What is their policy for dealing with businesses? Do they accept camera-ready copy? Prepare a list of questions before you visit the shop, and frame your questions as though you are interviewing the printer for consideration as a possible vendor for your corporation's production of documents.

CHAPTER
18

Managing Writing and Editing to Contain Costs

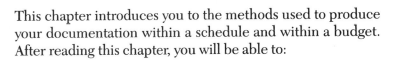

This chapter introduces you to the methods used to produce your documentation within a schedule and within a budget. After reading this chapter, you will be able to:

- Revise the present forms in your office, so they can be used for editing and production.
- Design new forms as needed to define your writing and editing tasks.
- Use forms to communicate with a supervisor about your writing.
- Use forms to direct the writing of your staff or employees.
- Use an editing system such as the Levels of Edit system to define your editing and production tasks.
- Prepare rough estimates for the production of your documents.

CASE STUDY   A Successful Writer Moves into Management

Jeff Beetham has joined Regional Environmental Interests (REI) in his local Atlanta suburb. As the fastest growing waste management corporation in the United States, REI has required large amounts of overtime writing to produce the proposals, manuals, environmental impact statements, and other documentation the expanding company needs.

Because Jeff has delivered many last-minute documents on time, he has been asked to manage, as the supervisor and lead writer, a team of individuals the company recently hired. Jeff's new position offers an attractive pay increase, more discretion in the management of his own time, and the levels of autonomy and responsibility he has hoped for.

However, Jeff, whose strength is his writing ability, now faces a unique challenge. He must plan and schedule an expanding collection of documentation, develop the mechanisms that will allow him to communicate with REI's financial employees, and present to his associates a writing and editing system that reviews their performance as writers and that evaluates the entire publications group's performance.

FORMS—EVIDENCE OF TIME MANAGEMENT

This chapter shows you how Jeff Beetham plans his first document review meeting, specifically how he budgets for the meeting. First, you need to understand why it is important to design and use forms for managing a written project, and how to design appropriate forms.

Why a Manager Uses Forms

Managers use forms to collect and distribute information about their projects. Forms enable managers to systematically collect and share information that other business associates can understand and respond to. These concepts of collecting and sharing information within a system appear at first to be simple, but they deserve some explanation.

Collecting Information. As you learned in Chapter 16, a publications manager collects, records, and distributes various kinds of information, including the schedules for writing and producing a document, the cost estimates for printing a manual, and a record of all the corporate and customer text reviews under production. If Jeff Beetham, for example, is to supervise 26 documents in their varied stages of development, he must develop a foolproof system for tracking each document's progress. As the supervisor, Jeff must translate his group's editing decisions into specific choices and specific corporate policies about the production methods of the documents. He must explain to his manager exactly how much each editing decision will cost once it is in production. Above all, Jeff must show that his group is cost-effective and that they earn profits for REI.

**FIGURE
18.1**

A Checksheet for the Levels of Edit.

EDITING EVALUATION FORM

Document
Name/Number _____

Author _____
_____
Editor_____
_____

Directions: Circle letter of edits which are applicable. Then check off areas that are incomplete or need attention. Space for evaluation and comments is provided on the second page.

Checklist

A. Policy Edit
    1_____Problems with required report elements.
    2_____Incomplete/unclear references.
    3_____I.S. system of measurement not used.

B. Integrity Edit
    1_____Text does not refer clearly/accurately to illustrations; illustrations do not refer clearly/accurately to manuscript.
    2_____Problems with letter and number sequences.
    3_____Contains duplicate captions/title tables.
    4_____Text refers to illustrations, pages, chapters, paragraphs that do not exist.
    5_____Other.

C. Screening Edit
    1_____Spelling errors.
    2_____Errors in subject/verb agreement.
    3_____Text contains incomplete sentences.
    4_____Illustrations and figures are not camera ready.
    5_____Other.

D. Copy Clarification Edit
    1_____Figures are not clearly explained.
    2_____Problems with subscript, superscript, or both.
    3_____Text/illustrations contain excessive detail.
    4_____Other.

FIGURE 18.1

Continued.

E. Format Edit
    1_____Improper margins/spacing for headings and subheadings.
    2_____Figures/tables incorrectly positioned.
    3_____Problems with top/bottom margins.
    4_____Other.

F. Mechanical Style Edit
    1_____Problems with capitalization.
    2_____Inconsistency in word compounding.
    3_____Inconsistency in acronyms and abbreviations.
    4_____Incomplete/improper bibliographic references.
    5_____Other.

G. Language Edit
    1_____Grammar.
    2_____Punctuation.
    3_____Usage.
    4_____Fluency, including transitional elements.
    5_____Parallelism.
    6_____Conciseness.

H. Substantive edit
    1_____Material not grouped and subdivided in a rational manner.
    2_____Apparent contradictions or inconsistencies between sections.
    3_____Inappropriate emphasis on various elements.
    4_____Tables not designed for easy reader comprehension.

Evaluation and Comments

Communicating Within a System. The forms allow Jeff to translate his writers' and editors' method into ideas and processes that Jeff's managers can understand: time, cost, supplies, travel, and other standard business language. Thus the forms offer the advantage of a systematic or standard approach to explaining the costs of researching, writing, editing, word processing, reviewing, illustrating, and the other very real labors that writers perform. In a sense, the forms illustrate to the managers how a group of writers work and how individual writers perform within their groups. Similarly, the forms allow Jeff to explain the workings of the corporation to the individual writers and editors in his group. By describing a writer's work on a form, Jeff can analyze that writer's performance, so the group can continue to improve its efforts.

Some Guidelines for Constructing Forms

As we show you how to construct forms for the writing, editing, and production of your documents, study this chapter's examples. These forms have evolved and been tested in various types of technical writing groups. Think about how you would change the forms to meet your needs as a technical communicator.

Most management information specialists discuss the forms for describing business processes in terms of their purpose, style, and organization.

Purpose. The forms you design may serve three purposes: to request an action, to create a record of a completed action, or to create a record of a long-term project.

An action form, such as a shop order or purchase order, places an order, or issues a written request for a product or service. For example, the Summary Bid for a Technical Manual is an action form because it lists the specifications for the printing job. It concludes with the signature and date that authorize the company to purchase the printing.

A memory form, on the other hand, confirms an action already approved informally or orally. For example, a confirming purchase order follows up a telephone order for goods and services. The Document Review Cost Estimate Jeff uses in this chapter is a memory form because he records his estimates on the form. However, the form itself is not a formal budget for accounting purposes.

A report form summarizes a completed process so a firm can maintain complete records. For example, a document history, which word processing software may record, is a report form. The Pre-Production Checklist, the Edit Evaluation Form, and the Editing Evaluation Form in this chapter are all report forms because the editor or reviewer completes these forms to explain to a supervisor that he or she has completed the tasks listed. The report form is the most commonly used form in technical publishing.

Style. Designers refer commonly to two styles—open and boxed—forms. Open forms— spreadsheets are examples—tend to have simple instructions and spaces where the user writes figures and comments. Closed forms usually structure the user's response by offering options the user can check, circle, or fill in. To some degree, closed forms limit the number and type of responses users can offer, but they are also easier to complete.

You may also specify that your editing forms be either cut or specialty. Cut forms— simple 8½″ × 11″, loose, boxed forms—are most common in business because they are inexpensive to print. Specialty forms may be continuous (boxed or on rolls, to be fed into an elec-

**FIGURE
18.2**

An "Open" Style Form for the Levels of Edit System.

EDIT EVALUATION FORM

Author: Editor:
Ms. Title: Date Received:
                                               Date Returned:
                                               Level of Edit:

Circle one number for each category below. You must explain scores of one or five in the comments sections.

1. Manuscript organization
1   2   3   4   5
Poor   Acceptable   Excellent

Comments:

2. Tone, Style, and Consistency

1   2   3   4   5
Poor   Acceptable   Excellent

Comments:

3. Clarity and Completeness

1   2   3   4   5
Poor   Acceptable   Excellent

Comments:

FIGURE 18.2

Continued.

4. Graphics Referencing and Sequencing

1	2	3	4	5
Poor		Acceptable		Excellent

Comments:

5. Institution's Policy Representation

1	2	3	4	5
Poor		Acceptable		Excellent

Comments:

6. Grammar, Word Usage, and Mechanics

1	2	3	4	5
Poor		Acceptable		Excellent

Comments:

7. Additional Comments

Strengths:

Weaknesses:

Suggested Additions or Deletions:

FIGURE 18.3   A Checksheet for a Document That Is Ready for Print.

PRE-PRODUCTION CHECKLIST

You must complete this checklist before your manuscript will be released for production and distribution. Attach this completed checklist to the front of your ms. when you submit your finished work for your supervisor's signature.

Front of the Book Checked For:

Date	Initials	
_____	_____	Correct title.
_____	_____	Correct edition and number.
_____	_____	Trademark.
_____	_____	Correct proper noun names of clients or divisions.
_____	_____	Conformity with corporate design regulations.
_____	_____	Signed approvals of all original cover lettering or designs.
_____	_____	Correct numbers and titles on title page.
_____	_____	Correct wording of copyright statement on title page.
_____	_____	Correct inclusion of copyright symbol on title page.
_____	_____	Correct date of publication on title page.
_____	_____	Correct inclusion on title page of corporate publisher, city, reproduction policy.
_____	_____	Inclusion of a preface.
_____	_____	Table of contents numbers agree with text.
_____	_____	Table of contents headings agree with text.
_____	_____	Table of contents starts on right-facing page.
_____	_____	List of figures is in order and included.
_____	_____	List of tables is in order and included.
_____	_____	Supervisory approval is obtained in writing if table of contents, list of figures, or list of tables is omitted.

Text Checked For:

Date	Initials	
_____	_____	Pages are all in order.
_____	_____	Correct document name and number appears in the footers.
_____	_____	Top, bottom, right, and left margins are correct.
_____	_____	Vertical spacing for illustrations is calculated correctly.
_____	_____	Accurate descriptions are written for artwork to be stripped in.
_____	_____	Correctly worded captions are attached to be stripped in with the artwork.
_____	_____	Every title from the table of contents appears in the text on the correct page.
_____	_____	Any extra materials, headers, footers, identification lines are included with type specs.

FIGURE
18.3

Continued.

Back of the Book Checked For:

Date   Initials
_____   _____   Special concluding chapter.
_____   _____   Appendixes included and listed in the front.
_____   _____   Complete index.
_____   _____   List of related manuals or publications.
_____   _____   Reader response questionnaire included.
_____   _____   Company name, address, telephone are on back cover.
_____   _____   ISBN is on back cover.
_____   _____   Price is on back cover.

Art and Photographs Checked For:

Date   Initials
_____   _____   Correct color codes marked throughout.
_____   _____   Instructions for special paper stock are marked.
_____   _____   Instructions for reductions or enlargements.
_____   _____   Backs of photographs and artwork are marked with pages for
                         stripping in.
_____   _____   Captions and labels on photographs and artwork are edited
                         for correctness and spelling.
_____   _____   All photographs and artwork are clean, ready for reproduction.
_____   _____   Printer's dummy is complete: all pages, composites, special
                         instructions.

Printer's Instructions Checked For:

Date   Initials
_____   _____   Color codes marked throughout.
_____   _____   Type sizes marked throughout.
_____   _____   Special instructions marked on dummy and on instruction sheet.
_____   _____   Composites included where needed.

Publication Inventory Checked For:

Date   Initials
_____   _____   Updated version of document stored on disk.
_____   _____   Current version of document on order for library.
_____   _____   Current version of document stored in publications depart-
                         ment's archives.
_____   _____   Usual order is placed for extra books for inventory.

_____   _____
Supervisor's signature   Date

tronic printer) or detachable (perforated for easy separation). Finally, you may decide on a single- or multiple-sheet form for easy distribution. The forms we show are all single-sheet, because we find it more cost-effective to make multiple copies after we use the form rather than to pay for printing multiple sheets and carbons.

Parts of the Form

All forms should contain five sections.

1. **Title.** Centered and in capital letters, the title of the form should appear at the top of the page. Keep the title brief, but use a title that adequately describes the form. Use descriptive titles, but do not even consider humorous titles.
2. **Heading.** Just below the title, the heading should label and ask the user to fill in all the identification data and general information you need. For example, you may need to know the date when the form is used, the job number it refers to, the user's name, address, and telephone number, or other facts. Our Editing Evaluation Form asks for the author's and editor's names because the form helps them to communicate with each other and with a supervisor. The Edit Evaluation Form asks for the date the editor received the document for review, the date it was returned from review, and the editor's choice of one or several Levels of Edit in the heading.

    Note, too, that some forms do not require a heading. The Pre-production Checklist has no heading because the signature and date appear on the second page as evidence of the form's completion by the editor.
3. **Instructions.** Below the heading, tell the user how to complete the form. Do you want check marks in the boxes? Do you want a circle around the correct answer? Do not assume the user understands these requirements. An exception applies in the case of forms that act as worksheets. For example, three of our forms, the Document Review Cost Estimate, the Budget for Slide Show, and the Summary Bid for a Technical Manual, lack instructions sections because they are self-instructing. The blank lines, the multiplication signs, and the marked lines for the totals tell the user how to complete the forms.
4. **Body.** In this section, the user states quantities, prices, and descriptions if the form is a purchase order. Note that on our forms, the users are asked to write very few comments. If we ask for the user's comments, we ask the user to make a decision first, as, for example, on the Edit Evaluation Form. We have found that, more often, users tend to complete closed, objective forms, and they tend to neglect open, subjective forms that rely on comments.
5. **Conclusion.** Include here spaces for approvals, signatures, or any summary information. For example, the conclusion of the Summary Bid for a Technical Manual has spaces for the prices of fewer or extra copies. If space allows, the conclusion may be used for the user's subjective or written comments that expand on the selections made in the body of the form.

Tips for Useful Forms

Which forms work best? The forms that users complete. Here are some tips for designing editorial and production forms:

1. Keep the design simple. Have the form typeset, if possible, because users tend to complete typeset forms more often. Use a simple roman typeface in capital letters and lower case

with occasional uses of boldface type for column or row headings. Keep the form clean and inviting.

2. Try to design a one-page form. If you must go to two pages, arrange the layout so that the user completes the most important sections first and the least important sections on page two.

3. Make the form compatible with your firm's system. For example, design your budget form so that its line items use the same company terminology. Also, design forms compatible with the company's computer system so you can review and update your information on a computer.

4. Do not ask for information you do not need. It is tempting to collect all the information pertinent to a project, but users complete more often the simple forms that address the task at hand. For example, do not ask editors to complete cost estimates as they complete their pre-production checklists.

5. Contain your costs. A multipage, multicolored production checklist may sound like a good idea, but can your production budget afford it? Complicated forms are expensive to print. As an alternative, consider maintaining and updating your forms on diskettes and printing high-quality batches with a laser printer.

6. Maintain a central file of all the forms for your publication group. Maintain a master list of the forms at the front of the file so you can conduct your inventory, update, and reorder effortlessly. Do not permit your master copies to leave the office; that way you can have a complete inventory both in hard copy and on diskettes.

AN EDITING SYSTEM—EVIDENCE OF YOUR SKILL AS A MANAGER

Why the Editor Needs a Classification System

Technical communicators like Jeff Beetham, who are successful, creative writers, are often dismayed when they must supervise a publications group, or even when they must coordinate the entire production of a lengthy, technical document. In fact, although his creativity and technical ability enabled him to write outstanding manuals, Jeff's success now also depends upon his ability to motivate the other members of the publication group. Jeff cannot hope to succeed as a supervisor by doing his group's work; he must, instead, use a communication system to define the objectives of his publication group for any document. Furthermore, he must use a system that both the writers and the managers can understand and respect.

Many publications groups use a form of the Levels of Edit system developed at the Jet Propulsion Laboratory (JPL) as a communication system for the writing and editing services they perform. The Levels of Edit system allows the supervisor to determine how a document will be edited as a matter of publications policy. Consequently, the writers and editors know exactly what they must do to prepare their manuscripts for publication.

An Overview of the System

How does the Levels of Edit system work? Simply, the system depends upon a concept called "controlled flexibility" for its success. The control results from the nine different types of routine editorial functions the system offers. The nine types of edits are then combined in five possible levels or decisions about the amount of detailed editing a document may receive, depending upon its importance to a corporation. Figure 18.4 shows how JPL combines the

**FIGURE
18.4**

A Table Showing the Functions Each Level of Edit Covers.

**TABLE 1
TYPES AND LEVELS OF EDIT**

Type	Levels of edit				
	1	2	3	4	5
Coordination	X	X	X	X	X
Policy	X	X	X	X	X
Integrity	X	X	X	X	
Screening	X	X	X	X	
Copy clarification	X	X	X		
Format	X	X	X		
Mechanical Style	X	X			
Substantive	X				

Source: Mary Fran Buehler, ''Controlled Flexibility in Technical Editing: The Levels of Edit at JPL,'' *Technical Communication* (Journal of the Society for Technical Communication), First Quarter, 1977, pp. 1-4.

types and levels of edit. Although JPL uses five levels of edit, a firm can create as many or as few levels as its managers desire based upon their firm's inventory of publications.

Figure 18.5 shows how the JPL managers apply the system to their hierarchy of publications. At JPL, three classes of publications exist:

Class A—These "premier" publications are typeset to a standard format using the highest quality production techniques.

Class B— High-quality word-processed publications use standard format but receive production techniques that allow for a lower quality of appearance.

Class C—Author-prepared, camera-ready copy in any understandable format receives lower-quality production.

The table shows that the supervisor may select the type of editorial treatment and the amount of editorial effort that an editor may expend on a document. Furthermore, the supervisor selects the level of edit as a matter of institutional policy rather than as an arbitrary choice. The supervisor can then use the system not only to produce documents of predictable quality, but also to predict costs, plan staffing, and draft budgets.

Before we discuss this system's advantages, we should review the types of edits that compose the system. (For an extended definition of these types of editing and of the system, see Mary Fran Buehler's "Controlled Flexibility in Technical Editing: The Levels-of-Edit Concept at JPL," in *Technical Communication*, First Quarter 1977, pp. 1–4.)

Coordination Edit. Takes the document through all the production processes to publication.

FIGURE
18.5

A Table Showing How You Can Use the Levels of Edit Concept to Create a Cost-Effective Hierarchy of Publications.

TABLE 2
TYPICAL USES OF THE LEVELS OF EDIT AT JPL

Type of report	Class	Levels of edit
Technical report	A	1, 2, 3
Technical memorandum	B	1, 2, 3
Technical memorandum	C	4
Special publication	A, B, C	1, 2, 3, 4
Meeting paper	—	1, 2, 3, 4, 5
Journal article	—	1, 2, 3, 4, 5

Source: Mary Fran Buehler, "Controlled Flexibility in Technical Editing: The Levels of Edit at JPL," *Technical Communication* (Journal of the Society for Technical Communication), First Quarter, 1977, pp. 1-4.

Policy Edit. Examines the document for conformity to corporate or institutional policies.

Integrity Edit. Reviews the document to be certain all the parts are there, numbered correctly, noted correctly, and in sequence.

Screening Edit. Review of camera-ready copy to make certain its language and illustrations reflect standard usage.

Copy Clarification Edit. Mark copy and illustrations for typesetters and other production personnel.

Format Edit. Mark copy so illustrator and compositor understand layout and illustration requirements.

Mechanical Style Edit. Review for consistent and standard usage of capitalization, abbreviations, references, and other elements.

Language Edit. Check for grammar, usage, word choice and flow, punctuation, and other aspects of language usage.

Substantive Edit. Reviews for coherence, internal logic, flow of ideas, and other rhetorical characteristics of strong writing.

Advantages of the Levels of Edit as a System

A systematic approach to the editing of publications, particularly the Levels of Edit approach, offers five advantages for the manager.

Controlled Flexibility. Why is controlled flexibility an advantage for a supervisor? Controlled flexibility means that the supervisor can make the best possible use of the editors and writers for meeting deadlines and client needs. Thus, although the publications group might not be able to complete every task its customers can imagine, the group can offer specific services such as mechanical editing and reviews for compliance with a company's policy. The customer, the manager whose budget must absorb the cost of the editing, can then select the type of service he or she needs most, and the supervisor can then arrange for the work to be completed.

Image Protection. A company may spend years and millions of dollars developing its image through specific marketing strategies, well-planned public relations campaigns, and carefully selected charitable or philanthropic activities. A firm simply cannot risk the misinterpretation or misapplication of its purpose or accomplishments. Through the Levels of Edit system, a trained editor can screen for the correct use of the company's image, the documents that are otherwise ready for distribution to the public. For example, a scientific paper may contain all the correct stylistic elements and equations for the journal in which the author will publish it. It may require, however, the "boilerplate" or standard description the company has approved for describing its research with human or animal subjects, which an editor may insert and then review for accuracy. This review, and the correct inclusion of boilerplate, may earn a firm or an institution substantial research subsidies and government support. Thus, the editors should review a company's or institution's printed matter for protection of its image.

Cost Analysis and Control. Supervisors such as Jeff Beetham learn quickly that they must predict each document's cost, and apply remedies when the documents cost more than estimated. An editing system enables the supervisor to define each type of edit, estimate the time needed for each edit, and predict the document cost in work hours. Supervisors use sheets like the examples shown in this chapter not only to estimate the costs of the goods for the production of a manual, but also to predict the labor costs for a document. The cost estimate enables the supervisor to discuss realistically the client's needs and requirements for writing and editing. The estimate also is necessary so the supervisor can hire extra writers or arrange for freelance assistance when deadlines are tight and the client has an adequate budget. Although cost estimating helps to create flexibility for the manager and supervisor as they decide how to do a job, a supervisor can complete an accurate estimate and draft a budget for a project only after translating the writing, editing, and production efforts into measurable work hours.

Clear Communication. As Chapter 17 explains, many individuals contribute to the manuals a company prints. So that all the differing expertise can be integrated and incorporated correctly into emerging drafts, the supervisor requires a standard means of communication that enables all the different people on the project to discuss the job in a standard vocabulary. The supervisor must assign the task and explain its requirements to the writers. They, in turn, must discuss their project with engineers, expert technicians, users, and editors. Finally, the supervisor must be able to explain to the manager why the group designs a particular kind of book, how the group proceeds with its efforts, what factors caused the efforts to conclude successfully, and what factors contributed to any problems or delays. Again, the importance of a standard vocabulary that reflects standard task definitions for the editors seems obvious, especially when it may be necessary for a supervisor like Jeff Beetham to communicate about all 26 projects his staff is managing, through all the various stages of the documentation cycle.

Clear Job Descriptions.   Although the writers and editors in Jeff Beetham's publications group may have studied technical communication at the best universities in the United States, they still need to learn their jobs as writers and editors according to the policies and publication goals of REI. Jeff must provide clear job descriptions, train the new personnel, and write a clear statement of each employee's achievements for that employee's quarterly or semiannual performance evaluation. Jeff's job descriptions must reflect REI's company policies for all of its employees as well. An editing system allows Jeff to make decisions about his employees' jobs: he can develop his editors as specialists in certain products, he can require all the editors to learn about the entire product line, he can delegate certain types of edits to some editors, or he can require each editor to become a master of all nine types of edits. The system helps Jeff allocate his staff according to the requirements and policies of the company and its publication needs.

Extraordinary Edits

No system is perfect. No matter how well-designed an editing system is, certain types of needs may present unique problems for the publications group. At JPL, for example, the staff has defined six areas of special needs that create unique editing challenges. They call these six areas, listed here, "extraordinary edits":

1. Researching references
2. Working with foreign language copy
3. Locating missing items and illustrations
4. Editing transcribed tapes
5. Incorporating more than one set of author's alterations
6. Handling multiple iterations of a manuscript

Certainly any substantial publications group has on its staff editors who can perform all six of the extraordinary edits. However, we focus in management of resources and personnel on the routine functions, the tasks editors perform most often to make the publications group successful. Less frequent tasks or extremely time-consuming endeavors are extraordinary in this system. By defining these jobs, or any infrequent tasks, as extraordinary, we make the best use of the editors' talents and time. Also, the supervisor can inform a client that he or she has several choices for obtaining the extraordinary edits. The client may hire freelance specialists, pay extra for the service, or perform that part of the edit in his or her own group.

COST ESTIMATING—SCHEDULES AND FORMS

Through learning about the documentation cycle, you understand that the creation of technical documentation for a corporation involves much more than just writing. You should also prepare to work on the planning and estimating phases of that documentation cycle.

Planning—The Least Expensive Phase of Production

By now, you may have realized that planning is a major theme of this text: planning as part of pre-writing, planning as you work toward an outline of headlines, planning through your illustrations before the illustrator begins to bill the company, and, most importantly, planning for your projects' resources, personnel, time and financial backing before starting to write.

Why does this book emphasize planning throughout? Because experts have stated that planning is the least expensive phase of the documentation cycle (Weiss, *How to Write a Usable User Manual*, p. 44). The cost of change increases at each stage of the documentation cycle and escalates dramatically once a manual enters the production phase. Thus, to create cost-effective documentation, a writer must concentrate the most effort in planning so that the actual drafting of the document occurs almost automatically. Throughout this text, we have asked you to design documents that can assist with the planning of your writing and the design of your finished product. It is now time to examine some of the documents that can help you to anticipate your documentation's costs.

Estimating—A Major Key to the Success of Your Plan

Of course, different companies and their publications groups use different methods to predict and record costs. For example, in one company, a supervisor may have to complete a budget for each document a publications group creates. In another firm, the supervisor may have to complete a budget for each specific product's collection of documentation. Yet another supervisor in a Federal agency must complete a zero-based annual budget, rather than an estimate for each project's cost. With any firm, you will go through an orientation period when you should learn how the firm expects its publications group to maintain financial accounts.

In this section, we will review generic forms such as for practicing cost accounting. As we teach you how to use and modify these forms, we will show you how Jeff Beetham anticipates his document's costs before a review meeting.

Scheduling a Document.　How long should it take a writer to complete a job? Of course, the completion of each task varies with the project's requirements, the writer's talent and experience, and other factors such as the writer's fatigue or the complexity of the manual under production. Nonetheless, Figure 18.6 offers a basis for the estimator who must price various writing and production tasks for a client.

For example, if Jeff Beetham wants to price the drafting and reviewing of an outline, he can determine the time, two weeks, on this list; multiply 80 hours (two 40-hour weeks) by $50 per hour, the standard rate for a writer in many parts of the United States; and obtain a cost of $4,000. Other factors may affect the costs of a writer's, editor's, or artist's time: the company's benefits (18 to 33 percent of cost), the degree of expertise required of the writer (do you require a medical indexer or an all-purpose, entry-level writer?), and the region of the nation in which you hire writers. Despite these variations, you can estimate your costs rather safe-

FIGURE 18.6	A List of the Typical Times for the Writing and Production of a 100-Page Manual.

SCHEDULING A DOCUMENT

How long should you allow for the phases of pre-writing, drafting, rewriting, editing and producing a document? Although the time and procedures may vary to suit your needs, you may borrow this schedule which we have used in planning a 100-page manual. This schedule relies on an outside shop for offset printing and an in-house artist for line art. Can you make a timeline for the writing and production stages listed here? How long will the whole book take?

Task	Time Needed
Pre-writing Through Editing	
Draft and review outline	2 weeks
Gather research	1 to 2 months
Write rough draft	1 month to 6 weeks
Revise and improve draft	2 weeks
Edit rough draft	2 weeks
Prepare first draft for review	2 weeks
Distribute first draft to reviewers	10 days to 2 weeks
Include reviewers' comments	1 week
Edit reviewed first draft	10 days to 2 weeks
Prepare illustrations	1 to 2 months
Prepare cover design	1 to 2 months
Prepare document design	1 to 2 months
Prepare second draft for review	2 weeks
Distribute second draft to reviewers	10 days to 2 weeks
Include reviewers' comments	1 week
Edit reviewed second draft	2 weeks
Prepare final draft	2 weeks
Distribute final draft to reviewers	10 days to 2 weeks
Include reviewers' comments	1 week
Prepare front matter	1 day
Prepare back matter	4 days to 1 week
Proofread the manuscript	3 to 5 days
Production	
Paste-up a dummy book	1 week
Review paste-up	3 days
Distribute galley-proof	3 days
Correct galley-proof	3 to 5 days
Return galley-proof to printer	2 weeks
Send completed camera-ready art to printer	2 weeks
Examine blue tint proof	3 days
Return blue tint proof to printer	2 days
Distribute finished document	6 weeks

ly if you remember that the geographic region seems to be the primary factor affecting the price of a writer. In the Midwest, most of the South or Southwest, and in the coastal Northwest estimate $50 per hour for a writer. In major metropolitan areas such as New York, Los Angeles, or Boston, your estimate should range between $75 and $100 per hour.

You might experiment with this schedule to practice your skill as an estimator. What will it cost to research a manual if that writer lives or works in New England? How do you feel about the costs of the research? How essential is that research to your firm? Can you think of ways to cut the cost of the research?

Request for Quotation.   This form might be used in any firm for obtaining the base prices of various printing jobs. You might modify the form to exclude the items your publications group does not purchase and to include specifications unique to your group. This form offers the advantage of asking for the price of printing in three categories. You might fill in quantities of 1,000, 2,000, and 3,000, or any category you need. Normally, a printer will expect to offer quotations on batches of 500 or more items. As the number of pieces increases, the unit cost should decrease.

Budget for Slide Show.   This form lists the equipment and the personnel involved in a typical slide show. As you scan the sheet, you can understand why audio-visual technology is so expensive. The company must pay for special equipment it must rent. To complete a professional slide show, the supervisor must hire one or several actors, musicians, and probably, technicians to work the equipment, although in some cases, the technician's fee is included in the price of the equipment rental. The sheet shows that the hour is the basic unit of measure. All hours are then multiplied by the base price for the service. Subtotals are required for each line item. You might enjoy estimating a slide show for a project that interests you. In the Midwest, short slide shows composed by professionals usually have a base price of $20,000 to $30,000.

Summary Bid for a Technical Manual.   Typically, a form such as this appears at the end of a lengthy document description that lists every detail of the manual to be published, including items such as the typeface, trim size, method of binding, number of line drawings, number of halftones, color specifications, and all the other facts that define the book as a product. You should be aware that composition, a labor-intensive process, adds greatly to your expenses. If your desktop publishing system allows you to supply camera-ready copy, modify this form to obtain prices both including and excluding composition.

Two other features are important. First, the estimator asks for the shipping price. Remember that the manuals must be stored in a climate-controlled warehouse for shipment with the firm's products. Second, the estimator asks for prices for a manual both "shorter" and lengthier than the 96-page estimate. Printers' estimates can vary greatly for overruns. Without a written commitment from the vendor, the estimator may end up well over the budgeted cost for a manual. Finally, note that the wise estimator specifies how to package the manuals when they are shipped and delivered, because this detail can save the company substantial costs in storing, packing, and shipping the manuals.

Document Review Cost Estimate.   Jeff Beetham uses this sheet to plan his intensive document review session for his radioactive waste disposal manual. At REI, company policy dic-

FIGURE
18.7

A Typical Form for Obtaining Prices from Vendors.

Request for Quotation

Job name _____   Date _____

Contact person _____   Date quote needed _____

Business name _____   Date job to printer _____

Address _____   Date job needed _____

Phone _____   **Please give** ☐ firm quote   ☐ rough estimate   ☐ verbally   ☐ in writing

This is a ☐ new job   ☐ exact reprint   ☐ reprint with changes _____

Quantity 1) _____   2) _____   3) _____   ☐ additional _____ s

Quality ☐ basic   ☐ good   ☐ premium   ☐ showcase   comments _____

Format   product description _____

    flat trim size _____ x _____   folded/bound size _____ x _____

    # of pages _____   ☐ self cover   ☐ plus cover

Design features ☐ bleeds   ☐ screen tints # _____   ☐ reverses # _____   ☐ comp enclosed

Art ☐ camera-ready   ☐ printer to typeset and paste up (manuscript and rough layout attached)

    ☐ plate-ready negatives with proofs to printer's specifications

    trade shop name and contact person _____

Mechanicals   color breaks ☐ on acetate overlays   ☐ shown on tissues   # pieces separate line art _____

Halftones ☐ halftones # _____   ☐ duotones # _____

Separations ☐ from transparencies # _____   ☐ from reflective copy # _____   ☐ provided # _____

    finished sizes of separations _____

Proofs ☐ galley   ☐ page   ☐ blueline   ☐ loose color   ☐ composite color   ☐ progressive

Paper   weight   name   color   finish   grade

    cover   _____

    inside   _____

    _____   _____

    _____   _____

    ☐ send samples of paper   ☐ make dummy   buy paper from _____

A Typical Planning Worksheet for the Cost of a Document's Review.

DOCUMENT REVIEW COST ESTIMATE

PERSONNEL	Days	× Daily Cost =	Total
Supervisor	_____	_____	_____
Technical Writer	_____	_____	_____
Technical Editor	_____	_____	_____
Assistant	_____	_____	_____
Engineer	_____	_____	_____
User/Customer	_____	_____	_____
User/Customer	_____	_____	_____
User/Customer	_____	_____	_____
User/Customer	_____	_____	_____
		Subtotal	_____

EQUIPMENT	Units	×	Days	× Daily Cost =	Total
Test Site/Room Rental			_____	_____	_____
Conference Room Rental			_____	_____	_____
Video Camera(s)			_____	_____	_____
Lighting Equipment			_____	_____	_____
Power Cables			_____	_____	_____
Microphones			_____	_____	_____
Audio Recording Equip.			_____	_____	_____
Review Copies of Manuals	_____		_____	_____	_____
Evaluation Forms	_____		_____	_____	_____
Other Supplies	_____		_____	_____	_____
				Subtotal	_____

tates that the personnel involved with a major document assemble for an intensive review session before the manual goes into production, the most expensive phase of the documentation cycle. Jeff must estimate some of the following costs:

Air fare from Seattle, lodging, and per diem for the nuclear engineer.

Rental car for the nuclear engineer for three days.

The costs of the local REI personnel who will bill the company only for their time, not their commutation. The technical writer and editors cost $50 per hour; the assistant, $25 per hour.

The costs of four users who are clients of REI. They will drive from Knoxville to Atlanta and stay in a hotel for two nights. REI is not covering mileage but will pay for their hotel rooms and all their meals (at above the per diem of $65).

FIGURE 18.8	Continued.

TRAVEL EXPENSES

Air Travel (Itemize all fares
        on separate form)   _____ × _____ = _____
                                No. fares   unit cost

Rental Car(s)   _____ × _____ × _____ = _____
                No. cars   No. days   unit cost

Lodging   _____ × _____ × _____ = _____
               No. rooms   No. days   unit cost

Meals   _____ × _____ × _____ = _____
             No. persons No. days   per diem

Mileage   _____ × _____ = _____
               No. miles $ per mi.

Cab Fares (Itemize all cab fares
        on separate form)   _____ _____
                           dates   subtotal

Gratuities (Itemize gratuities
        of more than $5.00
        on separate form)   _____ _____
                           dates   subtotal

Other Expenses
  (Explain on attachment)   _____
                                           subtotal

                              Subtotal Travel   _____

                                      TOTAL   _____

              Our margin for this contract (%)   _____

                                    TOTAL   _____

Jeff Beetham's Estimate

Jeff knows that he should "inflate" his estimate slightly for safety's sake. By estimating his costs at the more expensive figure, Jeff protects REI from a review session for which they might be under-funded.

How does Jeff complete an "inflated" but honest estimate? He creates his "margin"—that is, his allowance for emergency expenses—by estimating the variable items on the expensive side. For example, he must assign fixed costs, $50 and $25 to his REI personnel. He also must assign fixed costs to his supervisor and engineer. But he can inflate the user/customer category, because all REI has to do is provide meals and lodging. In fact, he feels pretty certain the customers from Knoxville will take him to dinner the second night they are in town, and he will recover some costs at that time. So he assigns a general fee of $70 per day to each

FIGURE
18.9

A Typical Budget Sheet for Estimating a Slide Show

BUDGET FOR SLIDE SHOW

Name of Project _____
Client or Division _____

Task	Hours	Cost
Script	_____	_____
Visuals		
Design	_____	_____
Pictorials	_____	_____
Storyboard	_____	_____
Typesetting	_____	_____
Mounting	_____	_____
Supplies	_____	_____
	Total	_____
Photography		
Studio Shots	_____	_____
Location Shots	_____	_____
Film	_____	_____
Developing	_____	_____
Duplication	_____	_____
Supplies	_____	_____
Special Lab Costs	_____	_____
	Total	_____

customer. Since Jeff knows prices in Atlanta, he thinks he can recover any surprise expenses in his budget by adding a bit extra to the user/customer costs he cannot predict precisely.

Similarly, Jeff inflates the cost of the manual's review copies, the evaluation forms, and the other supplies to create a margin for any extra expenses under line item two, the equipment category. Also, for line item three, travel expenses, Jeff creates an attachment explaining that the engineer's air fare may cost more than Jeff estimates because of last-minute delays that may cause the engineer to reschedule her flight from Seattle. By anticipating a possible problem with the engineer's ticket, Jeff prepares his manager for the worst possible expenses. Jeff and the manager can then adjust the budget to the actual figures that may be lower, but they will not be surprised and embarrassed by a deficit.

Finally, Jeff adds a surcharge of 25 percent in the second-to-last line of the budget, because the company plans to earn a profit of 25 percent on the documentation it produces for this client. Although Jeff's manager does not expect him to earn the entire 25 percent by over-billing for one meeting, an unethical practice, the budget requires Jeff to include this fee as a record of the profit REI plans to make through the entire process of creating the documentation the client has purchased.

FIGURE 18.9		

Continued.

Sound Track
   Actor's Fee   _____   _____
   Music/Musicians   _____   _____
   Permissions   _____   _____
   Studio Rental   _____   _____
                                          Total   _____

Production
   Mounting   _____   _____
   Programming   _____   _____
   Electronic Equipment   _____   _____
   Supplies   _____   _____
   Slide Duplication   _____   _____
   Audio Track Mixing/Duplication   _____
                                          Total   _____

Travel
   Location Shots   _____   _____
   Lab Work   _____   _____
   Sound Studio   _____   _____
   Other   _____
                                          Total   _____

                                       TOTAL   _____

_____   _____
Signature of Project Manager   Date

Tips for Estimating

Although estimating requires some exact inquiries such as telephone calls to verify prices or written request for quotation forms, it is also an art that requires the supervisor and manager to imagine every possibility or extreme situation that a budget must recover. In summary, remember these tips as you prepare your own writing, editing, and production estimates.

1. Nothing is free. Often, managers do not know how much it costs to print a page of documentation. Your budget must account for everything from personnel to paper clips. You cannot write the manual without the personnel and paper clips, so include them. Personnel and paper clips are often called "indirect costs" because they are not the raw materials of the manual. Bill for every indirect cost.
2. Include inflation factors in your budget. For example, by some estimates, printing inflates 26 percent annually. Personnel may inflate anywhere from 3 to 15 percent, telecommunications by 10 percent. How true are these estimates for the area in which you live and work? How much will postage inflate during the next year? How much will your office's utilities inflate next year? Your budget must "stretch" to include all these allowances.

FIGURE 18.10

A Typical Summary Bid Sheet for Printing Estimates.

SUMMARY BID FOR A TECHNICAL MANUAL

Title of Work: *Reference Guide for NP25 Ink-Jet Printer*

Summary: 3,000 copies, 96 pages each, shrink-wrapped, standard book cartons, manufactured to meet National Printers' specifications (available from the Purchasing Agent), shipped directly to the National Printers' Warehouse, Chicago, IL.

	3,000 copies
Composition: Original composition, page layout and make-up, reproducing and reading proofs, and other preparation and materials necessary for submitting camera-ready text for 96 pages.	_____
Preparation: All camera work, negatives, stripping and other preparation before blue tints.	_____
Proofs: Three complete sets of galley proofs, and one complete set of folded and trimmed blue tints.	_____
Presswork: All films, plates, preparation, and printing impressions.	_____
Paper: All paper for text and covers.	_____

3. If you must design a budget system for the publications of a small communication group, do not invent the whole system yourself. Use the models others created. Interview your firm's bookkeeper, and learn the method your company or institution uses. Then obtain sample budget forms from the order and form books of your public library. Consult the government documents section, as well, for budgets for specific occupational pursuits.

4. Study your costs during regular review sessions. List the direct and indirect costs that inflate the prices of your documentation. Create strategies for controlling or lowering these costs. This last factor, which is vitally important, can come as a real surprise. After studying your costs, you may find that an advanced tabletop publishing system can cut your composition costs significantly; then again, you may find that a new, fuel-efficient gas furnace may cut your publication group's utility bill more and save more on your general budget than the desktop publishing. Either way, you can cut the costs of your manuals.

FIGURE 18.10

Continued.

Binding: All binding operations, sheetwork, sewing, casing-in, jacketing, inspection, and packing. Also include dies, endpapers, and all other binding and packing materials.   _____

Shipping: To Chicago   _____

Total of Composition, Preparation, Proofs, Presswork, Paper, and Binding.   _____

Special Options for This Estimate.   _____

Name of paper recommended for this manual _____
_____

Prices for extra or fewer copies
   For extra copies, add this amount to base price (per copy)   _____
   For fewer copies, subtract this amount from base price (per copy)   _____

For additional signatures, list the additional
price per copy for   8pp_____   16pp_____   32pp_____

For fewer signatures, list the deduction per
copy for   8pp_____   16pp_____   32pp_____

Signed_____ Firm_____ Date_____

SUMMARY

To explain how corporations produce documentation, this chapter concentrated on the publications group. We discussed the need for student technical communicators to understand the corporate life and corporate structure. After reading this chapter, you should be able to:

- List the vertical structure of a typical publications group.
- Name the tasks the individual members of a publications group perform.
- Explain the differences between collegiate and corporate technical writing assignments.
- Define the terms "quality" and "accuracy" as they relate to corporate documentation.
- Describe four responsibilities of a publications group.

FIGURE 18.11	A Typical Planning Worksheet for the Cost of a Document's Review.

DOCUMENT REVIEW COST ESTIMATE

PERSONNEL	Days	× Daily Cost =	Total
Supervisor	3	500.00	1500.00
Technical Writer	3	400.00	1200.00
Technical Editor	3	400.00	1200.00
Assistant	3	200.00	600.00
Engineer	5	600.00	3000.00
User/Customer	3	70.00	210.00
User/Customer	3	70.00	210.00
User/Customer	3	70.00	210.00
User/Customer	3	70.00	210.00
		Subtotal	8340.00

EQUIPMENT	Units ×	Days	× Daily Cost =	Total
Test Site/Room Rental		3	125.00	375.00
Conference Room Rental		3	125.00	375.00
Video Camera(s)		N/A		
Lighting Equipment		N/A		
Power Cables		N/A		
Microphones		N/A		
Audio Recording Equip.		N/A		
Review Copies of Manuals	12	N/A	300.00	3600.00
Evaluation Forms	60	N/A	.50	30.00
Other Supplies	N/A	N/A	150.00	150.00
			Subtotal	4,530.00

- Describe the ten-step approach to production.
- Describe the process a typical firm may use to review its documents.

ASSIGNMENTS

1 Jeff Beetham has asked you to obtain some printing estimates so he can prepare a budget for his 98-page nuclear waste disposal manual. The manual will have:

Type: Text—10/11 Helvetica; Also chapter headings in 14 pt. and all running heads and footers in 8 pt.
Trim: 8½ × 11″
85 pages of text
10 full-page line illustrations

FIGURE
18.11

Continued.

TRAVEL EXPENSES

Air Travel (Itemize all fares
    on separate form) _____1_____ × _620.00_ = _620.00_
                        No. fares   unit cost

Rental Car(s) __1__ × __3__ × _74.00_ = _222.00_
              No. cars No. days   unit cost

Lodging __5__ × __3__ × _78.00_ = _1170.00_
        No. rooms No. days unit cost

Meals __5__ × __4__ × _65.00_ = _1300.00_
      No. persons No. days per diem

Mileage _100_ × _.28_ = _28.00_
        No. miles $ per mi.

Cab Fares (Itemize all cab fares
    on separate form) __N/A__   _3340.00_
                        dates   subtotal

Gratuities (Itemize gratuities
    of more than $5.00
    on separate form) __N/A__   _3340.00_
                        dates   subtotal

Other Expenses
    (Explain on attachment)   _3340.00_
                                    subtotal

Subtotal Travel _____

TOTAL _16,210.00_

Our margin for this contract (%) _4052.00_

TOTAL _20,262.00_

2 halftones (full-page)
Perfect binding
Paper cover with one color, Pantone #290
Paper stock: 50-pound, or suitable alternate
Cover stock: 60-pound, varnished, equivalent to Springhill coated cover stock

Jeff will supply camera-ready cover art and camera-ready art for all line illustrations. Find out what it will cost to prepare his book. Using the forms in this chapter, design your own cost estimate form for this project.

2   Now that you have contacted a printer about the cost of printing Jeff's manual, you will have questions about the information he failed to give you. Compose a memo to Jeff in which you list the technical information you need from him to complete a cost estimate form.

3 Prepare a production timeline for Jeff's manual. Assuming you and an assistant will work on the project, define the phases of research, writing, editing, and production. When can you deliver the product? What will the milestones be?

4 Obtain a paper that a classmate must write for this class or another class (if the instructor will permit you to collaborate). Design a Levels of Edit Checksheet for the paper based on the premise that you will spend no more than three hours editing and improving the copy. Which levels of edit will you choose? Why?

5 After you prepare the checksheet for your classmate, draft a memo in which you explain your selection of edits for the assignment and the time allotted.

6 Do you know what it costs to hire a technical writer in your geographic area? Find out and prepare a two-page report for the class. As you do so, consider the following questions. What level of professional education are technical communicators expected to have in your area? What types of projects do the writers work on? Do several types of industries dominate your area? What career opportunities exist for technical writers in your area?

7 Pretend you are Jeff Beetham, or that you are paid for your work. Prepare a budget for a term paper you are writing for another class. Include research time, planning, writing, editing, and production. Don't forget your indirect costs. Now obtain your technical communication instructor's approval for your budget by obtaining his or her signature and the date.

HANDBOOK

Editing Your Own Writing

The writing process has only begun when you have developed your ideas and written your first draft. Experts on the writing process estimate that successful writers spend 45 percent of their time on any one writing project revising their first draft. During this stage of the process, you should add any information you think your audience needs, and delete information that is irrelevant. You should check the structure and organization of your paragraph or essay. In addition, you should now begin checking your spelling, punctuation, and other mechanics in your sentences.

This section of the text gives you practice in proofreading and revising sentences from first drafts. After each discussion of a grammar rule or convention of punctuation, examples appear, both for illustration and to help you sharpen your own abilities at editing and revising.

The most successful writers almost always work with skilled editors who check their work for inaccuracies. Such problems as mispellings, poor punctuation, or vague pronoun reference make writing difficult for a reader to follow, and professional writers are anxious to avoid placing such barriers between themselves and their audience. Successful writing, published writing, always goes through several editings and revisions, beginning with the authors themselves. That's you.

1 Grammar and Sentence Structure

COMPARATIVE AND SUPERLATIVE FORMS

For short adjectives, use *-er* to form the comparative, and *-est* to form the superlative.

> heavier, lighter, leaner, easier
> heaviest, lightest, leanest, easiest

Use *more* to form the comparative and *most* to form the superlative of longer adjectives and adverbs ending in *-ly*.

> more easily, more malleable, more significantly, more established
> most easily, most malleable, most significantly, most established

Do not use *more* and *-er* or *most* and *-est* at the same time, as in *most easiest*.

DANGLING MODIFIERS

Introductory modifiers with *-ing* or *-ed* verb forms cannot be left "dangling." They must be clearly related to the word they modify, and this is done in one of two ways.

> Touring the plant, the new conveyor was in constant use.

The first tendency of most readers would be to link *touring* and *conveyor*, since conveyor is the first noun after the modifier. But conveyors don't tour; people do. The modifier must be somehow more clearly related to the person doing the observing. The sentence may be revised to put a person in the conveyor's place directly after the modifier.

> Touring the plant, I saw the new conveyor in constant use.

Or, the sentence may be revised to put the modified word even closer to the modifier:

> As I was touring the plant, the new conveyor was in constant use.

MISPLACED MODIFIERS

To avoid misunderstanding, all modifiers—adjectives, adverbs, prepositions, articles—should be placed as close as possible to the word they modify:

> We commissioned those plans from the architect that I sent to you.

Revising the sentence by moving the modifier closer to the word it modifies—plans—creates a clearer sentence:

> We commissioned those plans that I sent to you from the architect.

Unless the context makes it clear that the plans were not sent by the writer directly from the architect, the sentence should be revised again:

■ We commissioned, from the architect, those plans that I sent you.

Adverbs, such as *even, only, nearly,* and *hardly,* should be placed directly before the words they modify:

■ We have discovered one problem only in our parts inventories.

Revised:

■ We have discovered only one problem in our parts inventories.

PARALLELISM

When two or more parts of a sentence explain similar ideas and serve a similar purpose, the related parts should be in the same grammatical form:

> People in the communications department have *technical, professional,* and *liberal arts* degrees. [Three parallel adjectives]
>
> Molded plastic parts such as *glove-box lids, sun visors,* and *armrests* are all produced in the surrounding buildings. [Three parallel nouns]
>
> The butterfly operation joins *the sides and roof, which are lowered from an overhead conveyor,* to *the floor-pan and cowl, which is already mounted on the framing line.* [Parallel compound nouns and adjective clauses]
>
> You can contact me *by letter* or *by telex.*

Parallelism—similarity—such as this is easier to comprehend and is more emphatic than dissimilar, nonparallel statements of the same ideas, and is therefore usually preferred wherever related ideas can be stated in parallel fashion. It is especially necessary to make lists, headings for graphs and tables, outlines, and tables of contents as parallel as possible. It is also a grammatical requirement that items in a series be grammatically parallel.

■ Alongside the assembly personnel are electricians, tool-makers, service personnel, and other specialists who work on the line itself, not on the product.

PRONOUN AGREEMENT

Pronouns refer to nouns or other pronouns, so you must be sure that the reader is not needlessly confused. The pronoun you use must be similar to the word to which it refers, in both number and gender. Singular nouns are referred to by singular pronouns; plural nouns are referred to by plural pronouns. Feminine, masculine, or neuter nouns require that the gender of the pronoun match as well. Do not write:

Each woman wants their assignment immediately.

Make the pronoun agree. Write instead:

Each woman wants her assignment immediately.

PRONOUN FORM

Use the subjective case when the pronoun (*I, we, you, she, he, it, they*) is the subject of a verb:

We must look for ways to spread this cost over more products.

Use the objective case when the pronoun (*me, us, you, her, him, it, them*) is the object of a verb or of a preposition:

Give the report to her when you are ready to print it.

Use the possessive case when the pronoun (*my, our, your, her, his, its, their*) shows ownership or when the pronoun appears directly before an *-ing* form of the verb:

This is its weak point.

Our maintaining quality over the entire run is critical.

PRONOUN REFERENCE

When you use a pronoun, be sure the reader can easily determine its reference. Keep pronouns close to the words to which they refer, and revise any sentence in which the reader's choice of pronoun references is unclear:

The doctor asked the anesthetist to keep his office notified about his schedule.

After I read it, my respect for programmers grew.

Unless the context makes the pronoun references clear, sentences like these should be revised:

The surgeon asked, "Doctor Smith, can you keep your office notified about your schedule?"

After I studied computer science, my respect for programmers grew.

Readers habitually associate a pronoun with the noun that precedes it, so keep pronoun and antecedents close. Do not use *which, this, that,* or *these* to refer to a general idea expressed in the previous clause, sentence, or worse, paragraph. The simplest revision is to repeat the antecedent when use of one of these four pronouns alone would be vague.

Vague: A viscosity-index improver is a long-chain polymer which converts an ordinary single-grade oil into a multi-grade. That gives the cold viscosity of a light oil but the hot viscosity of a heavier one.

Revised: . . . The addition of the polymer gives the cold viscosity of a light oil but the hot viscosity of a heavier one.

Avoid using *it* and *they* for general and indefinite references:

Vague: *It* said in the morning paper that *they* don't expect the ticket surcharge to be popular.

Revised: The editorial in the morning paper stated that even the Civic Center Commission members don't expect the ticket surcharge to be popular.

The examples just above also imply another rule: make sure that the antecedent is actually in the sentence, or a reader can't possibly know who *they* are.

And, if you use the word *you*, as when writing directions, be sure the context means *you, the reader*. Do not use you when you mean *people, a person,* or *one*:

Vague: When I arrived at the dock, you could see dozens of safety violations.

Revised: When I arrived at the dock, I could see dozens of safety violations.

SHIFTS

Do not unnecessarily shift from one *number* to another:

Each writer should proofread their own work.

Or from one *person* to another:

It is impossible to predict what you might discover once we start investigating.

Or from one *tense* to another:

I looked in every manual we own, but I fail to find directions for this procedure.

Similarly, do not shift unnecessarily from active to passive *voice*:

The contractor writes the estimate, but the cost is determined by the subcontractor in such a contract.

Some shifts may be necessary for clarity, but random shifts of focus make reading difficult. Decide on the simplest way to convey your meaning; then be consistent with *number, person, tense,* and *voice*:

Each writer should proofread his or her own work.

It is impossible to predict what we might discover once we start investigating.

> I looked in every manual we own, but I failed to find directions for this procedure.
>
> The contractor writes the estimate, but the subcontractor determines the cost in such a contract.

SUBJECT-VERB AGREEMENT

Subjects and verbs in the same sentence must have endings which "agree." If the subject of a sentence is a singular noun or a third person singular pronoun (*he, she, it*) then the verb must be the third person *s* form, too, for all verbs in the present tense and *to be* in past tense:

> He *is* trustworthy and dependable.
>
> She *does* all specialized work with the CAT scan.
>
> It *grows* at a geometrically increasing rate.
>
> Dr. Fuller *wishes* he had remained in Augusta, I believe.

If the subject of a sentence is a plural noun or third person plural pronoun (*they*), then the verb must be in its plural form as well:

> They *are* on emergency standby.
>
> They *do* our emergency repairs on a retainer contract.
>
> Red-leaf maples *grow* fast enough to create shade in five years.
>
> Fuller and Richards both *wish* they had stayed, we suppose.

When proofreading your drafts, be careful of inverted sentences, in which the subject will follow the verb. Check carefully for prepositional phrases which separate subjects and verbs; the subject of the sentence cannot also be the object of a preposition. Be equally careful with compound subjects. If the compound subjects are joined by *and*, the verb must be plural; if the compounds are joined by *or* or *nor*, then the verb must agree with whichever of the compound subjects is nearest to it.

> There are four new policies that apply to this purchase.
> The policies of our board are not necessarily binding on us.
>
> The auto industry and the steel industry have strong lobbies in Washington.
> Neither the textile industry nor the garment industries have strong lobbies.
>
> Neither the government industry nor the textile industry has been successful in influencing legislation in Washington.

Finally, remember that the pronouns *several, few, both,* and *many* are singular, along with *each, either, neither, one, anybody, somebody, someone, everybody, everyone, no one,* and *nobody,* all of which are singular, and thus require the singular *s* form of the verb. In the past tense, the same rules apply, but only one verb, *to be*, changes its past tense forms to match the number and person of the subject of the sentence. *Were* is the third person plural form of *to be*, and *was* is the third person singular form of *to be* in the past tense. (Note that *was* is still an *s* form.)

2 Punctuation

Apostrophe
Capitalization
Colon
Commas
Hyphen
Parentheses and Dashes
Quotations and Quotation Marks
Run-On Sentences, Comma Splices,
    Sentence Fragments
Semicolon
Slash (Virgule)

APOSTROPHE

To form the possessive case of nouns and indefinite pronouns, add only an apostrophe if the word ends in *s*:

■ Jess', days', weeks', employers', companies'

Add *'s* to words ending with other letters or sounds:

■ Ford's, mask's, week's, month's, company's, someone's

Add the appropriate possessive marker to only the last element of a compound or to nouns of joint possession:

▌ brother-in-law's, no one else's

▌ Chevrolet and Toyota's joint venture

The possessive pronouns *its, whose, hers, his, your, yours, ours,* and *theirs* do not require the apostrophe.
　　Use an apostrophe to indicate a contraction or the deletion of letters and numbers:

▌ you're, it's (for *it is*), who's (for *who is*), don't, shouldn't

▌ '78, o'clock, *Singin' in the Rain*

The apostrophe is also used to form the plural of words used as words, of letters, numbers, and other symbols such as abbreviations:

▌ Don't use too many *but*'s and *and*'s.

▌ Dot your *i*'s.

▌ People expected greater growth in the 1980's.

▌ How many DA's does a large city have?

CAPITALIZATION

Capitalize all proper nouns used as names, whether the name of a person, group, race, language, organization, holiday, month, or day, or name of a place.

■ Larry, Hindu, League of Nations, English, Mother's Day, March, Tuesday, Times Square

Capitalize all widely used abbreviations and historical events.

■ FBI, CORE, AARP, CIA, Vietnam War, Bay of Pigs

Capitalize the first word and all other words except articles, conjunctions and short prepositions in the titles of books, plays, documents, magazines, newspapers, articles, poems, and songs.

Magna Carta, Bill of Rights, Forbes, The Chronicle of Higher Education, Journal of the American Medical Association

Capitalize all titles that precede a proper noun and all common nouns used as a part of a proper name.

the Reverend John Black, Dr. E.B. House, General Richardson, Clemson University, Sea of Galilee, Benson Gymnasium, Madison Avenue

In addition, the pronoun *I*, the first word in a sentence, the first word in a quotation, and sacred religious words are always capitalized. Words that are *never* capitalized, unless they form part of a proper noun or a place name, include the names of directions — *north, east, west, south* — family relationships — *aunt, uncle, father*, etc. — and the names of seasons.

COLON

Some uses of the colon established by convention include its use in divisions of time, between titles and subtitles, in biblical citations, and in introduction of a formal quotation.

4:30, John 3:16, Genesis 1:2

The Second Time Around: Remarriage in America

Fowler explains the use of the colon with a metaphor: ''It promises to deliver the goods that have been invoiced in the preceding words.''

A colon is used to introduce a list or series of examples to explain the more general material in the sentence before the colon. Sometimes such a colon is preceded by phrases like *as follows* or *the following*.

Marbleizing a surface requires only simple equipment: paints, brushes, and scrap cloths.

Consumers generally find the following techniques easiest to use at home: marbleizing, combing, and spattering.

It is not required that you use a colon after *are* or *including*, unless these words and the colon introduce a long list you intend to center or emphasize with graphic aids. Even then, some style sheets suggest that you revise the sentence to avoid the linking verb or the word *including* before the colon.

The benefits of universal term insurance include: . . .

Revised:

Four major benefits of universal term insurance will immediately change your cash flow: . . .

COMMAS

Use a comma to separate the name of the person to whom you are speaking:

Steve, would you like to serve as chairman?

Use a comma to separate the major items in dates and addresses:

Your oil will be shipped Tuesday, March 18, 1986.

The lawyer's office is at 2327 Laurel, Augusta, Maine.

Use a comma to separate independent clauses linked by *so, for, or, nor, and, yet,* or *but:*

We can finish the overhaul of the pump motor in two days, but removing and re-installing will require a total of 12 hours' more work.

Use a comma to set off an introductory sentence modifier—a word, a phrase, or a dependent clause—from the main idea of the sentence:

Actually, we keep those pump parts in stock here.

During the first two hours, we plan to finish the inventory.

If I had one more assistant, we could finish the inventory today.

Use a comma to separate every item in a series, including adjectives in a series:

We are responsible for removal, repair, and transportation of the motor.

Old, oily, or unadjusted plugs affect engine performance.

Use commas to separate parenthetical elements, including non-restrictive adjective clauses:

Cummings, the manufacturer, publishes over 20 manuals about this engine.

The globe valve housing, the useful housing, is the subject of this report.

The valve, as a matter of fact, has one simple application in this design.

The designing architect, who retired last year, will have to be consulted about this change.

Sulfur, which produces a distinctive kind of corrosion, is a problem in all our pumps.

The pump vault, which is near a utility line, may remain there.

Finally, do not overuse commas.

HYPHEN

Use a hyphen to join two or more words which are being used as a single adjective before a noun:

an ill-conceived idea

a well-formulated proposal

an evil-smelling preparation

As the rule implies, do not use a hyphen if the first word is an adverb ending in *ly* or if the two words form a predicate adjective following the verb:

a calmly presented proposal

a poorly conceived idea

That idea is ill conceived.

Your proposal is well formulated.

Hyphens are also used to form compound words that are not yet accepted for use as single words (check your dictionary to determine any compound word's status) and to avoid an awkward combination of letters or sounds:

well-lift

ex-champion

all-consuming

self-important

President-elect Brown

Hyphens are also used in spelled-out fractions and numbers twenty-one through ninety-nine:

forty-three, one-half, two-thirds

PARENTHESES AND DASHES

Parenthetical elements interrupt the grammatical order of a sentence to explain or offer comment that is in addition to the main idea of a sentence. Parenthetical elements are not *necessary* to the understanding of the sentence; therefore parenthetical elements are similar to non-restrictive elements, which are set off by commas.

If, however, the parenthetical information might be confusing if set off by commas, use parentheses or dashes. Parentheses seem to imply that the parenthetical information is somewhat unimportant, while dashes do just the opposite, emphasizing the importance of the information between them.

ROBOGATE (developed by the Comau Group) was first used at the Fiat factory in Turin.

Dashes are preferred to set off parenthetical elements that represent a major shift in thought or sentence structure. Dashes also set off an introductory list or summary.

> ROBOGATE has no fixed production line—trolleys are guided by magnetic cables under the floor—to take the body shell from station to station.
>
> Conventional, robot, and laser-aimed—all three types of welders are used in the ROBOGATE complex.

Parentheses are also used to enclose numbers or letters labelling items listed in a sentence:

> ROBOGATE offers very important advantages such as (1) speed, (2) lower labor costs, (3) the ability to produce several body styles on one line without retooling, and (4) amortization of equipment cost over several flexible model lines.

Don't overuse this last technique: if your list is long, complicated, or important, you should probably "break out" the list into a parallel, centered list, with white space and graphics to emphasize it. Don't rely on numbers in parentheses to emphasize your list for you.

QUOTATIONS AND QUOTATION MARKS

Use quotation marks to enclose direct quotation of writing or speech.

> David Owen says, "The methodological forefather of the modern cult of mental measurement was Alfred Binet."
>
> "Binet was reluctant," Owen claims, "to extend his method beyond the narrow, diagnostic purpose for which it had been designed."

As in the examples, use commas to set off "tag lines" such as *he said* or *he claims* unless the quotation ends in a question mark or exclamation point.

Commas or periods that occur at the end of the quotation should be placed inside the quotation mark; colons and semicolons are placed outside the quotation marks. Dashes, question marks, and exclamation points are placed inside quotation marks when they are part of the quotation, outside when they are part of the larger sentence.

> "Can coaching help your performance on this test?" asks Owen.
>
> "Of course!" Owen answers immediately.
>
> For years, the tests' publishers have claimed, "Of course not"!

If the quotation is closely related grammatically to the sentence which introduces it, the commas are sometimes deleted.

> Owen snarls "How much aptitude can dance on the head of a pin?" to show his opinion of the whole idea of aptitude testing.

In typed or printed final drafts, very long quotations are often formally introduced by a preceding comma and single spaced and centered on the page. In such cases the quotation marks are deleted.

Quotation marks are also used to set off the titles of articles, short stories, poems, essays, and songs. Punctuation which is part of the title goes inside these quotation marks, and punctuation which is part of the larger sentence goes outside.

> "Eye of the Tiger" was a popular motivational song for a while.

> Most of this data appeared originally in "Sense or Nonsense?" in *Readings in Educational and Psychological Measurement*.

Quotation marks are also used to set off words being used in a special way.

> What do the psychologists mean by "developed ability" in a high school senior?

RUN-ON SENTENCES, COMMA SPLICES, SENTENCE FRAGMENTS

Two sentences (two independent clauses) joined together with no punctuation and no conjunctions between the clauses form a run-on sentence:

> Dr. Porsche designed the Porsche he also designed the Volkswagen.

A run-on sentence can be corrected in several ways. You may separate the two clauses with a semicolon:

> Dr. Porsche designed the Porsche; he also designed the Volkswagen.

Or you may separate the clauses with a comma and a coordinating conjunction such as *and, or, for, nor, yet, but,* or *so*:

> Dr. Porsche designed the Porsche, and he also designed the Volkswagen.

Or, a preposition or a subordinating conjunction may be added to one of the clauses, or the clauses may be separated to make two complete sentences:

> Before he designed the Porsche, Dr. Porsche designed the Volkswagen.

> Dr. Porsche designed the Porsche. He also designed the Volkswagen.

A comma splice is similar: two independent clauses joined by only a comma:

> His first design was a gasoline-electric, it was called the Lohner-Porsche.

A comma splice may be corrected with a semicolon:

> His first design was a gasoline-electric; it was called the Lohner-Porsche.

Or insert a coordinating conjunction after the comma:

> His first design was a gasoline-electric, but it was called the Lohner-Porsche.

Or by inserting a subordinating conjunction, making one of the clauses a dependent clause:

■   His first design was a gasoline-electric, which was called the Lohner-Porsche.

Or the comma splice can be separated into two complete sentences:

■   His first design was a gasoline-electric. It was called the Lohner-Porsche.

There are two types of sentence fragments, the first of which is a group of words, punctuated as a complete sentence, but lacking a subject or a verb.

▨   A technical writer using her handbook.

The second type is a dependent clause which is mistakenly punctuated as a sentence.

▨   Which is the only certain way to proofread your work.

▨   If shift C comes in an hour earlier, for example.

Both types of fragments are occasionally used by professional writers for emphasis, but technical writers should probably avoid sentence fragments, which are often ambiguous or awkward.

SEMICOLON

Place semicolons between independent clauses that are not linked by *so, for, or, nor, and, yet,* or *but* (Contrast this to the use of commas between main clauses; see the handbook entry "Commas"):

■   We cannot tolerate errors; everyone suffers for mistakes.

■   Many durable goods manufacturers are merging; however, more modern production techniques ensure that this will not force the consumer to choose among fewer model lines.

Use semicolons to separate items in a series if the items themselves have commas within them:

■   We will be joined by J.H. Smith, their lawyer; F.T. Wharton, their accountant; and C.B. Wills, their agent.

SLASH (VIRGULE)

This punctuation mark (/) represents the word *per*; for example, the symbols "90,000 revs/min" mean the phrase "ninety thousand revolutions per minute."

The virgule also represents the word *or* when the use of the word itself would be awkward; the symbols "and/or" mean "and or or." This use indicates alternatives such as in, for example, the listing of several part numbers, any of which is correct, as in "122/124B/125C." The virgule is also used by convention to separate the denominator and the numerator in fractions. In addition, some style manuals use the virgule to separate day, month, and year in numerical representations of dates.

3

Spelling and Diction

Diction
Spelling

DICTION

Technical writing is formal writing, even when you are sending a hand written memo to a co-worker down the hall, so you should not use words which your dictionary indicates are non-standard, slang, colloquial, dialect, obsolete, or archaic. (These labels vary from one dictionary to another, so check the front matter, the first few pages, of your desk dictionary to ensure that you understand its labelling system.) Words that bear one of these usage labels should never be used unless you use them in a quotation, and in technical writing very few quotations you'll use will contain such non-standard diction.

Here are some examples of informal and colloquial expressions you should avoid:

most for *almost*, as in "Most everyone has worked here for two years or longer."

anyways for *any way*, as in "Anyways, we've all been trained on each machine."

be sure and for *be sure to*, as in "Be sure and check your invoice."

but what for *that*, as in "I'm not certain but what it might work."

could of for *could have*, as in "I could of made a mistake."

in regards to for *in regard to*, as in "In regards to your fee, we feel it's reasonable if it includes your expenses."

irregardless for *regardless*, as in "Irregardless of your fees, we want to schedule two meetings."

kind of a for *kind of*, as in "What kind of a consultant is she?"

nohow for *not at all*, as in "We don't keep her calendar that full, nohow."

off of for *off*, as in "Can you pull that seal off of the shaft for me?"

plenty for *very*, as in "I'm plenty busy with all these reports."

real for *very*, as in "That was a real rough assignment."

somewheres for *somewhere*, as in "My report is around here somewheres."

sort of a for *sort of*, as in "What sort of a spline will we need?"

sure for *surely*, as in "The engineers sure specified the wrong valve."

try and for *try to*, as in "I'll try and get the specs right this time."

used to could for *used to be able to*, as in "You used to could order direct from the factory."

ways for *way*, as in "It's a long ways from the pump vault to the discharge sump."

where for *that*, as in "I read where these old piston pumps were being replaced."

A, An

Use *a* before any word beginning with a consonant or a *y* sound:

a Eurodollar, a filter, a university, a vault

Use *an* before any word beginning with other vowel sounds:

■ an entomologist, an incompetent, an armrest, an interesting idea

Get

Try to avoid overusing the word *get* in your writing; substitute more specific synonyms as *order* or *purchase* whenever appropriate, and avoid informal uses of *get* such as *I don't get what you mean* and *We'll get finished in an hour.*

Every day, Everyday

Every day means "always;" *everyday* (one word) is an adjective meaning "common, routine, or mundane":

He follows the same schedule every day.

This is my everyday suit.

Lie, Lay

Lie (lay, lain) means to "assume a horizontal position or to be horizontal":

Temporary orders often lie around all day waiting to be delivered.

The plans lay around for almost a year.

Lay (laid, laid) means "to place an object":

Where did you lay the blueprints?

We laid the problem at the feet of the people responsible.

Raise, Rise

Raise (raised, raised) means "to lift or to cause a lifting motion":

■ If we can raise those speakers, the sight lines will be improved.

Rise (rose, risen) means "to lift oneself up":

The Genie lift rises on its self-contained compressed air.

Your grocer rose before dawn to travel to the wholesale market.

Sit, Set

Sit (sat, sat) means "to take a seated position":

We sat at the airport for over two hours.

Our officers may sit there for similar periods unless we lease a plane.

Set (set, set) means "to place an object":

Set the flange in place, behind the spline.

We set the matter before the board of directors.

Very

Writers often overuse *very* as a modifier. While you revise your drafts, try substituting synonyms such as *intensely, much* (before past participles used as adjectives), or use a precise term or figure whenever possible.

Who and Whom

Who and *whoever* are used as subjects of verbs; *whom* and *whomever* are used as objects of verbs or objects of prepositions:

> Who has the contract for managing this timber?
>
> Who received the shipment?
>
> To whom was the contract given?
>
> They chose the same manager whom they chose last year.

There are no exceptions to this simple rule, even if the sentence is inverted, and even if the sentence structure is extremely complex. Even in a sentence as complex as this, the rule holds true:

> We will cooperate with whoever wins the contract.

Here, the verb *wins* requires a subject, so subjective case *whoever* is used, even though the pronoun follows a preposition. (*Whoever* is not the object of *with*; the object of *with* is *whoever wins the contract*.)

SPELLING

There are five rules that explain *most* English spelling.

Rule 1. When adding a suffix (except *-ing*) change any final *y* to *i*, if the *y* follows a consonant.

> | copies | amenities |
> |---|---|
> | companies | applies |
> | utilities | carries |

Rule 2. Usually, *i* precedes *e*, except immediately after *c*, or when the combination *ei* is pronounced as *a*. Use your dictionary to check the exceptions.

> | believe | weigh |
> |---|---|
> | receive | neighbor |
> | deign | reign |

Rule 3. When adding a suffix that begins with a vowel, double any final consonant that follows a single vowel, as long as the accent is on the final consonant.

coping	willing
benefited	cooling
submitted	

Rule 4. When adding a suffix that begins with a vowel, delete a silent final *e*. If you add a suffix that begins with a consonant, leave a silent final *e* in place.

coping	management
hiding	judgement
purely	insurance

Rule 5. When adding prefixes or suffixes to a word, never delete any part of the word unless Rule 4 specifically demands it.

taxed	janitorial
deployed	finished
hangars	turbocharges

4 Manuscript Form

Manuscript Form
Abbreviations
Numbers
Underlining

535

MANUSCRIPT FORM

Unless you have been instructed otherwise, assume that you will use a standard format for the preparation of manuscripts.

☐ Use 8½ ″ × 11 ″ bond paper.
☐ Type on one side only.
☐ Use black ribbon.
☐ Leave 1 ″ margins at the bottom and right.
☐ Leave 1½ ″ margins at the top and left.
☐ Always double space a final draft.
☐ Indent the first lines of a paragraph 7 spaces.
☐ Put page numbers in the upper right of each page.

Make sure that any last minute corrections are legible; make these corrections with black ink. Remember that the appearance of your manuscript is the first impression your audience receives of you and your work. Sloppiness often implies a lack of serious attention, and is rewarded by the reader's lack of serious attention.

ABBREVIATIONS

In technical writing, use abbreviations as little as possible unless you are absolutely sure your audience will understand them. Never abbreviate common nouns such as *university* or *company*. The names of government units, organizations, or business firms may be referred to by abbreviations or initials—IBM, NATO, PTA—as long as you know your audience will require no further explanation. If you use abbreviations which are unique to a technical field—BTU, PTO, ET—define them in the text when you first use them, or consider including a glossary of terms for readers who may need such an aid.

NUMBERS

Technical writing often includes dates, addresses, precise measurements, percentages, references to time, and other uses of numbers. You should be consistent in your use of numbers and follow the conventions of the discipline or an applicable style sheet.

Numbers that begin sentences should be spelled out; if an extremely long number begins the sentence, you should revise so the number can appear in figures later in the sentence.

■   Twelve men using laser alignment can perform 24,000 inspections each hour.

In technical writing, figures are preferred in virtually every other instance, and especially when the figures are accompanied by symbols or are meant to be exact:

3:15 P.M., 4:30 A.M.   32°
$12.17   9½
10" × 12"   67%
55 mph   3.14159

This preference for figures holds true for identification numbers, such as:

Channel 12   Part .111A03597-A36
Volume 6   1957 Chevrolet
Scene 2   Flight 1189
Page 3   Train 12, Track 6

Some cautions: many technical writers follow the "rule of ten," which requires that numbers of ten or less be spelled out, and larger numbers appear as figures:

I wrote the proposal in 30 minutes, but I planned it for six months.

In addition, avoid using *st*, *nd*, *rd*, and *th* after dates, and avoid sentences which spell out a number and then repeat the number as a figure in parentheses. While common in legal writing, such practice is unnecessary in technical writing if the number has first appeared in figures.

Legal: This note will be paid in fifty (50) equal installments.

Revised: This note will be paid in 50 equal installments.

UNDERLINING

Underlined words in a manuscript, written or typed, are set in italics when the manuscript is printed. Underline titles of magazines, books, newspapers, plays, and movies:

I thought The Encyclopedia of the Motorcar was too general for my research.

Excellence Was Expected might be a rewarding book to start your Porsche research.

Underline names of ships or aircraft, underline foreign words that are not yet part of the English language, and underline words used as words.

Mullah is an honorific title.

Index